Earth in
Human Hands

Also by David Grinspoon

*Venus Revealed: A New Look Below the Clouds of
Our Mysterious Twin Planet*

Lonely Planets: The Natural Philosophy of Alien Life

Earth in Human Hands

Shaping Our Planet's Future

David Grinspoon

GRAND CENTRAL
PUBLISHING

NEW YORK   BOSTON

Grand Central Publishing
Hachette Book Group
1290 Avenue of the Americas; New York, NY 10104
grandcentralpublishing.com
twitter.com/grandcentralpub

Printed in the United States of America

LSC-C

First Edition: December 2016

10 9 8 7 6 5 4 3 2 1

Grand Central Publishing is a division of Hachette Book Group, Inc. The Grand Central
Publishing name and logo is a trademark of Hachette Book Group, Inc.

The publisher is not responsible for websites (or their content) that are not owned by the
publisher.

The Hachette Speakers Bureau provides a wide range of authors for speaking events. To
find out more, go to www.hachettespeakersbureau.com or call (866) 376-6591.

Library of Congress Cataloging-in-Publication Data

Names: Grinspoon, David Harry, author.
Title: Earth in human hands: shaping our planet's future / Dr. David Grinspoon.
Description: First edition. I New York, NY: Grand Central Publishing, [2016] I Includes
    index.
Identifiers: LCCN 2016025817 I ISBN 9781455589128 (hardcover) I ISBN 9781478908906
    (audio download) I ISBN 9781455589135 (ebook)
Subjects: LCSH: Human ecology. I Human beings—History. I Nature—Effect of human
    beings on. I Global environmental change. I Environmental protection. I Earth (Planet)
Classification: LCC GF47 .G75 2016 I DDC 304.2—dc23 LC record available at
    https://lccn.loc.gov/2016025817

For Jennifer, who always makes me feel that
I am in good hands.

With all the will in the world
Diving for dear life
When we could be diving for pearls.
—Elvis Costello, "Shipbuilding"

Contents

Introduction

A Planetary Perspective on the Human Predicament

Gazing over the countless fluctuations and transformations in Earth's multibillion-year history, I am struck by the unique strangeness of the present moment. We suddenly find ourselves sort of running a planet—a role we never anticipated or sought—without knowing how it should be done. We're at the controls, but we're not in control. This book is my view of how we got into this situation, and where that leaves us now.

A child of the space age, I grew up captivated by the romance of planetary exploration. My timing was right to become a NASA research scientist working in the new field of astrobiology, the scientific study of life in the universe. My participation in the spacecraft exploration of other planets has informed my view of our presence on this one. In these pages I'll describe how we humans fit into the long-term story of Earth, and how I believe this knowledge can help us to navigate our current time of environmental stress and uncertainty about the future.

Although climate change is the most obvious, it is only one of a large number of interconnected ways in which we have suddenly begun to modify the planet we inhabit. The scientific community is now converging on the idea that we have entered a new phase, or epoch, of Earth history—one in which the net activity of humans has become a powerful agent of geological change, equal to the other great forces of nature that build mountains and shape continents and species. The proposed name for this new epoch is the "Anthropocene" or the age of humanity. This concept challenges us to look at ourselves in the mirror of deep time, measured not just in decades or centuries or even in millennia, but over hundreds of millions and billions of years.

The realization that humans may have initiated a new geological phase also disrupts and rearranges all our systems of timekeeping. Think of these three timescales: human history with its long procession of civilizations, migrations of people, and waves of technologies, ideas, and modes of thought; geologic time with its ages and epochs marking fluctuations in environment, climate, and sea level, recorded in the layers of rocks; and, closely related, biological time, the evolving stages and forms of life logged in the fossil and molecular records. The Anthropocene represents a braiding together of these into one inseparable narrative. This is "the end of history," and of biology and geology—at least as separate stories. Have they become *irreversibly* entwined? It's conceivable that there may never again be geological change without human influence.

We are witnessing, and manifesting, something unprecedented and still completely unpredictable: the advent of self-aware geological change. As an astrobiologist, I study the possible evolutionary relationships between life and the planets that may host it. I see the Anthropocene as a tricky new step in the long, intricate dance between Earth and its biosphere that has been going on for four billion years. There are those who object to the name *Anthropocene* as being too self-aggrandizing and serving a destructive, human-centered viewpoint. But this epoch is well-named because it represents a recognizable turning point in geological history brought about by one species: *anthropos.* And our growing acknowledgment of this inflection can be a turning point in our ability to respond to the changes we've set in motion. I believe that, more than the extreme and undeniable physical changes to the planet being caused by human influence, it is this dawning self-recognition that is really fundamentally different and, ultimately, promising about the Anthropocene. Many species have changed the planet, to the benefit or detriment of others, but there has never before been a geological force aware of its own influence.

Popular treatments of this rich subject have too often focused on the ongoing argument about whether or not the Anthropocene should officially be recognized as a new epoch within the geologic timescale, and at what point it began. Yet that debate has spurred a fascinating re-examination of the various stages of encroaching and accelerating human influence on Earth since the end of the last ice age twelve thousand years ago. And the Anthropocene has now garnered the attention of people thinking far afield of geologic timekeeping. The sharp human torqueing of Earth's landscapes, habitats, and global cycles presents a range of challenges that go well beyond the physical and biological sciences.

In 2012, I saw an announcement that opened up a door for me. The John W. Kluge Center at the Library of Congress, in partnership with the NASA Astrobiology Institute, was advertising a new position, a chair of astrobiology. They would support a scholar in residence to pursue "research at the intersection of the science of astrobiology and its humanistic and societal implications." I fantasized that this might give me the opportunity to write the book that you are now reading. My proposal was accepted and I was selected as the inaugural chair. I was psyched just to have some time to read and write, but I soon discovered how truly fortunate I was. The Kluge Center is a scholar's dream: a supportive place to study, converse, and create, set within the ornate, infinitely stimulating palace of learning that is the Thomas Jefferson Building. I initially wondered if, as the only physical scientist there, I would be off playing in my own corner of the sandbox. I quickly learned that many of the humanists—historians, literary theorists, and theologians among them—were also keenly interested in exploring the changing relationship between humanity and the planet. My conversations with them[1] were great fun, and I hope that some of the broadened perspective they provided has found its way onto these pages. Also among the joys of writing in Washington is the group of local scholars I've come to know, who are studying the Anthropocene from many different points of view. Our informal "Washington Anthropocene Group" started meeting at the Library of Congress, and has since migrated to several museums and universities, often continuing the conversation at local bars. I'm the only astrobiologist, but we've had geologists, historians, geographers, anthropologists, paleontologists, materials scientists, and others. At conferences and on the Web, I've met philosophers, ethicists, artists, and economists, all wrestling with the concept of humanity as a planet-changing force. Whatever its fate as an official geological time period, the concept of the Anthropocene has sparked a thousand worthwhile interdisciplinary conversations about the human role on Earth.

I don't believe in "great person" theories of science history. Ours is a team sport, a transpersonal endeavor. No individual is essential, and all worthwhile ideas will emerge eventually. But the story is also full of brilliant heroes and colorful characters who carve out the specific, somewhat random path science takes toward understanding. Without bombarding you with names, I do want to share some tales of a few of the people I've worked with or been influenced by. In looking over my manuscript, I notice that a disproportionate number of these people are men. Among other things, this reflects a tremendous loss to science, of potential contributions by those who have not been as welcome in our field. Of the five scientists I think of as my closest mentors, all are men. This is not unusual for a scientist who started out in my field in the 1980s. We are making progress, and it would be much more unusual for a scientist starting out today. Though the history of science is dominated by men, the future of science is not.

One of the characters who winds his way through this book is astronomer Carl Sagan. My own personal connection with Sagan began when he and my father, a professor of psychiatry, bonded over the fact that, at one point, they were among the few Harvard faculty opposed to U.S. involvement in the Vietnam War. They became best friends and from the age of six I grew up around "Uncle Carl." When I was a kid, his tales of space exploration and speculations on human evolution cast a powerful spell. In college I worked summers in his laboratory at Cornell, zapping mixtures of gas to simulate the organic goop on Saturn's moon Titan and mathematically modeling potential alien signals. Later still, we worked together on a study of Earth's early climate that became part of my doctoral dissertation.

Carl also introduced me to Jim Pollack, who had been his first grad student at Harvard, and who decades later became my postdoctoral adviser at NASA's Ames Research Center, where many of the scientific breakthroughs I describe in this book took place. In the 1970s, '80s, and '90s, Sagan, Pollack, and other planetary scientists modeled runaway greenhouses, asteroid strikes, and nuclear winters, along with the possible "terraforming" of Mars, Venus, and other worlds. Long before the current debate on whether or not we should consider "geoengineering" our planet, purposefully tweaking it to compensate for our clumsy climate modification, Sagan and Pollack were considering how humans might transform planets both accidentally and deliberately. The research these scientists did on planetary climate catastrophes informs my view of Earth's history and the new kind of human-induced catastrophe it is now experiencing.

It may seem circuitous to begin a book about the human role on Earth by talking about the exploration of other planets. I will try to make the case, however, that looking homeward from the vantages we've gained through our interplanetary journeys gives us valuable perspective for navigating the planetary-scale changes we are now facing—and causing. We need to learn all that we can about how planets work, so we can make the transition from inadvertently messing with Earth to thoughtfully, artfully, and constructively engaging with its great systems.

The planetary perspective allows us to step away from the noise of the immediate present, to see ourselves from a distance, in time-lapse. When we do so, what we see is not just a problem facing our civilization but an entirely new evolutionary stage in the development of life. In seeing ourselves as a geological process, we also see the planet entering a phase where cognitive processes are becoming a major agent of global change. Earth's biosphere gave birth to these thought processes, which are now in turn feeding back and reshaping its changing planetary cycles. A planet with brains? Fancy that. Not only brains, but limbs with which to manipulate and build tools. We are just beginning to come to grips with this strange new development. Like an infant staring at its hands, we are becoming aware of our powers but have not yet gained control over them.

It's a challenging moment for human civilization. The great restless cleverness of our species has gotten us into a tough spot. Our collective actions, over which we often seemingly have little control, threaten the well-being of many of our fellow humans,

not to mention vast numbers of our more distant biological rel-
atives. Our very survival may be threatened. Paradoxically this
comes at a time, and even largely as a result, of unparalleled
advancement in our scientific and technological prowess. But
if we're so great at figuring things out and inventing solutions
to survival problems, how come we're in this mess? Part of the
reason we are, so far, stumbling through this transition is that
we have not yet seen it clearly for what it is. I think our fundamental Anthropocene dilemma is that we have achieved global
impact but have no mechanisms for global self-control. So, to
the (debatable) extent that we are like some kind of global
organism, we are still a pretty clumsy one, crashing around with
little situational awareness, operating on a scale larger than our
perceptions or motor skills. However, we can also see our civilization, such as it is, becoming knitted together by trade, by satellite, by travel, and instantaneous communications, into some
kind of new global whole—one that is as yet conflicted and inco-
herent, but which is arguably just beginning to perceive and act
in its own self-interest.

We have, unconsciously, been making a new planet. Our
challenge now is to awaken to this role and grow into it, becoming conscious shapers of our world. We have to "human up" and
accept the responsibility we've stumbled into. We didn't ask for
this. And we may not be up for the challenge, but at this point
we have no choice. However we got here, we find ourselves domi-
nating many of our world's systems and needing to learn, under
duress, how to handle that. It's a complex task, and we have to
learn how to do it without a manual and on the fly. Fortunately,
we may have a leg up. There are some ways in which our evolutionary history and our unique plasticity as a species may equip
us for the job.

Some of the most amazing things are very easy to take
for granted. How cool is it that I am writing this and you are

reading it? By thwacking away at these dirty little plastic keys in some coded pattern, I'm sending you a detailed message over malleable expanses of space and time. What is this magic? Our fingers were not evolved for this. They were made for grasping and throwing, touching and feeling, making tools, making dinner, making love. And yet here we are. These human hands and the nervous systems they are wired to are so flexible, seemingly made to be rewired as needed. Time and time again our species has escaped existential threats by reinventing ourselves, outsmarting the toolkit evolution gave us, finding new skills not coded in our genes to survive new challenges not previously encountered by our forbearers. We've bounced back a few times from the edge of extinction, and ultimately thrived due to our abilities to communicate, work collectively, adapt creatively to changing environments, and solve problems through technological and social innovation.

Now we need to do so again. To do the Anthropocene right, we'll have to use our innate skills at cooperative problem-solving and innovative tool use to become more effective global actors. We would so like to avoid this reality, this responsibility, but having seen what we've seen, and what we've done, we can't go back to being ordinary members of the animal kingdom. We also can't afford to ignore natural constraints, or imagine we are so clever we can just invent our way right past them.

I think we're junior apprentice planetary engineers. We can't shrink from this role and to do it well will require finding a much deeper understanding of both our world and ourselves. We have, without knowing it, thoroughly reworked our planet. Now that we are realizing this, what should we do? We cannot un-rework it. So we need to rework ourselves into the kind of creatures who can successfully play the role we've unwittingly assumed.

It sometimes seems to rub people the wrong way to say

anything sympathetic about humanity, positive about our potential influence on Earth or hopeful about our future. How could you not be shocked and alarmed by our jarring, accelerating influence on this planet? We rightfully feel some deep regret, and some shame, at how we have (not) managed ourselves.

However, our obligation now is to move beyond just lamenting the job we've done as reluctant, incompetent planet-shapers. We have to face the fact that we've become a planetary force, and figure out how to be a better one. By seeing our role clearly, we take the first step toward assuming our responsibilities. We need to pause and look up at the distant horizon to see where we really are. A planetary view of the human journey, where we take in the wider timescape, suggests that we are not stuck, just disoriented, not evil, just confused, struggling to find our way in a world increasingly of our own making, and confronting aspects of our existence for which we are not yet fully equipped. We've been building an expanding, rapidly changing civilization on a finite world with no long-term plan. Our challenge is to acknowledge, with clear eyes, the tough predicament we're in, and not to succumb to toxic fatalism. Our most valuable resources—creativity, communication, invention, and reinvention—are in fact unlimited.

Our oblivious stumbling into ecological and climate danger could be just a phase, characterized by inadvertent, clumsy human interaction with planetary systems. There is another way to do this. I'll describe some examples of a more thoughtful mode of engagement that, while still in its infancy, is clearly something we are capable of. I believe the true Anthropocene, what I call the "mature Anthropocene," characterized by intentional, deliberate interactions with the planet, is something that should be welcomed. Though it is only in its infancy, it can already be glimpsed. Awareness of ourselves as agents of geological change, once propagated and integrated, could provide us with the capacity to avoid doom and to take our future into our own hands.

Here I've been speaking of the human species as one thing, and human civilization as one thing, when obviously both are a great many things with diverse and complex histories. Yet since the problems we face are global, and to some degree our solutions must be as well, this begs the question of who we are really talking about when in this context we say "we." This is a dilemma I will revisit.

The planetary perspective provides a kind of out of body experience for us—hovering in orbit and watching ourselves sleepwalk through a slow disaster of our own making. Now, can this experience help us to shake ourselves awake? For virtually all of its history Earth has evolved without us, and we have always seen ourselves as autonomous actors on a passive planetary backdrop. But now we are beginning to see that our futures—those of humanity and of planet Earth—are tightly conjoined. If human civilization is to persist and thrive we will need a completely different view of our planet, and of ourselves, in which we acknowledge both our deep dependence and our increasing influence. We need visions of a future in which we have applied our infinite creativity to the task of living on a finite world, where we have embraced our role, become comfortable and proficient as planet-shapers, and learned to use our technological skills to enhance the survival prospects not just of humanity but of all life on Earth. My name for this vision is *Terra Sapiens*, or "Wise Earth."

A recent scientific breakthrough enriches this story: the exoplanet revolution. As we long suspected and have now confirmed, this universe is full of planets, orbiting nearly every star. It is now very close to inconceivable that we could be the only life, and only technological intelligence, in the universe. An interplanetary perspective on Earth's current dilemmas incites us to wonder whether parallel dramas may have unfolded on distant worlds. Do other planets also grow inventive brains that end up causing themselves problems? Do other species develop technology and build civilizations that create dangerous instabilities on their planets? How do they cope? Do planetary biospheres become self-aware? The Anthropocene leads us to a new way of looking at SETI—the search for extraterrestrial intelligence—which in turn illuminates changing notions about ourselves, how we fit into our planet, and what kind of future we dare imagine.

One hundred million years from now, what will our time have been? A brief climate spasm that Earth shrugged off and largely forgot, leaving a thin layer infused with bizarre plastic objects? Or the beginning of a lasting new phase when the biosphere finally woke up and adjusted its grip on the planet?

In 1758, when Carl Linnaeus—the Swedish botanist who invented modern biological taxonomy and classification— needed a name for the human species, one that distinguished us from others in the genus *Homo*, he called us *Homo sapiens*, or "wise apes." Is this a good name for us? Or was this wishful thinking? Linnaeus saw that we were good at problem-solving, and he was himself an exemplar of the scientific revolution in which the human intellect was rapidly teasing apart the mysteries of the universe like no other animal could. We had learned to "tame nature" in so many ways. We had reason to be proud of ourselves. But at that time Linnaeus could scarcely have conceived the kind of wisdom that we need now. We've been so successful at solving problems of survival in many different local environments that we've exponentially increased our numbers and global influence. We've spread ourselves so widely around the planet that we're now confronting a new environmental factor that may not be so easy for us to tame: ourselves.

What does wisdom mean for a species with the power to

change its home planet, affecting the fortunes and futures not only of themselves but of all life? Certainly it requires us to comprehend our role in the physical workings of the planet and to act in ways that are not obviously self-destructive. It is strange that geology and planetary science, investigations we began out of simple curiosity, have now become crucial forms of self-knowledge. Although science has helped us stumble into the Anthropocene trap, it also provides the tools with which, armed with wisdom, we could spring ourselves from it.

Can we live up to our name and create a wisely managed Earth? We have no choice but to try, because events we have already, unwittingly, set in motion are leading us inevitably toward a branching point between calamity and wisdom. Yet once we overcome the fear and embrace this new reality, then thrilling new future possibilities for our planet and ourselves open up before us.

Earth in
Human Hands

LISTENING TO THE PLANETS

And you may ask yourself, well, how did I get here?
—Talking Heads, "Once in a Lifetime"

Sidewalk Wisdom

For the past two years I've been living in a carriage house on Capitol Hill, seven minutes' walk from my office in the Library of Congress. It's a leafy, gentrified neighborhood of colorfully painted restored Victorian brick houses, and on sunny weekend days you can score all manner of quality household items on the wide cobbled sidewalks, choice stuff left free for the taking by residents too generous or busy to dispose of them otherwise. The boxes of books that appear weekly along East Capitol Street (a stack of used baby manuals, a pile of tomes on population dynamics from a year at an NGO, discarded cookbooks from someone's vegan phase) provide irresistible glimpses of the lives in these handsome Washington homes, and occasional serendipitous enlightenment.

Once, on Seventh Street, I spotted a sky-blue book lying on the redbrick curb. Upon approach, I saw it was *Anatomy of the Sacred: An Introduction to Religion* (sixth edition), by James C. Livingston, conspicuously folded open to a specific page. A message of some sort? I brushed off the dirt and read:

> Whatever the cause of our unease and strife, few thoughtful persons would deny that we humans have, over the millennia, sensed a tragic flaw, a falling short of our potential, a missing the mark of life as it was meant to be or should be. Something, we feel, is "out of joint."

"How true," I thought. "What is it about humanity?"

The very next day, I found a true treasure, containing one possible answer. In a carton of mostly forgettable, fluffy self-help books, out-of-date travel guides, and other discards, an old textbook caught my eye: *Psychology Today: An Introduction* (second edition, 1972). Reflexively I glanced at the list of contributors on the inside cover. At the very bottom, it read, "Chapter Introductions by Isaac Asimov." Say what? Are you kidding me? I took that one home. Each of the thirty-four chapters is preceded by a short, informal essay by the great science-fiction maestro. The writing is playful, irreverent, personal, and insightful. Asimov introduces each topic of human psychology by riffing on free will, the nature of consciousness, intelligence, and morality, and his famous "laws of robotics." These constituted Asimov's science-fictional device for ensuring that conscious robots could not behave dangerously. If we invent powerful, capable machines with smarts, awareness, and autonomy, we might, he imagined, build into the very fabric of their minds inviolable ethical precepts to be obedient, cooperative, and altruistic. Asimov was prescient. Today his laws come up frequently in discussions about how we can keep artificial intelligence from becoming threatening. Yet his robots were also foils for us and our troubled relationship with our own runaway cleverness. His stories explored the conflicts inherent in complex moral minds, whether evolved or engineered with technologically amplified capacities.

In one of these microessays, Asimov summarizes the way in which life has organized itself into a hierarchy of structures, with organisms at each level being built from collectives of simpler organisms. He looks at where we ourselves fit into this scale-spectrum of complexity. First there are the viruses, which are almost too simple to be alive. They seem more like molecules than creatures, each not much more than an encapsulated coil of RNA or DNA. The ultimate freeloaders, they hijack the machinery of more fully developed living cells in order to function. Then there are the simplest, undifferentiated cells, the bacteria. More complex cells are made up of various component parts, each of which is itself rather like one of those simple bacterial cells. These complex cells can exist as free-floating individuals or can be loosely bound with others in various colonial arrangements. Finally, as Asimov describes,

> cells can drown their individuality and abandon their free-living abilities in order to form a multicellular organism, which may be as simple as a flatworm, or as complicated as a giant sequoia, a whale or a man.

He points out, however, that the hierarchy does not end there. A multicellular organism by itself is often useless. It needs others to survive and propagate. All but the simplest reproduce sexually. Many are dependent for their survival on more complex social arrangements: a herd, a school, a flock, or, in our case, a tribe or society. Some (for example, the social insects) have such tight interdependence with other individuals that they form what might be called superorganisms, and in those cases, it may legitimately be questioned whether individuality resides in the organism or the hive.

Just as we individual humans are multicellular organisms, each an exquisite arrangement of forty trillion cells, with the whole being greater than the sum of its parts, so we cannot fully manifest our humanity, or survive for long, without joining together to form larger associations. Asimov sums it up:

> As individual multicellular organisms, however, we would be less willing to agree that a complex society or state is greater than the sum of the individual organisms making it up. We would be less ready to judge that it is a cheap price to give up our individualism to become part of a society.
>
> Yet the tug is there. It is as though we are at some stage of evolution between the multicellular and the multiorganismic.

It strikes me that the tension Asimov identifies here is at the heart of many of our political, economic, spiritual, and environmental struggles. We are trying to work out this question of how to thrive as individuals who also cannot exist without some larger cooperative order. Ever since the time most of us gave up hunting and gathering in the clans with which we roamed for almost all human history, and instead domesticated into villages, cities, and nations, we've been trying to figure out how to organize ourselves. Our initial success at banding together in groups to solve problems and invent technologies allowed us to change our world without knowing it. Now that we see what we've done, we face new challenges that require us to develop new cultural technologies, new civilizational solutions that allow for coherent action among much larger groupings. Yet we're not insects, and we cannot subsume ourselves to the hive. We need our individual freedom and creativity, but now, as never before, we are confronted with the need to make smart, coherent collective technological choices on a global scale in order to survive. It would not surprise me one bit if someday we learned that intelligent species on other planets also have had to struggle with some version of this same evolutionary dilemma.

This tension between our individual and collective selves is evident in the trouble we're having grasping the depth of our role in the changes now enveloping our planet. Somehow our minds and our inventions, our cognitive systems and their material extensions, have become part of the workings of this world. We got here because we are so good at communicating, cooperating, inventing, and making plans. But we never planned for this.

What a strange journey this planet has been on. Born as an inanimate orb, it hatched and became animated, with life soon spreading everywhere around and through it, and embedded deeply in its mechanisms. Later, life generated mind, which now is also becoming integral to the way the world works. What a thing for a planet to grow brains, sprout opposable thumbs, build machines, and start to rework itself. If we worry about our own creations getting out of hand and running roughshod, how do you think the planet, if it feels anything, feels about us? Well, this planet does feel, because we feel—and as the unwitting agents of this latest transformation, discovering it already in progress, we feel confused, like sleepwalkers awakening to find ourselves in the middle of rewiring our home without a manual.

How did we get here?

A Curious Anti-Accretion

The birth of Earth was fast and rough. In thirty million years, tops, nearly every bit of our planet came crashing down from space, falling together in a violent coalescence we call accretion.

In that same brief flurry, all the planets of our solar system self-assembled, congealing out of the dusty, spinning disk swaddling the infant Sun. In the surrounding blackness the swirling hot mess radiated and cooled. The temperature plummeted, and it started to snow. Sticky flakes of ice, rock, and metal clumped together into dirty snowballs. These grew slowly and randomly, until they were big enough to feel one another's presence. With the encouragement of gravity, the growth quickened. Pebbles gathered into cobbles, then boulders, then "planetesimals," little planets up to a thousand kilometers across. As these more massive bodies threw their weight around, the smash-ups gathered speed. Ever more furious collisions annihilated most planetary contenders and added their remains to the few survivors. Soon only a handful of new worlds were left, traumatized, orbiting the Sun amid a dwindling swarm of leftover debris.

With the accretionary scrum seemingly complete, four rocky worlds remained in the "habitable zone," the band of space favorable to surface water, where the Sun is close enough to melt ice but far enough away not to boil oceans. Of these final four, two were Venus-size and two Mars-size. Yet the orbits were not entirely settled, and there was still to be one very nasty encounter. With a furious, glancing blow, one of the smaller, Mars-size orbs smashed into one of the larger, Venus-size ones. Earth was made in this last apocalyptic collision, which spat an incandescent shower of melted and vaporized rock into an orbiting ring of fire that soon condensed into our freakishly large and close moon. After this final calamity, there remained Venus, Earth, and Mars, three newborn planets where oceans might gather and linger.

Largely molten from their fiery formation, thoroughly pocked with craters, and blanketed with hot steam, these newborn sibling worlds began to solidify and cool. The inner solar system was largely cleared of planet-forming bodies, but for another half billion years a diminishing fusillade of late hits would randomly trigger relapses to the primordial, molten state. In the calmer interregnums, various worldly activities commenced. Volcanoes erupted, storms raged, rivers flowed, and oceans filled. For a time, each of these three planets enjoyed a nearly identical childhood, with warm water lapping promising chemicals onto virgin rocks under the light of a faint young sun. Yet each would soon suffer a catastrophic change and head down its own unique path.

And on this one world, the one in the middle with the looming oversized moon, organic molecules at play along the shores and vents of the new oceans stumbled into configurations that could replicate themselves. That replication changed everything. These carbon copies enabled chemical memory, so that good designs persisted and better ones prevailed. This planet enmeshed itself in an evolving sequence of changing forms, enfolding and transmuting the entire surface and atmosphere. This planet came to life.

Eventually, four billion years later, again came something completely new: novel, stark, inorganic geometries began quickly remaking the surface. Suddenly, the nightside lit up in bright, spreading webs. Then there was a curious anti-accretion: some pieces of Earth, in seeming defiance of the laws of gravity, started launching themselves back out into the surrounding space from whence everything had once quickly fallen.

To Distant Climes

I don't know if space junkies are born or made, but my timing was good. Humans have, for fifty-three years at this writing, been launching small probes packed with instruments to investigate the other planets of our solar system. My life fits neatly into this new age of exploration. Space travel and planets were my obsessions from a very young age. *Mariner 2*, the first success-ful interplanetary spacecraft, arrived at Venus a week before I turned three, in December 1962. I don't remember that, but I sure remember Uncle Carl (Sagan) showing up at our house when I was nine with glossy eight-by-tens of the new pictures sent back from Mars by *Mariner 6* and *7*, depicting abundant cra-ters and none of the fabled canals. And I recall, from when I was eleven, the saga of *Mariner 9*, in which the true, haunting, faded glory of Mars was finally revealed.

I don't believe our destinies are written in the stars or plan-ets, but these early exposures to interplanetary mysteries had a strong effect on me. Despite the occasional temptation to put aside scientific instruments for musical ones, I was irresistibly lured by the siren songs of Titan, Iapetus, and Aphrodite, and I followed them from childhood fixation to adult profession as a comparative planetologist. It is a young field, a by-product (more than we like to admit) of the Cold War development of launch vehicles that allowed us to start lobbing scientific instru-ments toward the other planets in the 1960s.

Before we actually visited our neighboring worlds, prevail-ing viewpoints held that their climates were not so different from that of Earth. Many scientists argued that Venus was a water world, that the thick clouds enshrouded a lush, tropical, and possibly verdant planet. Mars was believed to be somewhat colder than Earth, with a thinner atmosphere, but pre–space age descriptions of the Red Planet often hinted or stated that the seasonally shifting surface features seen through telescopes were signs of vegetation.

Spacecraft data woke us rudely from these sweet dreams of almost-home. Early results from other planets carried sobering hints about the extremes that climate change can take. The very first thing we learned from any spacecraft at another planet, the first result from *Mariner 2* at Venus, was that Earth's sister is absurdly hot. Venus was radiating a frightening amount of heat from its surface. It took years, and several more missions, before we realized the true extremity of the Venusian climate. The Soviet Union had a remarkable string of fruitful Venus missions in the 1970s, including the only successful landings on that planet (still, to date), which left no doubt that Venus is an oven world, with a surface where no liquid water or living matter could exist. Heat is the enemy of complex organic molecules, and the stuff we're made of doesn't stand a chance anywhere within twenty-five miles of that searing surface. Yet early spacecraft results also hinted that Venus is a planet with a past, one that was likely cooler and wetter. We began to see Venus as a place where planetary climate had started off like Earth's but had gone completely off the rails, into the hot zone.

In July 1965, *Mariner 4* became the first Earth craft to fly by Mars. The pictures showed a lunar-like, barren landscape of craters and—not much else, just craters. Again, naïve expectations of a place where our kind of life could thrive were found wanting. The exploration continued, and after several flyby missions that only snapped pictures of small areas, *Mariner 9* was launched to Mars in 1971 with a promise to become the first spacecraft to orbit another planet, allowing us systematically to photograph the entire surface.

Mariner 9 made it there and entered orbit successfully—a bold new feat. Its global view had long been eagerly anticipated. So when the camera was finally turned on and the first images beamed down to Earth, scientists were amazed to see ... absolutely nothing: just a bland, featureless, fuzzy disk.

There was nothing wrong with the camera. Rather, Mars was in the throes of a global dust storm. Every Martian year, when it is Southern Hemisphere summer, large storms erupt. Afternoon winds stir up thick clouds of dust, much as they do in

Arizona. Dust absorbs sunlight, which further heats the summer air, driving faster winds, which whip up more dust. A Martian dust storm can quickly grow into a vast regional tempest, visible from orbit or even from Earth. Once in a while, every few years, a storm grows and grows until it becomes a planet-shrouding monster, engulfing and obscuring all of Mars.

We didn't know about this when *Mariner 9* showed up at the peak of one of these events, much to the astonishment of the mission scientists. They watched in wonderment and relief as, over a period of several weeks, the dust settled, revealing first the tops of a few giant volcanoes and, gradually, the rest of the Martian surface. This episode demonstrated dramatically that weather and climate on other worlds can be complex and changeable. As we'll see, the study of dust storms on Mars soon proved invaluable for understanding some big mysteries of Earth history as well as some troubling changes that human technology might yet cause on our planet.

Mariner 9's orbital mapping mission revealed a much more storied and mysterious world than earlier missions had suggested. Seen in its global fullness, Mars is varied, wind-whipped, ice-capped, and carved by ancient dried-up river channels, giant extinct volcanoes, and vast antediluvian eroded canyons. These all spoke of a long and dramatic evolutionary history, and hinted at ancient oases, wet and fertile, lost across the red sands of time, a promising but vanished Martian past that, half a century later, we are still seeking.

We began to understand that Mars, like Venus, had suffered through a climate catastrophe that long ago turned a once-more-Earth-like planet into an entirely different and more forbidding kind of place.

In comparison, Earth emerges as the true oddball of local space. Forested continents, rippling streams, and a flagrantly oxygenated atmosphere are, we have learned, far beyond the norm in this solar system. What happened here?

A New Science

Suddenly the planets were no longer just wandering lights in the sky, but diverse and mysterious locales. They were not just pixels but places, and the information was pouring in. This data explosion created a problem. Who was going to interpret it and figure out what it all means? Who was going to do the science? Nobody had ever studied other planets up close. It was not something astronomers did. They were good at using telescopes and studying stars and galaxies. Interpreting ancient Martian rivers or Venusian clouds would require—what? Geology? Meteorology? Chemistry? Geologists didn't think about Venus and Mars. Nobody was trained to work on these questions.

Making matters worse, science in the twentieth century had exploded and splintered into a heap of separate fields, each with its own specialized knowledge, techniques, culture, and language. Yet here was a task that required minds ready to bridge these conceptual fences. This challenge (and the new funding available from NASA) attracted a small group of intellectually adventurous, broadly educated young scientists. Gradually the new hybrid field developed a unified identity and took on a lasting name: planetary science.

It was still a young field in the early 1980s when I showed up in Tucson, one of the early hotbeds, to start work on my doctorate. Arizona's clear, dry, star-studded skies had drawn Dutch astronomer Gerard Kuiper, the founder of modern planetary astronomy and one of Carl Sagan's mentors, to establish his Lunar and Planetary Laboratory there, where in 1973 the

University of Arizona started the first Department of Planetary Sciences in the United States.

None of my professors there had degrees in planetary science. They were the first generation: chemists, physicists, meteorologists, and geologists; veterans of Apollo and the audacious first missions to the planets. They were still figuring out what planetary science was. Now, a generation later, most professionals in the field have planetary science degrees, so it seems to have become a real thing.

Arizona was a magical place to study planets. In addition to those profoundly deep skies, the geology is endlessly rich. A highlight of grad school was the weekend geological field trips spent camping out in the volcanic fields, steeply faulted mountains, and vast erosional canyons; learning to connect the outcrops, debris flows, and cliffs along which we hiked with the bigger, hidden picture of underground structures, geological maps, rock types, and planetary histories.* Having lived my whole life in flat, forested New England, for me these were like visits to a raw, exotic planet, or a whole series of them. I've spent most of my subsequent career poring over spacecraft data on computer screens, writing code, and running models of distant worlds, but I always feel grounded by these formative experiences clambering over Arizona rocks and dirt.

Comparative planetary geology[1] has allowed us to recognize and make sense of myriad Earth forms found on other planets: volcanoes, faults, landslides, frost heaves, folded mountain belts, and braided streams. These comparisons also sometimes yield fresh insights about Earth.

In the last half century, we've had several major conceptual breakthroughs in understanding our home planet, several big "aha" moments for science where the picture abruptly comes into

* Yes, we drank a lot of beer, too. It's what geologists do.

focus. It's not a coincidence that these were the same decades when we took our first tentative forays out into the unknown darkness beyond the terrestrial village. Yet note that "terrestrial" has two opposites, the other being "marine." The paleo-space age of the 1960s was also the decade when we completed much of the initial exploration and mapping of the deep ocean floor,* revealing a previously hidden half of Earth's surface. In visiting these terrae incognitae above and beyond the surface and the land, we first saw Earth whole, and could begin to see the path that planetary evolution took here as only one of many possible paths. This unleashed a burst of self-discovery. From this convergence of new knowledge and perspective emerged three new "big picture" insights into the nature of our world.

The first of these big ideas was the theory of plate tectonics, a once-fringe concept that has become *the* key to understanding how Earth works. Before we had this unifying vision, geologists studying disparate parts of our planet were like the proverbial blind men puzzling over an elephant. Now we see it as one beast, its seemingly separate mountain ranges, canyons, and ocean ridges, and its apparently independent patterns of earthquakes, volcanoes, uplift, and erosion, all revealed to be connected in one global system. The outer skin of our planet is broken up into about a dozen rigid pieces: the *plates*, the shifting shards of a broken sphere. These slowly drift around the planet, colliding, jostling, sliding against and elbowing into one another. Most geological activity can be explained by these interactions, the *tectonics*.

At the dawn of the space age, when *Sputnik* spooked America into jumping skyward, plate tectonics was still a suspect idea, considered controversial and largely rejected by mainstream geologists. Just over a decade later, by the time Apollo

---

* A massive global undertaking in which the contributions of the Soviet Union were key. As with the space race, ocean mapping was motivated largely by the quest for Cold War advantage but left a priceless legacy for all humankind.

astronauts were first driving buggies and birdies over lunar landscapes, and sending back the first whole Earth selfies, the idea was rapidly taking root as the unifying theory of the earth sciences. Like continental drift, which any child can see explains the neat puzzle fit of Africa and the Americas, once you get it, it seems obvious. You wonder how generations of brilliant scientists could have missed or doubted it. Not only did plate tectonics explain, under one theory, the history and geographical distribution of devastating earthquakes in Turkey, California, and Japan; the conical volcanoes and steep ocean trenches facing off across the west coast of South America; and the continuing, trembling uplift of the high Himalaya, but it went much deeper, showing how all these surface activities manifest hidden forces from Earth's insides. The tectonic plates, these sluggishly gliding rock rafts on which we ride out our hurried lives, are pulled along by currents arising far below in Earth's mantle, a vast realm of rock that is solid but squishy like butter. These inner rock flows are the convection pattern of Earth's interior, the original lava lamp where hot continent-size blobs rise and cool slabs of ocean floor sink into the mantle. Now we see that all Earth's major landscapes and their accumulated changes are part of one coherent, if chaotic, system, driven by the heat emanating from Earth's marrow.

A second big new idea involved the deep, integral role of life. The more we study the entangled history of Earth and its biosphere, the more we see how many features of our planet result from a complex relationship between life and the "nonliving" world. Our planet has been brought to a strange, anomalous state by a force that is (as far as we can tell) absent on the neighbors but that has come to dominate here: the life force. The interplanetary perspective helped us to see the deep, pervasive, planet-altering role of life. Perhaps life can even be best defined as a kind of transformation that might happen to some planets. This is the topic of the next chapter. But first...

The third big new idea about Earth, born of the space age, was the realization that our planet has not been so isolated from the rest of the solar system as earth scientists had long assumed. Ancient extraterrestrial collisions watered our world and seeded it with organic molecules. Occasional large impacts have continued to disturb and prod the evolution of Earth and life. As was the case with plate tectonics, this discovery encountered fierce resistance before being widely accepted. Once we broadened our perspective to include other worlds, and widened our temporal view to include the immense swaths of time laid bare on the many more dormant and aged surfaces we found out there, we realized that Earth, with its restless activity and eternally youthful surface, had been hiding something.

Worlds in Collision

One of our best grad school geology field trips was visiting Meteor Crater, the well-preserved, fifty-thousand-year-old, mile-wide hole in the Northern Arizona desert that has played a key role in allowing us to connect the geology of Earth with that of other planets. It's a surreal and sensational place to visit, but what made it especially memorable for us was scrambling through the crater with Eugene Shoemaker, the brilliant, amiable man who had unlocked its mysteries. As Kuiper was to planetary astronomy, Shoemaker was to planetary geology. He foresaw that the space age would revolutionize earth science, and he led the charge. He knew that crater as if it were his backyard, which it sort of was. As a grad student at Princeton in the 1950s, Shoemaker did the pioneering fieldwork that proved that, yes, the giant hole in the ground had been caused by an object crashing down from space, and not by a volcanic steam explosion. Later he established the astrogeology branch of the U.S. Geological Survey in nearby Flagstaff, and he recruited (or simply attracted through his magnetic, genial intellect) many of the first generation of American planetary geologists. What a treat to visit the crater with the man himself. Gene seemed familiar with every cobble and shrub. He had the geologic map burned into his brain, and could effortlessly point out the telltale signatures of the ancient impact. He showed us the upside-down sequences of sedimentary and volcanic rock, where the explosion had blasted out layers of bedrock and folded them back over the desert like a sheet. Belying the magnitude of his accomplishments and influence, he was an unassuming, humble, gentle, and extraordinarily nice guy, always listening to students and young scientists and offering helpful, constructive responses to their ideas. One of his intellectual legacies is a wider appreciation of the continuing influence of impact explosions on Earth and other planets. His loss in a car crash while out investigating a crater near Alice Springs, Australia, in July 1997 was deeply mourned by our community.

Gene Shoemaker's proof that Meteor Crater was the scar from a space impact provided the crucial link between Earth's surface and the craters we see on the Moon. This helped us realize the extent to which Earth has been repeatedly hit and changed by such events. The exploration of the Moon and planets made it obvious that all worlds endure these insults and that Earth could not have escaped. So geologists began to scour orbital and aerial photos and maps for circular features on Earth. Craters, it turns out, are not that rare, but many are no longer obviously visible from space. On a planet this geologically restless, craters do not stay in their pristine bowl shape for long. Many are squished, warped, scoured, partially filled, or fully buried. After a while, planetary geologists became adept at recognizing the signs of craters that had been altered almost beyond recognition. Often satellite imagery gives hints of possible craters, but it takes field expeditions on the ground to identify the mineral and geologic markers that Shoemaker and others discovered to discriminate an impact from volcanic processes that can also make circular features.

It wasn't until the 1980s that we started to fully recognize the important role of impacts in Earth history. The watershed was the "Alvarez hypothesis"—the proposal that a large asteroid struck Earth sixty-five million years ago, causing the "end-Cretaceous extinction." This was a controversial new solution to the long-standing mystery of what killed off those most famous of all extinct organisms, the dinosaurs. Yet it wasn't just the dinosaurs who got snuffed out at that moment in Earth's history. More than 70 percent of all species suddenly went extinct. Something happened to Earth that wiped out nearly everything. The mystery was solved by physicist Luis Alvarez, with his geologist son, Walter, and a team of other scientists. All around the world there is a thin layer of clay, about a centimeter thick, separating the older Cretaceous rocks, in which dinosaur bones are plentiful, from the younger Paleogene rocks, in which dinosaur bones are absent. The Alvarezes examined samples of that clay layer from a site in Gubbio, Italy, and found them to be heavily laced with the element iridium, which is known to be a marker of extraterrestrial origin. They proposed that this layer of sediment spiked with iridium was actually the fallout of dust thrown around the world by a huge impact explosion that caused the mass extinction. In June 1980 they published a paper in *Science* entitled "Extraterrestrial Cause for the Cretaceous-Tertiary Extinction." This idea crashed into the established geological worldview like a hypersonic rock out of the blue.

Earth, it seemed, was less isolated than we thought. Maybe

large impacts from space had had a repeated and significant influence on the evolution of our planet. People started referring to this as the "new catastrophism." This phrasing implied a reversal of the dogma of "uniformitarianism," the geological principle that holds that features on Earth result from the slow accumulation of gradual changes. The uniformitarian mind-set was itself the result of an earlier liberating, revolutionary triumph in the nineteenth century: the discovery of deep time. Geologists back then realized that the major features of Earth could be explained without biblical floods or other cataclysms, but rather as the accrued result of many smaller events and changes (storms, earthquakes, volcanoes, erosion, and subsidence) occurring over thousands of millennia. The uniformitarian mantra is "The present is the key to the past." You don't need abrupt or miraculous events to explain the origin of mighty mountains and valleys. You just need previously unimaginable expanses of years, not the hundreds or thousands of years of recorded history but the millions and billions of years recorded in rock layers and the nuclear clocks hidden inside minerals. Once we learned about deep time, then, we no longer needed catastrophic changes to explain the world, and catastrophism, when it was mentioned at all in our college courses, was taught as a relic of ancient, ignorant, pre-Enlightenment thinking.

Some of the initial resistance to the Alvarez hypothesis was surely due to this training. Yet as we connected the dots between Earth, the heavily cratered surfaces of other planets, and the stray asteroids and comets populating interplanetary space, we realized that this picture was incomplete.

One thing we discovered in exploring the solar system is how very young Earth's surface is compared to almost everywhere else. Most of the planets have been much more inactive than our oddball, hyperkinetic Earth, and their beyond-ancient histories are laid bare, billions of years unburied and raw on their surfaces waiting to be observed and interpreted.

When geology discovered deep time, it seemed to eliminate the need for catastrophic explanations. Then, however, space exploration uncovered deeper time. This expansive view revealed that, on even longer timescales, sudden, catastrophic events actually *have* played a big role. Of course we missed this before we looked beyond Earth. The typical interval between very large impacts is not only greater than the life expectancy of a geologist, and not only greater than the whole expanse of human civilization, it is also much longer than the few million years during which there have been creatures anything like human beings. There are, we started to realize, important events shaping Earth that come around only every few tens of millions of years. To understand this, we needed to expand our horizons and realize that, geological dogma notwithstanding, the present is *not* the key to the past—or, at best, it's an incomplete key. Guided by the flawed uniformitarian doctrine that what we can observe occurring now represents the complete set of geological processes, we had missed a major element of Earth's story.

Yet the Alvarez hypothesis remained controversial. Many old-school geologists and paleontologists strongly resisted the idea. There was a fair amount of scientific mud wrestling, with some insisting that the mass extinction had been caused by sea-level changes or a sudden outpouring of volcanic lava that flooded the climate with CO_2. Eventually, as usually happens, the net effect of this conflict was to move science along. Those defending the Alvarez hypothesis came up with better and better evidence for it. Then, in the 1990s, the hole left by the impact was finally found, a 190-mile-diameter crater called

Chicxulub buried along the coast of the Yucatán Peninsula in Mexico. Today there is little doubt that a large impact was the major cause of the end-Cretaceous extinction.[2]

When I started grad school in 1982, large-impact events were all the rage. The Alvarez hypothesis was the source of much buzz and debate at several of the first conferences I attended. Also around that time, another huge mystery was solved invoking a giant collision. The origin of Earth's huge moon* was still unexplained. This was an embarrassment—kind of pathetic, really, given our two decades of exploring the solar system, our collections of Moon rocks, and our (we thought) sophisticated understanding of planetary origins. How come we still couldn't explain our moon? We had theories, but none of them really made sense. Then, in 1984, scientist/artist/Renaissance man Bill Hartmann from the Planetary Science Institute in Tucson proposed that the Moon was formed in the last act of accretion, when proto-Earth suffered a collision with a Mars-size protoplanet. The idea works. It explains why Moon rocks, chemically, are like Earth rocks that have been thoroughly cooked. Sophisticated new computer simulations showed that such a collision would have surrounded Earth with a ring of molten material that, in a matter of months, would have coalesced to form a giant moon. Problem solved.

Also at that time, yet another wild idea involving worlds in collision was causing a stir: meteorites from Mars. About a dozen strange meteorites had turned up in collections around the world. Based on their rock types and ages, some researchers suggested the outlandish hypothesis that these stones came from the Red Planet. The community reacted with appropriate skepticism to what seemed like a marginal idea. Then it was

---

* Almost unique in the solar system as being a significant fraction of the size of the planet it orbits. "Almost unique" because Pluto, too, has a giant moon.

found that some of these rocks had little bubbles of trapped air that did not seem like Earth's atmosphere but that *did* exactly fit the atmosphere measured on Mars by the *Viking* landers. That clinched it. They really did come from Mars! Only, how did they get here? It was proposed that they were blasted off the Martian surface in large impacts, drifted through space, and eventually crashed to Earth. Computer models projecting the orbital path of shrapnel ejected from Martian impacts confirmed that it was not only possible but likely that some scraps of Mars should have ended up here.

The implications were enticing. The planets were not isolated. They had, over their lifetimes, been occasionally splashing material onto one another. Could this possibly mean that living organisms might have passed between them, seeding life across the void? Especially when you consider that the frequency of impacts was much greater when Earth and Mars were young (something we know from counting craters on different surfaces) and that some microbes and spores are impressively hard to kill, this seemed to change the equation when it came to figuring the odds of life on neighboring planets. Life might be something that could spread, naturally, between worlds!

When I was figuring out what I would do for my thesis work, all this was in the air: giant collisions had caused mass extinctions on Earth, had formed the Moon, had perhaps determined the nature and timing of the origin of life on Earth, and had blasted rocks, maybe even living organisms, between Earth and Mars. It was an exciting time to be studying planetary science, and Tucson felt like a place where important new ideas were being generated and discussed. One day Walter Alvarez stopped by to talk to my PhD adviser, John Lewis, about the atmospheric effects of large impacts, and I got to join in the conversation. The idea of disruptive invaders from outer space had stirred up ideas about planetary evolution, and the dust had not settled.

So I pursued my dissertation research on "when bad things happen to good planets"—my actual nerdy title was "Large Impact Events and Atmospheric Evolution on the Terrestrial Planets." This led me to a career in studying the (often catastrophic) ways that planetary climates and environments can change.

Building a Greenhouse

One of my thesis chapters examined how giant impacts might have changed the very early environment of Earth. I was learning about the evidence that such impacts were much more frequent four billion years ago, when life was first trying to get a cell hold. I wondered how that would have affected climate during that formative chapter, and realized that I could simulate this.

First I had to learn the basics of climate modeling, build a working model of Earth's early climate, and then see what happened when you disturbed it with multiple impacts that repeatedly filled up the atmosphere with massive clouds of dust.[3] So I built "baby's first climate model." The starting point was learning the equations representing the passage of different kinds of radiation through atmospheric gases, and translating them into some lines of computer code. My first model was pathetically simple by today's standards, but there is something very empowering about building a model that works. It was a rush when the numbers came out, I graphed them up, and they made a temperature profile that mimicked the actual atmospheric temperature structure of Earth's atmosphere. The model atmosphere got colder with increasing altitude, with the same slope as the actual real-world atmosphere, and then leveled off at the right height. When your model fits the data, the satisfaction has its own specific flavor. You feel that you have really figured out some small part of nature, not conquering it, but learning to hear its music, and to sing along. Now, having had that experience, whenever I see data representing temperature structure on any planet, they are forever slightly less mysterious to me. I simulated a physical process that was then obscure but has since become part of everyone's vocabulary. I produced a simple model of the greenhouse effect, the process by which an envelope of atmospheric gases heats up a planet's atmosphere.

An airless planet—and by now we've been able to study many of these—absorbs energy from visible sunlight, so it heats up until it is emitting the same amount of energy back into space in the form of infrared radiation. If you surround that planet with an atmosphere, it has more trouble radiating away the same amount of energy, so its temperature will rise by an amount that depends on both the overall thickness of the air and what portion of it is composed of "greenhouse gases" that are relatively opaque to infrared radiation. What makes a good greenhouse gas? Molecules such as CO_2, CH_4, or H_2O that are composed of three or more atoms. These more complex molecular structures have the arms-a-flying, twirling-hippy-dancing-to-Phish kinds of vibrational motions that get excited when hit with infrared light. Diatomic gases (those with just two atoms), such as O_2, N_2, and H_2, don't dance like that, don't absorb infrared, and are not good greenhouse gases. If a planet has a sizable atmosphere with a considerable portion of the big, floppy gas molecules, its climate will be warmed by a significant greenhouse effect.

The greenhouse effect is a blessing for us, one of the crucial factors that make life on Earth possible. Without about 90 degrees Fahrenheit of greenhouse warming, provided mostly by traces of carbon dioxide and water vapor, our planet would be permanently frozen, and quite possibly lifeless. The problem, of course, is that we've been loading the atmosphere with carbon dioxide (one of those flippy-floppy triatomic molecules) faster than it can be absorbed by the carbon cycling of Earth, and thus increasing the magnitude of the effect, with consequences that we can't fully predict.

I first learned about greenhouse warming as a subset of the knowledge we needed to understand planetary evolution. It's one of several aspects of climate physics that have (somewhat strangely for us students of planetary science) become quotidian. Not too long ago this was merely an arcane topic tackled by grad students learning the basics of planetary climate, but in recent years it has become familiar to anyone even remotely plugged in to the cultural and political issues of our time. I have marveled at how some of the esoteric concepts I absorbed in grad school, when it seemed nobody cared, have now become commonplace in the cultural lexicon. Terms such as *polar vortex* (a whirlwind pattern found around the poles of Venus, Mars, Jupiter, Saturn, and Earth), *albedo* (the total fraction of light reflected by a surface), and *cloud forcing* (the ability of cloud formation to magnify or dampen a change in climate), terms that seemed then to be part of our secret nerd language, are now bandied about in newspapers and online political forums. Now the greenhouse effect is everywhere. Community activist groups hold teach-ins about it. Schoolkids learn it in their science lessons. Pundits shout about it on television. This is all for the good. Despite the polarization, the noise, and the fury, this diffusion of science concepts from the priesthood to the masses is necessary and long overdue.

Today, climate change is on everyone's mind and screens, but few realize how our fundamental understanding of climate and its potential for change or stability has been enriched through comparative planetology. As we look around the solar system, we see a range of planetary atmospheres, each with its own version of the greenhouse effect, some more extreme than Earth's and some less. Historically, our gradual discovery of conditions on the neighboring planets has been intertwined with our increasing ability to understand, reconstruct, and predict climate change on Earth.

The Discovery of Global Warming

The first to predict that the burning of fossil fuels would warm Earth was Svante Arrhenius, a Swedish chemist who also tried to work out climate conditions on Venus and Mars. In 1896 he crudely calculated the magnitude of global warming that would result from human CO_2 emissions, and concluded that it would likely be a wonderful thing for life and civilization. After winning the Nobel Prize in chemistry in 1903, Arrhenius felt free to publish his wide-ranging theories about extraterrestrial life and the environments of other planets. His best-selling 1918 book, *The Destinies of the Stars*, is an erudite romp through mythology, cosmology, astrophysics, climate theory, terrestrial ice ages, environmental history, and comparative planetology, and includes discussions about the likely climates on Mars and Venus, possible life on those worlds, and the future climate and habitability of Earth. He concludes that

> a very great part of the surface of Venus is no doubt covered with swamps, corresponding to those on the Earth in which the coal deposits were formed, except that they are about 30°C. warmer...analogous to conditions on the Earth during its hottest periods. The temperature on Venus is not so high as to prevent a luxuriant vegetation...The vegetative processes are greatly accelerated by the high temperature. Therefore, the lifetime of the organisms is probably short. Their dead bodies, decaying rapidly, if lying in the open air, fill it with stifling gases.

Arrhenius's vivid description of a swampy, tropical, vegetated Venus proved very influential on scientists and science fiction writers throughout much of the twentieth century. It was pretty much the state of the art until better data, from telescopes and then spacecraft, started to give us hints of a much more extreme climate.

The first good clue came in the 1930s, when American astronomer Theodore Dunham Jr. was observing stars with an infrared spectrometer, which can measure the intensity of light at different colors, revealing the chemical composition of the source. One night, out of curiosity, he turned his telescope toward Venus, and was surprised to find that the bright disk was dark at two specific colors corresponding to absorption by carbon dioxide at high pressure. Realizing that Dunham's discovery would have big implications for climate on Venus, planetary astronomer Rupert Wildt, in 1940, did the first calculations of greenhouse warming in a thick atmosphere loaded with CO_2. The results showed that the temperature must be above the boiling point of water. Romantic visions of Venus as a nearby, moist jungle world began to evaporate.

In 1952, a remarkable Symposium on Climatic Change was held by the American Academy of Arts and Sciences in Boston, with contributions from a wide range of experts in earth, space, and life sciences. The gathering was organized and chaired by Harvard astronomer Harlow Shapley, a sort of Carl Sagan of the 1950s, who combined rigorous astronomical research with daring cross-disciplinary studies, humanitarian and political activism,* and the popularization of science. Shapley's greatest scientific contribution was to do for the solar system what Copernicus and Galileo had done for Earth: he showed that the Sun is not at the

---

* Shapley was accused of being a Communist Party member by Sen. Joseph McCarthy and investigated by the House Un-American Activities Committee.

center of the Milky Way galaxy but is orbiting at its outskirts. He published numerous popular books explaining astronomy and cosmic evolution for the lay reader, speculating on the possibility of extraterrestrial life and waxing poetically about the cosmic meaning of human existence. His popular 1963 book, the Saganesque *View from a Distant Star*, begins with the phrase "Mankind is made of star stuff." The impressively multidisciplinary proceedings of his Symposium on Climatic Change reveal that before the space age, some astronomers were already thinking about Earth's climate evolution as an extension of space science requiring collaboration with other fields. A round table discussion was held on "Climatic Conditions Required for the Origin and Continuance of Life on This and Other Planets." In Shapley's own account,

> The session was serious, deep, cheerful, and inciting, for on the inner circle of chairs we had representatives of biochemistry, paleontology, astrophysics, meteorology, geophysics, physical chemistry, geology, and astronomy, and they were not amateurs.

In his talk, Shapley discussed what was then known about the climates of Venus and Mars, calculated the likelihood of life-supporting climates on planets around other stars (extremely pessimistic compared to today's best estimates), discussed the climatic requirements for the origin and sustenance of life on Earth, pondered whether the extinction of the dinosaurs had been caused by climate change, and reviewed the history and theory of ice ages. Finally, he speculated on the causes of the recorded warming of Earth during the first half of the twentieth century. In his conclusion, he stated that

> the growth of industry, and the increasing population of fuel-burning inhabitants of the earth, have in the past seventy

EARTH IN HUMAN HANDS

years put enough additional carbon dioxide into the atmosphere to affect (perhaps quite slightly) the atmospheric control of climate.

The speakers recognized that climate is a result of a planetary radiation balance that is susceptible to change by a variety of provocations. The fact that the species doing all the talking was the same one starting to perturb that balance was mentioned, but only barely.[4]

In 1963, a decade after Shapley left Harvard, there arrived there a young, charismatic astronomy professor who was also fascinated by the biological implications of climate evolution on Earth and other planets. Carl Sagan was in many ways Shapley's successor at Harvard, and even taught some of the same courses to a new generation of undergraduates. Sagan was handsome, hyperconfident, and intellectually audacious, to a degree that struck some of his colleagues and mentors as reckless. In addition to his research on planetary science, he pursued collaborations with biologists in search of the origins and cosmic distribution of life, helping to establish what became known as exobiology and later astrobiology. He played a leading role in the fledgling community of SETI (the search for extraterrestrial intelligence) and was politically active, opposing the war in Vietnam and becoming faculty adviser to the leftist Students for a Democratic Society. It was these activities that likely got him booted out of Harvard when he came up for tenure in 1968.

As part of Sagan's PhD thesis at the University of Chicago, he had revived and extended Rupert Wildt's theory of climate warming on Venus, adding in water vapor as a second greenhouse gas, one that absorbed additional infrared radiation and produced higher temperatures than CO_2 alone. He defended his thesis in 1960, two years before the age of planetary exploration began with the launch of *Mariner 2*. Sagan arrived at

Harvard right after *Mariner 2* had arrived at Venus, reporting back that conditions there were hellishly hot. Building on these results, his first graduate student there, James (Jim) Pollack, under Sagan's guidance, built the most sophisticated Venus greenhouse model to date. Sagan and Pollack would become lifelong collaborators, and during their three years together at Harvard, they produced a string of important papers on the atmosphere, clouds, and climate of Venus.

By the time Sagan was ejected from Harvard's orbit and captured by Cornell in 1968, he probably knew more than anyone alive what the heat-absorbing properties of CO_2 and water vapor could do to climate on an Earth-like planet. He was also becoming more visible—as a popularizer, activist, and public spokesman for science—and concerned about widespread public ignorance of science in a democratic society where, increasingly, scientific literacy was key to understanding important issues. Once the science of climate modeling became good enough (in the 1970s) to indicate that global warming from industrial CO_2 was cause for serious concern, Sagan became the first person to speak about the problem effectively to the American public; he brought it up often and with urgency in the 1980s. He inspired Al Gore to focus on the topic, and in this way can be said to have helped plant the seeds for our current noisy cultural debates on the issue.

The Cold and the Dark

Shortly after Sagan left Harvard, Jim Pollack took up residence at Ames Research Center, a sprawling NASA facility housed in a former naval air base built on the bay-fill flats jutting into San Francisco Bay, a few miles north of San Jose. Geographically, it would be accurate to describe it as being nestled within Silicon

Valley, but Ames was there first, since the founding of NASA in 1958, long before anyone imagined such a thing as a personal computer or an industry built around it.

Much of our knowledge of planetary climate has come from Ames, and this can be traced largely to the outsize influence of Jim Pollack. Pollack was knowledgeable about seemingly everything in planetary science. He was ubiquitous at conferences, where he always sat in the front row (like his mentor Sagan) and seemed to ask the first question after every single talk.

Jim Pollack in his office at NASA Ames.

One of the most thrilling intellectual experiences of my life was, after grad school, going to work at Ames as a postdoc, with Jim Pollack as my adviser. Pollack was a classic nerd in the most loveable way, one of the quirkiest people I've ever known. His office looked like it had been decorated in 1970, on the day he moved in, and never altered since. It had the feel of a place where minds ranged freely across the universe, paying scant attention to the here and now. Atop his filing cabinet was a little plastic diorama of an Apollo lunar landing site, complete with astronaut and flag, and tacked on the walls were several large posters, yellowing with age, of cute lions and tigers lounging like pussycats. I don't know if Jim identified with these big, mellow felines, but I thought he was one of them: he was at the top of the scientific food chain, but I never saw him bare his claws. Once, Carol Stoker, another planetary scientist at Ames, told me, in shock, that Jim had complimented her on something she was wearing. "I always thought that I could walk in there completely naked," she explained, "and he wouldn't notice." Jim had a habit of absentmindedly tearing off pieces of Scotch tape and sticking them on his desk as he spoke to a visitor or on the phone. His desktop, showing years of tape accumulation, had a semitranslucent sheen. He spoke so slowly and deliberately— sounding a lot like Cheech or Chong without the cannabis haze—that you were always tempted to finish his sentences for him. This, you quickly learned, was a mistake, because his response to any interruption was to begin the sentence over again. You had to let him proceed at his own pace, but if you listened, you always learned. I can still picture his diminutive handwriting on the dusty blackboard in his little office as he drew diagrams of cloud layers and wrote equations describing the clouds' response to radiation.

Building for decades on his thesis work with Sagan, Jim had become a leading expert in "radiative transfer," a key tool for calculating the magnitude of the greenhouse effect and other aspects of planetary climate. Radiative transfer is the mathematical treatment for how different kinds of radiation pass through different kinds of atmospheric gases and particles (a word that, in Jim's quirky pronunciation, rhymed with "sparkles"). His numerous students and mentees spent hours imitating his singular mannerisms and patterns of speech. This was done for amusement but also with great love. Maybe my memory is hazy or rosy, but I think I can say, as much as anyone I have ever known, that everybody loved Jim Pollack. He was also openly gay at a time when that was nearly unheard of in our field.

One of Jim's intellectual passions was figuring out general principles of climate evolution on Earth-like planets. Fortunately for me and many others, he was an enthusiastic and patient mentor. There was usually a line of acolytes outside his door, as his assistants, students, and postdocs on the many projects he juggled queued up for an audience with the master. Once you made it inside the room, there was no hurry, though. Learning radiative transfer from Jim felt like training with a Jedi master. Together we worked on a number of problems involving the influence of clouds, dust, and impact events on climate and planetary evolution. Jim was a "great attractor" of planetary climate. Through him I met many of the people at the interface of planetary exploration and climate studies and saw firsthand the cross-fertilization between earth science and planetary climate science that has helped us to expand our ideas, gain confidence that we haven't missed anything important, and sharpen our tools as we continue to study the phenomena that all these worlds have in common.

At Ames for three decades, until his untimely death to spinal cancer at the age of fifty-five in 1994, Jim led, trained, inspired, and guided a small army of researchers at the forward edge of interplanetary studies. His mentorship touched an impressive number of careers. Twenty years after his death, many leading voices in Earth and planetary climate studies count him as a major influence.

During the 1970s, Pollack and his colleagues integrated the results coming back from the first wave of interplanetary space-craft into new models of planetary climate. They followed a path of discovery that ultimately led through Martian dust storms, historic volcanic eruptions, dinosaur extinctions, nuclear war, primordial climate history on Earth and Venus, a smoggy moon of Saturn, and current debates about "geoengineering."

Comparative planetary studies have increased our ability to model atmospheres in physical condition not found on Earth today. This helps us understand possible past and future environments on Earth. Sometimes what we find on other planets provokes us to ask questions about Earth that might not have otherwise occurred to us. Good science often starts with "what if." What if massive amounts of light-obscuring dust suddenly filled a planet's atmosphere?

In 1971, Jim Pollack arrived at Ames, and *Mariner 9* arrived at Mars amid that intense global dust storm. The dust was an annoyance for the geologists who were hungry to study the surface features of Mars. For Pollack, though, it was an irresistible scientific mystery. He wanted to understand the genesis and evolution of Martian global dust storms. Bringing to bear his great insight into radiative transfer, he teamed up with one of Sagan's first grad students at Cornell, Brian Toon, whose 1975 thesis was on "Climate Change on Mars and Earth." Toon is a master of microphysics. That's the part of climate modeling where we consider populations of aerosols, that is, tiny particles suspended in a planet's atmosphere. A microphysical model simulates all the things that can happen in aerosols' interactions with one another (colliding and merging, dissolving in raindrops, falling to the ground, etc.) and with radiation, and reveals how, in aggregate, all these minuscule events affect the overall climate. Toon had done a lot of early development work on a model called CARMA (Community Aerosol and Radiation Model for Atmospheres). If you want to include clouds, hazes, or dust in your climate model, you need a microphysics code, and CARMA, which has Toon's fingerprints all over it, is still one of the most widely used for climate modeling on Earth and other planets.

Pollack and Toon added large amounts of dust into a Mars climate model to see what would happen. What happened was that, as dust absorbed sunlight and reradiated into the surrounding air, the atmosphere heated up. At first this stirred up winds, kicking up more dust, and feeding back to grow the storms into gargantuan size. Yet, as the dust got thicker and sunlight was blocked from penetrating into the depths, the lower atmosphere cooled off, muting and then eliminating the greenhouse effect. Eventually, when the global pall got thick enough, this shadowing evened out the surface temperature, which calmed the winds, which allowed the dust to settle, causing the model storm to come to an end in much the same way the actual Martian storm had.

In elucidating the mysteries of Martian global dust storms, Pollack and Toon learned a lot about the effects of dust on planetary climate, and further developed modeling tools that soon came in handy for some problems closer to home. They used CARMA to study the effects of large volcanic eruptions on Earth's ancient and modern climates. In the late 1970s they teamed up with their mentor Carl Sagan to study how humans, over the history of civilization, had changed the reflectivity of Earth and thus the climate. Using the techniques, perspective, and language of planetary exploration, Sagan, Toon, and Pollack published a paper in *Science* in 1979, long before climate change became the issue it is now, entitled, "Anthropogenic

Albedo Changes and the Earth's Climate," in which they discussed how changing land use practices by human societies (starting with fires set by hunter-gatherers, expanding with the Agricultural Revolution, and accelerating with the Industrial Revolution) had likely been influencing our planet's climate for a very long time.

Then, in 1980, Walter and Luis Alvarez published their earth-shattering proposal for the end-Cretaceous mass extinction. The evidence was solid that there had been a massive impact at just the right time in Earth's history to coincide with the extinction event. It stood to reason that such a large impact would have caused calamitous environmental changes, but what exactly would such an object have done to the planet and its life? What was the kill mechanism? One effect of such an impact would have been to throw massive quantities of dust into the atmosphere, which would quickly have spread around the world by winds. We know this happened because the centimeter-thick layer of clay that marks the end of the Cretaceous sequence of rocks (the same layer where the Alvarezes found the extraterrestrial "iridium anomaly") is the remnant of this pall, the globally distributed fallout left when the dust settled.

Could the impact-generated dust cloud itself have brought about enough climate change to cause the mass extinction? Pollack and Toon, armed with knowledge and models from Martian dust storms and terrestrial volcanoes, were well positioned to attack this problem. They teamed up with meteorologists Tom Ackerman and Rich Turco, computer code specialist May Liu, and a young Ames postdoc named Chris McKay, who had already, as a grad student in Colorado, made a name for himself as an expert on Mars and an activist for future Mars exploration.

To start, they needed to know how much dust would actually have been blasted into the atmosphere and spread around by such a large explosion, and how finely ground that dust would have been. The best data came from nuclear test explosions. So Toon and colleagues immersed themselves in studies of bomb-generated dust plumes and figured out how to scale those numbers up for the much larger explosions caused by giant asteroid impacts.

In 1982, they published their results: impact dust would have drastically reduced sunlight reaching Earth's surface, dramatically cooling and darkening our planet for several years. In a suddenly dim and wintery world, many organisms would have frozen to death. Most of those left would have starved because photosynthesis would have been shut down. This work established that an impact-generated dust cloud was a likely cause for the mass extinction, and stands today as the seminal work on the climate effects of a large-impact-generated dust cloud.

All that immersion in nuclear test data and visions of a world suddenly gone dark fed into another project the team had started. Remember, at this time the Cold War was still raging. This conflict had provided the rocketry and impetus for the first wave of planetary missions, which had given us the tools to imagine and model the climates of past and future Earths. The world had become used to the superpower standoff. Yet the hidden machinery of global mass destruction—the underground silos with their thousands of missiles, armed with multiple independently targetable city-incinerating bombs; the nuclear submarines stealthily prowling the deep; the bomber squadrons at the ready, practicing for a day nobody wanted and everyone feared, when they might, through accident, escalation, miscalculation, or madness be pressed into service—still sat and waited, occasionally snapping to heightened attention when the United States and the USSR squabbled or their proxies came to blows.

Turco, Toon, Ackerman, and Pollack realized that with their studies of Martian dust storms, ancient mega-volcanoes, and Cretaceous impacts, they had developed the tools needed to model the climate effects of a nuclear war. Looking into the problem, they realized that smoke would cause a more severe climate effect than the dust from the bomb explosions. The smoke from burning cities would rise to the stratosphere in giant thermal plumes and quickly spread around the entire Earth. When they included the massive petrochemical fires that would be ignited in such a conflict, they found that the nuclear destruction of as few as one hundred cities would spread enough soot and dust into the stratosphere to plunge the world into a deep freeze, similar to the aftermath of a large asteroid impact. They named this "nuclear winter."

This work proved to be problematic. In addition to the technical challenges of ensuring they had done the modeling right for what would clearly be an incendiary result, they realized their results had huge implications for defense policy. They basically showed, scientifically, that current U.S. defense postures were suicidal. They had followed their intellectual muse into political territory that NASA, the hand that fed them, might perceive as a bite. Could they get away with publishing this work and still retain NASA support? Toon and Pollack enlisted Sagan's help. Not only did their old mentor have a valued big-picture perspective on this type of scientific problem, but he had also become, by that time (post *Cosmos*), a big shot within NASA and beyond. Sagan added his gloss to the science, but mostly he helped navigate the NASA politics and prepare for the global cloud of controversy they knew would be raised by the publication of this work. They added Sagan's name to the paper, which subsequently and forever more became known as the TTAPS study (pronounced "Tea Taps"), for Turco, Toon, Ackerman, Pollack, and Sagan. In late 1983 the TTAPS paper was published in *Science* with the title "Nuclear Winter: Global Consequences of Multiple Nuclear Explosions."

As predicted, the work was controversial. Hawkish politicians denounced it as politically motivated, often in combination with personal attacks on Sagan. Some scientists attacked the work on technical grounds, nitpicking various model assumptions (which is a normal and healthy part of the process). Sagan went on the offensive, arguing in lecture halls and TV studios that nuclear winter theory demanded deep rethinking of the strategic postures of both superpowers. He debated Edward Teller in front of Congress and led a delegation to meet with the pope.

Whatever one thought about the details of their physical models, the TTAPS study and the wider debate it ignited helped drive home the absurdity of nuclear strategies dependent on massive deterrence. The United States and the USSR had created a situation where even a limited nuclear conflict would cause a climate disaster that could quite possibly, among other things, collapse global agriculture, dooming civilization as we know it. With these weapons, there was no destroying your enemy without also destroying yourself. It brought to mind Stanley Kubrick's brilliant Cold War dark comedy, *Dr. Strangelove*, in which the Soviets create a "doomsday machine" that will detonate if a nuclear war starts, rendering the entire world uninhabitable. The TTAPS nuclear winter study revealed that we had, unwittingly, built such a machine. These results were widely discussed in the security communities of both superpowers, and are often cited as helping to motivate the partial disarmament that both sides undertook as the Cold War wound down.

Anti-Greenhouse

In all these studies, Pollack and his collaborators were discovering variations that can be induced, by changes in quantities of gases or suspended particles, in a planetary greenhouse. Then, by extending this work to the outer solar system, they discovered that planets can also have an anti-greenhouse.

A billion miles from here is another world where cold rain falls through nitrogen skies. Out in the realm of gas giants and ice dwarfs orbits Titan, Saturn's strangely Earth-like moon. It's the only other world we know of with a thick atmosphere made mostly of nitrogen. Complex climate feedbacks seem to have played a central role in Titan's evolution, and understanding all the competing, interacting processes is an irresistible challenge for comparative climatology. While I was at Ames, Pollack's research group was modeling the hazes of Titan, and as so often happens when we study alien atmospheres, this work shed new light on a phenomenon that is important for understanding the past, present, and possible future climate of Earth.

The second most abundant gas on Titan is methane (CH_4, otherwise known as "natural gas"). It plays the same role there that water plays on Earth. It's so cold on Titan that methane condenses into liquid and rains out on the surface. There it carves steep river valleys, erodes desert plains with occasional flash floods, pools in great lakes, and evaporates into clouds, forming weather fronts and storm systems that, again, bring rains to the icy plains. On Titan, this "methalogical cycle" (an analogy to the hydrological cycle that defines so much of Earth's character) shapes surface landforms that seem dreamlike to our earthly senses—strange yet oddly familiar. Until recently, however, when NASA's *Cassini* spacecraft showed up with infrared and radar eyes, these details were completely hidden to us. They were obscured by another feature that also results from all that methane in the air. You'll never see that interesting surface from orbit or through a telescope. When you look at Titan, all you see is an orange-brown, fuzzy, featureless ball, not unlike Mars in the throes of its worst global dust storms. Also, if you lived on Titan, you would never see the stars, or even giant Saturn hanging in the sky. The upper atmosphere is permanently shrouded in a thick brown haze that, we've learned, is smog made up of organic molecules. (And you thought mid-twentieth-century Los Angeles was bad!) This organic haze is produced when the ubiquitous methane molecules are ripped apart by ultraviolet sunlight* and the desperately unstable molecular fragments find one another, eagerly recombining to make various organic molecules. Titan's upper atmosphere is a nonstop factory of complex organics, which both shroud this world in its permanent smoggy haze and snow down on the icy surface. There they gather in vast dune fields that blow around in the nitrogen winds and dissolve in the methane lakes. The presence of all these organics on Titan resembles our picture of the primordial Earth and the conditions that led to the origin of life. It seems to present a freeze-dried portrait of a crucial lost phase in our own biological origin story. This is one reason we astrobiologists are obsessed with Titan. Another is the fascinating and complex climate balance.

Like water on Earth, methane on Titan is a strong greenhouse gas that evaporates and condenses, amplifying climate change and causing strong feedbacks. In studying the climate balance of Titan, Pollack's group of researchers discovered some interesting and seemingly paradoxical effects. Titan, by our standards, is really cold, at -290 degrees Fahrenheit. Without any methane greenhouse, it would be much colder still, by about 22 degrees. Yet, if we put all that methane into a basic climate model, we find that there should be about twice the level of greenhouse warming that is actually observed. What's missing from the model?

This question led to the discovery of the "anti-greenhouse effect."[5] It has to do with all that orange organic haze suspended

---

* And also by energetic particles caught in Saturn's strong magnetic field.

in Titan's upper atmosphere. It turns out that the passage of radiation through this haze is having an effect exactly opposite from that of a greenhouse gas: it blocks visible light but allows infrared light to pass through. Such a haze will prevent sunlight from warming a planet yet will allow the planet to cool efficiently into space. The effect on Titan's climate is to negate about half the value of the greenhouse warming caused by methane. The Titan anti-greenhouse effect turns out also to be a pretty good match for what happens to Earth's climate in the immediate aftermath of a huge asteroid impact or giant volcanic eruption and what would happen to it in the aftermath of a nuclear war. So the term *anti-greenhouse effect*, first used to describe climate processes on Titan, is now commonly invoked in modern descriptions of nuclear winter and other aspects of climate change on Earth.

Now, when you hear some people advocating or warning against "geoengineering" Earth by spraying sun-blocking aerosols into the upper atmosphere, they are proposing to induce a process that is constantly at work on Titan. I'll return to the physics, and the wisdom, of such an anti-greenhouse project in chapter 4.

Climate Catastrophes in the Solar System

The interplanetary approach to climate allows us to ask a basic question: why is Earth like this and the other planets so different? The deep-time perspective afforded by planetary exploration lets us pursue this question all the way back to the origin of the planets, and the divergent evolution of their atmospheres and climates.

Spacecraft studies have provided plenty of evidence that the extreme climates we see today on our neighboring planets are completely different from their earliest climates. Venus, Earth, and Mars have each experienced catastrophic climate change, and as is often the case with siblings, their personality differences have become accentuated over time. Now each seems almost a caricature of a different planetary type: one hot and dry, one cold and dry, one water-soaked and thoroughly infested with life.

Climate is complex, messy, and hard to predict, in large part because planetary climate systems are full of feedback mechanisms, which tend to dampen (negative feedback) or amplify (positive feedback) any initial change. Positive feedbacks will destabilize a system, meaning that small changes can grow and knock it into a completely unrecognizable state. Negative feedbacks do the opposite: resist change and add stability.* Examples of both negative and positive feedback can be found everywhere in our daily lives. Just this morning a friend with an infant child joked that he was too tired to remember how to make coffee. This is an example of a negative feedback that in theory (but, fortunately, not in practice) could keep him in a stable state (tired and uncaffeinated) all day long. Now, suppose he made one pot of coffee and started to wake up. If this made him restless, he might get up and make another pot. The more nervous he got, the more coffee he'd make and consume, which would only make him even more hyper. This is an example of positive feedback, which leads to instability and a rapidly changing state. Yet suppose he reached a point where his hands were so jittery that he could no longer operate his coffee machine. Then another negative feedback would kick in, which would stabilize him at the maximum level of caffeination at which he

---

* As a result, it is usually positive feedback that has the most "negative effects," as in "undesirable," and vice versa.

could barely make coffee. No doubt tomorrow morning he'll again be tired.

Earth's climate has fluctuated through the ages. Ice ages arrive every few hundred thousand years, and are themselves punctuated by multithousand-year periods of slight warming, called interglacial periods. Right now we're in one of these interglacials, which started about twelve thousand years ago.

A few times our planet has entered much more prolonged and severe global freeze-overs during which it looked like the ice world of Hoth from *Star Wars*. In other intervals it has warmed to a globally ice-free, tropical state. Somehow Earth has always bounced back from these extremes. There are many processes that can doom a planet to fire or ice. As we learn about the history of Earth's fellow travelers, and become more sophisticated about planetary evolution, what really stands out as unusual about Earth is how stable its climate has been. This is especially remarkable given that the Sun has slowly been getting hotter. The amount of energy entering Earth's atmosphere from the Sun is 1.4 kilowatts per square meter. We call this the "solar constant," but it is not. The Sun is steadily getting brighter, and has been since its birth 4.6 billion years ago. When our planet was born, our star was only about 70 percent of its current brightness. With so little solar heating, Earth should have frozen solid. Apparently it didn't, and therein lies a paradox.

There's something puzzling about early Earth. Climate models show that, *all other things being equal*, when warmed only by that wimpy young sun, the planet should have been completely iced over. Yet geological evidence (for example, the presence of ancient water-formed sediments) shows us that young Earth was not frozen. This contradiction was first pointed out by Australian geologist Ted Ringwood in 1961, and was discussed in 1972 by Carl Sagan and George Mullen, who named it the "faint young sun paradox" and, with Sagan's flair for branding and popularizing a topic, brought it to the wide attention of the scientific community.

The obvious way out of the paradox is that "all other things" were not equal: Earth's atmosphere has evolved, and the greenhouse effect was much greater in the past. The answer may be simply that early Earth more closely resembled its two triplet siblings, Venus and Mars, and had an atmosphere that was loaded with CO_2, supporting a global greenhouse strong enough to compensate for the feeble early Sun, keeping Earth's climate warm, wet, and habitable.

The earliest climate of Earth is still somewhat of a mystery, but it is clear that although the atmosphere has changed radically over time, the climate has generally remained stable, within the right temperature range for liquid water and "life as we know it" over billions of years. How has this happened? It turns out Earth has a natural thermostat, resulting from the strong negative feedback between two major parts of the Earth system: weathering reactions and volcanoes.

Weathering reactions are the way in which carbon atoms are pulled from atmospheric gas molecules to be locked into solid mineral crystals. They happen when rainfall flows over and trickles through silicate rocks on the continents. Atmospheric CO_2 dissolved in this water makes carbonic acid, reacts with minerals in the rocks, and eventually ends up in the ocean, where the carbon is deposited in sedimentary carbonate rocks. The process is aided on both ends by various organisms, bacteria that increase the absorption of CO_2 in the soil and ocean creatures such as corals that secrete carbonate shells. The net effect is almost alchemical: air turning into rock. Weathering cools the climate by sucking CO_2 out of the atmosphere.

Yet carbon doesn't stay sequestered in carbonate rocks forever. Volcanoes eventually return it to the air as CO_2. Over millions of years, in the slow, unceasing roil of plate tectonics, every part of the seafloor is eventually subducted, pulled deep into Earth's mantle. Carbonate rocks don't survive under that kind of heat and pressure. The carbon is again turned to gas that, dissolved in fresh magma, returns to the surface. Throughout Earth's history, volcanoes of every sort (from gently oozing seafloor pillows to wide, gushing basaltic floods to devastating Krakatoan mega-eruptions) have always produced, along with fresh rock and new lands, massive volumes of CO_2.

Volcanoes adding carbon to the atmosphere, and weathering removing it—that's the cycle. Yet what turns this into a self-regulating thermostat is the fact that the two parts of this cycle respond quite differently to climate changes. Volcanoes barely react to climate. The rate of CO_2 production by volcanism is ultimately determined by the interior heat flow of Earth. So volcanoes deliver carbon like the Pony Express: through any kind of weather. When it comes to climate, volcanoes just couldn't care less. Through ice ages and hothouse eons, Earth exhales volcanic CO_2, sometimes stuttering or hiccupping, but never stopping to worry about surface conditions.

By contrast, the weathering reactions removing CO_2 from the air are highly sensitive to climate. During hotter epochs, chemical reaction rates speed up and the hydrological cycle accelerates, fueling more rainfall and faster weathering, all of which sucks CO_2 from air and into rock. Yet when the continents freeze over, and rock is buried under solid ice, these weathering reactions grind to a halt and the removal of atmospheric carbon ceases. Thus, a prolonged ice age will always be self-limiting because it will eventually cause a buildup of warming CO_2, and a hot phase will always bring itself to an end by increasing the rate of weathering, drawing down CO_2 and cooling our planet again.

So, we've got negative feedback, which stabilizes the climate.

Over the ages, the planet's internally driven volcanic cycle and solar-powered hydrologic cycle conspire to adjust the CO_2 level such that surface temperature remains comfortably in the range of stable liquid water. This seems fantastically convenient for us, and for Earth's water-based biosphere. Also, as the Sun slowly brightens over billions of years, the thermostat responds, lowering the CO_2 content of the atmosphere. Earth slowly self-adjusts its level of greenhouse gases to balance against the rising influence of its warming star. Given enough time to respond to any provocation, the climate will return to a stable state.

Only when we compare Earth to other planets do we see how many factors have contributed to creating and maintaining this remarkable life-sustaining planetary thermostat. Venus and Mars, it seems, also started out with warm oceans, and when they were young worlds, a similar carbonate thermostat was likely once operating on all three planetary siblings. Yet on each of our neighbors, for different reasons, the thermostat broke down and the climate veered off toward an uninhabitable state.

Venus shows us what would happen if Earth had been born too close to its star. The essential factor is that water vapor, like CO_2, is a strong greenhouse gas. This can cause a positive feedback, which is like having your thermostat wired the wrong way, so that when it gets too hot, the heat turns on. When a planet warms, more water evaporates. The water vapor increases greenhouse heating, which causes still more water to evaporate, which increases heating, ad infinitum. Unchecked, this will lead to a "runaway greenhouse." Venus, 30 percent closer to the Sun, gets twice as much sunlight as Earth. As the faint young Sun brightened, Venus passed a point where it got too much sunlight and fell unavoidably into this runaway greenhouse state. All the water boiled off and seeped out into space. The volcanoes kept pumping out CO_2, and as a result, Venus today has an almost pure CO_2 atmosphere that is nearly one hundred times as thick as Earth's, and it's hot as hell there: hot enough to cook all forms of earthly life and to fry our spacecraft.

Size Matters

What about Mars? Why did the climate thermostat break down there? Of the three siblings, Mars was the runt and shows us what would have happened to Earth if it had been made too small. The more we study with rovers and orbiters, the more we see signs of an early, more Earth-like time on Mars, with vigorous rainfall, rushing rivers, and wind-lashed lakes under a sky thick enough to be an earthly blue. Mars seems to have had a billion-year spree of warmer, wetter climate and possibly even life. Then it all stopped, and not much has changed since, in more than three billion years. What happened? With only one third of Earth's gravity, Mars couldn't hold on to its atmosphere. An early pummeling by asteroids and comets created explosive impacts that repeatedly blasted air off the planet. Held only by the weak gravity, more gas was stripped off by the solar wind, or speeding air molecules simply flew into space. The Martian greenhouse grew ever feebler. Today the thin CO_2 atmosphere is so cold that it is partially frozen into polar caps of dry ice. Any water that didn't get swept into space or destroyed by the ultraviolet light that penetrates the remaining thin air lies frozen in the ground or locked up in the polar caps.

It's even worse than that: little Mars was always destined to be a frozen desert; the planet is doubly damned by its small size.

As we've explored and begun to develop a science of comparative planetology, a key question has been whether other planets also have plate tectonics, with all that it implies on Earth (continual renewing of the land and perpetuation of geochemical cycles with their stabilizing effect on climate and enabling potential for life)? It turns out that, to a large extent, planetary character is determined by size.

So far we have not found another planet with plate tectonics. Mars shows some hints that it may have had such a system early on, during a brief, vigorous phase of internally driven geologic activity. Yet, early on, Mars seized up, its crust thickening, its convective heat engine losing steam and its plates annealing into one solid, unbroken, immobile sphere. This, we've come to understand, is to be expected on smaller worlds. They lose heat quickly, and thus don't maintain the vigorous internal convection that has sustained plate tectonics on Earth. Their internally driven geological activity ceases and they end up covered with craters, largely devoid of active or recent geological features.* When it was still a young world, Mars lost most of the internal heat that once drove massive volcanoes, replenished the air with CO_2, and kept the carbonate thermostat working.

Size is key. Below a certain size, worlds will end up like Mars. Larger planets will hold in enough heat so that their deep interiors will remain molten for billions of years, giving them more active geology and younger surface ages. Given this emerging understanding, hopes were high for finding plate tectonics at Venus, which, at 95 percent Earth's diameter, is nearly a twin in size. Yet, for those interested in the global geology of Venus, the history of exploration has been marked by delayed gratification. The planet is surrounded in a thick acid fog that makes photographing its surface from orbit impossible. Long after we had global geological maps of Mars, for Venus we had only the foggiest, only that fog. This finally changed when orbiting spacecraft

---

* When we explored the outer solar system, we discovered an important way that worlds can defy this rule. Many moons of giant planets are more active than we expected, and we learned that the gravitational tussle among bodies orbiting a Jupiter or Saturn can provide enough heat to melt interiors, ignite volcanoes, renew surfaces, and wipe away ancient craters.

were sent up with radar imaging equipment that could see through clouds. The Soviets did this first with *Venera 15* and *16*, which in 1983 managed to map part of the Northern Hemisphere. Then, in 1990, NASA's *Magellan* spacecraft got to Venus and, over its four-year mission, made detailed radar maps of the entire surface. We saw thousands of volcanoes of a bewildering range of types and sizes, steep canyons, vast plains, towering mountains, mysterious river channels apparently carved by lava, and nearly one thousand impact craters scattered around the planet. What we didn't find was any evidence of plate tectonics. If you look at a global map of Earth at the same level of detail, you can easily see a planetwide network of structures defining the boundaries of tectonic plates: steep trenches where plates are being dragged down into the mantle, and seafloor ridges where plates are growing and spreading. On *Magellan*'s maps of Venus, we see no global pattern suggesting the push-me-pull-you dynamics of plate tectonics.

If size is key, why did a planet so similar in size to Earth evolve in such a different way, absent the global tectonic system that seems key to the character of our planet? This remains a mystery. The difference is likely related to the extreme dryness of Venus compared to Earth. On Mars, when the geology died it drove the climate to extremes. On Venus it may have been the other way around: I suspect that the extreme climate change that befell Venus at some point in its history doomed plate tectonics and condemned the planet's surface to remain a lifeless inferno. After Venus lost its oceans to a runaway greenhouse, the interior also would have started to dry out, and this might have shut down plate tectonics. The question has forced a closer look at how and why plate tectonics works on Earth. We've learned there are many ways that plate tectonics is aided and lubricated by the presence of our planet's pervasive hydrosphere. Venus could have started out with Earth-style plate tectonics and then lost its ability to recycle its surface and interior, as it lost its water to a runaway greenhouse, and the interior of the planet was slowly wrung dry. This seems plausible, but without further missions to Venus we can't really claim to understand the divergent evolution of these sister planets. Also, as I'll discuss in the next chapter, we are still learning of all the ways the ubiquitous presence of life itself has deeply altered Earth.

We still don't know what critical factors determine a planet's destiny as one of these very different types of worlds. The fact that we don't really understand the Earth-Venus difference in tectonic evolution shows that we can't yet predict whether a given planet will have plate tectonics. This is currently one of the greatest weaknesses in our ability to predict which planets elsewhere in the universe are good candidates for life. We are understandably obsessed with finding other Earth-size worlds, but we still don't really know the history of our nearby twin. Future exploration of Venus will help us here. We know the secrets are sitting there, on and below the surface of Venus, in that crushing atmosphere, beneath that shroud of acid cloud. We can go and find them. It won't be easy, but we'll do it. The answers may clue us in to some of the deep connections among climate, geology, and biology on worlds like Earth, and prepare us to make sense of the new harvest of rocky planets we are just starting to discover around other stars.

Billions and Billions of Worlds

For at least as long as we've been human, we've looked at the night sky and wondered what the stars were and how far away they were. For hundreds of years we've known they are distant suns, and wondered if they shine on worlds of their own. Now, finally, we know. When you look at a night sky studded with stars, most of them are orbited by unseen planets.

Planetary scientists long assumed that our sun was not unusual in having a family of orbiting planets, and that many of the hundreds of billions of other stars in the Milky Way, and the trillions and trillions in other galaxies, were likely centers of their own extrasolar systems. Indeed, this is what I was taught in grad school, in the late 1980s, long before we knew it was true. Our professors told us that the formation of the planets was a predictable by-product of the messy birth of the Sun, an inevitable fusing of orbiting bodies from the leftover debris, and therefore something that should have happened around most stars. Yet, at that time, this belief was a kind of informed faith. We believed other planets were there, but we had no direct evidence.

Science marches onward. Better instruments were built, and some clever, dogged astronomers refused to take "We don't know" for an answer. The breakthrough came in October 1995, when astronomers in Switzerland observed that 51 Pegasi, a star similar to the Sun and fifty light-years away, was wobbling, responding to the gravitational tug of a small, unseen companion. A week later this observation was confirmed by American astronomers who caught the same wobble from a telescope in California. The gates were cracked open, and a flood of discovery began. More and more extrasolar planets (exoplanets, for short) were found by this and other techniques. Now we've acquired the solid data to transform our longtime belief in exoplanets from a well-justified hunch to a known fact.

The watershed came from NASA's *Kepler* spacecraft, the brainchild of Bill Borucki, another longtime denizen of Ames Research Center. Bill had been advocating for this planet-hunting mission for decades, seemingly forever. I remember him doggedly pushing the concept when I was a postdoc at Ames, twenty years before *Kepler* became a reality. His concept was to launch a small telescope into orbit just to obsessively stare at one little area of sky. He proposed that if we could precisely monitor the brightness of a large number of stars in one random area of the galaxy, watching for any flickering, we could tell if planets ever passed in front of any of them. A lot of people thought it was clever but kind of "out there." Things changed when observers started to find planets from ground-based telescopes. Bill Borucki finally got his mission. *Kepler* was launched in March 2009 and began revealing to us the approximate number and demographics of planets in one tiny (and presumed typical) patch of the Milky Way, covering one wingtip of Cygnus the Swan. Now we know planets are ubiquitous. They're a normal thing for a star to have, and they exist in a stunning diversity of size, density, and orbital arrangement.

The question of what these newly discovered planets are really like as worlds, as places, is both overwhelmingly interesting and incredibly difficult to answer. When we started exploring the solar system, we knew about only nine planets. Now our effort to put Earth in context takes place against a backdrop of "billions and billions" of them. Planets are diverse and quirky, so the handful of local examples may not be sufficient for us to learn universal patterns of planetary evolution. We need many more, and fortunately we are about to get them as we slowly learn more details about exoplanets. Only extrasolar planets are all so ridiculously far away, and each is orbiting close to a star that is billions of times brighter, making it a ridiculously daunting technical task to observe the planets themselves directly. Detecting them and determining size, density, and orbit has become relatively routine, but finding out much more about them is going to be really, really hard. How many of them are similar to Earth in the ways we most care about? Do they have tectonic cycles, stable climates, watery surfaces, and conditions otherwise copacetic for our kind of life?

The only way we have right now of gaining any hints is through painstaking analysis of data sets at the hairy edge between noise and meaning. Yet a fantastic amount of cleverness is being displayed by the community of scientists, many of them young, who are now bravely trying new ways to use the tools we do have to glean information about these distant new worlds.

For a very long time, perhaps forever, we will know much, much more about the handful of planets within our solar system. To understand exoplanets, we'll always rely heavily on our detailed knowledge of the local bunch—because planets are incredibly complex. Climate and atmospheric evolution are intricately bound up with interior, and tectonic (and biological? and technological?) evolution. Our progress will be limited until we explore our own solar system much more deeply. If we want to know what makes planets tick, we have to do a much more thorough job with the ones we can get to.

Exoplanets and solar system planets—we won't be able to understand one without the benefit of exploring the other. Fortunately, we live at a time when we can proceed with discovery in both realms, and the net effect will take us a great way toward learning how planets like ours function. The modelers are starting to map out the possible climate states of exoplanets. I've participated in some workshops about this and have found it fascinating watching astronomers and terrestrial climate modelers try to talk to one another. There is a huge gulf in scale and perspective. Our knowledge of exoplanets is so sparse. Each of these worlds is known to us as, at best, a few numbers: mass, distance from a star, and in some cases vague inferences about temperature or atmospheric composition. In contrast, our Earth climate models are supplied with dense grids of data: millions of points of temperature, humidity, and wind velocity. The mismatch in perspectives and techniques makes communication challenging. Yet the questions raised are vital: increasingly, we are banking on our climate models for the future well-being of human civilization on this planet. If our models really are any good, shouldn't they also be able to predict the climate on any other world as well? At these workshops, I've felt as if I were seeing the future, a new era of climate understanding that will come when we can study our own planet's qualities in the context of thousands of its peers.

The discovery of exoplanets has been a transformative watershed for planetary science. These planets will test our theories and deepen our insights in so many ways that were impossible during the phase of human history (which ends right now) when we knew nothing about virtually all planets in the universe. Earth climate modelers, so mired in the details of their difficult and urgent work, may be slow to realize it, but they need exoplanets, too.

Another Earth

A half century into the space age, we've learned that we really can't discover all that we need to know about our planet without looking beyond it. Planetary exploration has become crucial for understanding and protecting Earth. Astrobiology is closely tied to the topic of climate change on Earth because the starting point for thinking about planetary habitability is to look at the qualities that make our own planet habitable.

We're still just beginning to take the census of planets in our galaxy. A statistical study of *Kepler*'s harvest suggests that one in five Sun-like stars has Earth-size planets in its habitable zone.[6] This is a pretty loose estimate, with some arbitrary definitions.

"Earth-size" here simply means having a diameter between one and two times Earth's diameter, and "habitable zone" is defined very crudely. Yet this study succeeded in showing that, however one chooses to define these things, the news is good: there are planets in abundance, including plenty of the types of worlds where we can most easily imagine life evolving. In fact, in August 2016, we learned that Proxima Centauri, the *nearest star* to the sun, apparently has a planet that is slightly larger than Earth orbiting in its habitable zone!

Still, what does an Earth-size world really imply? Just because a planet is Earth-size, does that mean it is Earth-like? This might seem a silly semantic question, on a par with "is a dwarf planet a planet?" but it has generated some heated and interesting debate. What do we really mean by Earth-like? Perhaps a good criterion would be the continuous presence of both stable surface water and vigorous geological activity over billions of years, because the most meaningful way for a planet to be Earth-like, to have the quality we really care about, would be to favor our kind of carbon-based, water-immersed life.

Finding "another Earth" has long been an obsession of the exoplanet science community, often referred to as the Holy Grail. We are primed for such an announcement. We can expect many exciting discoveries of worlds that are tantalizingly similar in enticing ways. We may even soon find one with an atmosphere strange enough to hint at the presence of life. Yet, despite the sex appeal of "another Earth," I don't believe we will find one. Why? Because planets are more like people than protons. Particles such as protons are interchangeable. See one, you've seen 'em all, and it is likely that we can learn all the types of elementary particles. We may never learn all the types of planets. Each planet is the result of a complex and contingent history. Like people, they are individuals, and no two will be exactly alike.

What is it really that makes Earth Earth? The surface, where we live, exists at the interface between two giant heat engines: below us the churning mantle, above us the restless, windy troposphere. All rocky worlds large enough to hold an atmosphere will have some combination of these, and the dynamic possibilities of each realm have been substantially illuminated by the variations we've found on other worlds. We live on the convoluted, shifting shoreline between cycles of earth and sky that are incessantly driven by the Sun above and the heat below. So much about our world can be understood as the interplay between these inner and outer cycles. Life thrives at the boundary, enabled and sustained by the great cyclic flows of carbon, nitrogen, oxygen, sulfur, and phosphorous.

Still, something else is going on here. It's a two-way exchange. These cycles feed, but also feed off, Earth's biosphere. Gradually, we've come to realize that there is another great force at work on Earth: life itself. Life is so enmeshed in all the cyclical workings of Earth, and the recycling of the elements, that these geochemical cycles are now more commonly referred to as *biogeochemical cycles*, a word whose multiplicity of prefixes reveals the causal complexity of our world.

During the formative eons when Venus and Mars were undergoing their climate catastrophes, Earth was also in the earliest phases of its own radical transformation. Had anyone been there to observe its humble beginning, it would have seemed innocuous at first: a strange kind of chemical scum that started forming somewhere around a hot vent on the ocean floor or in some warm tidal pond. This tiny disturbance had a peculiar property. It was self-perpetuating, and thus marked a propitious branching point, the lowly start of a major transition that forever changed our planet's fate. Not only was this stuff self-propagating, but its imperfections made it adaptable. It changed with the planet, and then it changed the planet. The phenomenon persisted and spread and eventually globalized, taking over Earth.

2

CAN A PLANET BE ALIVE?

Wake up to find out that you are the eyes of the world.
—Robert Hunter

Life Disturbs

The search for life elsewhere presents some juicy puzzles. Do we know what we're looking for? What should we assume that we and aliens have in common? Despite our cultural saturation in Spielbergian imagery, we don't really expect little green men with big heads and teardrop eyes. Yet must life everywhere be built of the same molecules? Are DNA and proteins universals or just locally frozen-in accidents? Must other life, at least, be built of carbon molecules in liquid water, or might there be some entirely different basis for complex biochemistry that would operate better in an alien environment? What *is* life, anyway? Would we know it if we found it?

We know of only one biosphere, and we're in it, so we have no perspective. On Earth we're all related, stemming from the same beginning, so we have only one example of life, the opposite of what you need for scientific perspective on a complex system. How can we try to define it before we've found multiple cases? Yet how well can we search for something we haven't defined? As space exploration began in the early 1960s, scientists in the nascent field of exobiology were aware of this philosophical conundrum, but it became a practical problem in the mid-'60s, as NASA began to plan the search for life with the first Martian landers, a process that would culminate in 1976 with the landings of *Viking 1* and *2* on Mars. Out of this effort came an idea that changed the way we think about life on Earth.

NASA hired James Lovelock, an iconoclastic, independent British scientist/inventor,[1] to help devise instruments for finding life on Mars. Lovelock realized that a quality of life on Earth, one that should indeed be truly universal, is that it radically changes the global environment. He proposed that searching for such anomalies on another planet was a better idea than trying to find actual organisms. It should be easier to spot these large-scale deviations than to observe living creatures, which might well be microscopic and/or unrecognizable. Such a search strategy doesn't assume anything specific about alien life, only that life interacts chemically with its surroundings, extracting energy and leaving its mark. He concluded that life would always knock an atmosphere out of chemical equilibrium, leaving a strange brew of gases that would differ recognizably, detectably, from the air of a lifeless world. He began to suspect that to find life on Mars, we merely had to make very precise observations of the atmosphere. This could be done from Earth with sophisticated telescopes and spectrometers. This advice didn't really help NASA with their question of what instruments to put on a lander to find life. Jim Lovelock is definitely not the type to give you the answer he thinks you want to hear.

He worked on this problem at the Jet Propulsion Laboratory

in Pasadena, where he shared an office with Carl Sagan. In 1965, he learned of new infrared observations of Mars, from the Pic du Midi telescope in France, showing that its atmosphere, like that of Venus, was made almost entirely of CO_2. This is fundamentally different from Earth, and what you would expect if the atmosphere were in perfect equilibrium, completely undisturbed by any biological activity. As far as he was concerned, these measurements showed that there was no life on Mars. Lovelock concluded that a lander mission was no longer needed to determine if anything was alive and breathing on the Red Planet. Of course his officemate vehemently disagreed. Sagan's mentor, astronomer Gerard Kuiper, had earlier argued that the "regenerative dark areas" seen on the surface of Mars were caused by vegetation, and Sagan had developed his own speculative hypotheses about the kind of life we might discover there with surface landers.

I would like to have been a fly on the wall for that exchange. Lovelock is an erudite provocateur who seems to love tweaking the establishment. To the degree that there *was* an establishment line of thought in exobiology, still a small and barely reputable effort at that time, it was Sagan's. Carl relished a good intellectual argument and loved kicking around new ideas, but he was also heavily invested in sending a lander to Mars to look for the microbes or spores he suspected just might be crawling or hibernating in the dirt there. NASA ignored Lovelock's advice and proceeded to design, build, and fly the *Viking* orbiters and landers—two of each.

The *Viking* landers were magnificent, among the most complex and elegant machines ever built. They both arrived on Mars in 1976 and gave us our first vivid portraits of the Martian surface. As a teenager, I was enchanted by these missions, the drama of their landings, and the first pictures of that rock-strewn, alien desert where *Viking* would search for life.

When I went to college, *Viking* provided my first taste of real scientific research, in the form of a summer job working for Tim Mutch, my freshman planetary geology professor at Brown University, who led the camera team for the *Viking* landers. My first task was recording the size and position of every rock at both *Viking* landing sites, which would allow us to investigate the history of the site by doing statistical comparisons with different kinds of rock fields on Earth. (Yes, my job was counting rocks, which sounds like a parody of a boring occupation. But these rocks were on Mars!) I spent months immersed in the dusty plains of the Red Planet. By the end of that summer, I felt I knew what it was like to tromp around on Mars with the crimson dust blowing around my ankles in thin, frozen breezes.

The landers functioned nearly flawlessly, and provided the foundation for much of our modern understanding of the environment and history of that neighboring orb. But the centerpiece of the mission was the biological experiments, and these taught us more about ourselves and our wet-behind-the-ears approach to life than about any critters in the Martian soil. NASA took the opposite approach from that recommended by Lovelock, searching for very specific signs of organisms in the soil around the landing sites. For reasons that seem a bit questionable with the clarity of hindsight, the Viking experiments were designed to look for life that would thrive under not current Martian conditions, but warmer, wetter conditions more like those found on Earth today. This decision was largely justified by a wishful idea about life on Mars that Carl Sagan had proposed and promoted in the early 1970s. According to his "Long Winter Model of Martian Biology,"[2] the Martian climate might flip-flop between its current frozen, dry state and occasional wet, warm spells. Martian organisms might have evolved a "cryptobiotic," or dormant, phase that would allow them to survive the long cold snaps. Such sleeping beauties might spring to life when the *Vikings* landed and fed them "chicken soup," as the mix of water and generic organic nutrients carried on board the spacecraft was called. No such luck. The Viking biology experiments squirted this Purina Martian chow into the dry soil and waited to see if any thirsty creatures would try some and belch, betraying their presence.* If there were Earth organisms living there, we would have found them, but apparently nobody was home.

At first, one experiment did seem briefly to signal a biological response, generating momentary excitement. Yet, over the hours and days, the results followed a pattern much more indicative of reactive soil chemistry than native life. It wasn't strange Martian bugs but strange chemistry, something in the dirt that was not like Earth dirt, and was extremely reactive with water.† Today the *Viking* experimental approach may seem naïve, an attempt to find life as defined by simple geocentric assumptions. Yet perhaps, in our ignorance of how alien aliens might be, it was worth a try to grab some alien dirt and give it a squirt of water just to see. None of our landers and rovers since has included any instruments actually meant to test directly for living organisms. We realized that, short of seeing something in a camera or microscope that is unmistakably alive, we don't yet know how to design such an experiment. Our search has largely shifted toward finding signs that Mars was once habitable billions of years ago, when it and Earth were young worlds.

Yet we'd still love nothing more than finding possible environmental disturbances of the kind that Lovelock suggested decades ago, such as strange gases that "shouldn't" be there.

---

* Okay, in detail, the experiments were more sophisticated than this, but the approach was very much predicated on the idea that organisms similar to those found on our planet might be lying dormant in the soil, waiting for some drops of water to revive them.

† Quite possibly these reactions were caused by the perchlorate compounds that were later found in the dirt around the 2008 *Phoenix* lander.

One recent discovery falls into this category: the *Curiosity* rover has sniffed a possible trace of a strange gas, methane (CH_4) that should not be there according to our best understanding of Martian geochemistry, and might conceivably come from living organisms. Right now it's not clear if the methane is really there, or present in enough quantity to be a possible indicator of life. By the time this book is published, that may become clear.* Either way, the debate underscores Lovelock's prescience about what a sign of life could look like.

This line of thought about Mars also led Lovelock to look anew at how life has changed Earth. What if Earth had never detoured down the path of life? What would its atmosphere be like? The answer is that without life, our planet would be drastically different from its current state, in ways that would be obvious to the casual alien observer. All this oxygen, along with healthy traces of methane and several other gases, forms a bright signpost of disequilibrium. That is, it's an unstable mixture, a chemical house of cards that is actively propped up by the ongoing activity of abundant life. To any extraterrestrial astronomers examining the inner planets of our solar system, this condition would shout, "Hey, look at *this* one!" If it is life they were seeking, the aliens wouldn't give airless Mercury a second glance, and they'd likely, after a quick spectral examination, pass on Mars and Venus, finding their stale CO_2-dominated atmospheres unpromising. But Earth would stand out as strange and perturbed, with something in the air that geology and chemistry alone could not explain, with some huge ongoing, active chemical disturbance. That disturbance is life, and were it to disappear, or had it never evolved on Earth, the molecular oxygen (O_2) would largely vanish from the air. A promiscuous element if ever there was one, oxygen never stays single for long. Without life's continuing input, oxygen would have long

---

* Note added in proof : Still don't know!

ago hooked up with whatever elements it could find, and the carbon would have reverted to its more stable form, carbon dioxide. In other words, the air of a lifeless Earth would closely resemble the nearly pure CO_2 we see on Mars or Venus today.

His efforts and insights in the area of life detection led Lovelock to a new view of the role of life on Earth. The full realization of this vision came about when the maverick chemist teamed up with the brilliant renegade biologist Lynn Margulis.

Mother of Gaia

The last time I saw a roomful of scientists cry was at an Indian restaurant a few miles south of the Kennedy Space Center, in Cape Canaveral, Florida, where a two-hundred-foot rocket sat fueling up at Launch Complex Number 41, preparing to blast off two days later, carrying the *Curiosity* rover to Mars. It was a small gathering of colleagues who had just learned of the untimely death of biologist Lynn Margulis on November 22, 2011. We would all miss her funeral because of the imminent launch of *Curiosity*, so we congregated that evening over curry and saag to tell stories about Lynn, laugh, cry, and feel that bond we humans share when a person crosses over from sharp individual existence to the more diffuse persistence of collective influence, memory, and love, her microbiota shuffling off in search of other mortal coils, new living collectives to animate for a while. At such moments, we glimpse something that Lynn helped us learn: that our own individuality is a brief illusion transcended by love and biology. The handful of women and men who convened that evening, NASA übernerds all, had our lives enriched by Lynn's incandescent intellect and nurturing heart. She seemed timeless and tireless, and she died suddenly, felled by an aneurism at the age of seventy-three. So we shed those aqueous, saline tears for Lynn, who in some way had changed all of us, and who had done so much to shape our ideas about planets and life.

Lynn Margulis was a giant of modern biology, but she was also more than that: she braved the boundary between science and philosophy, fearlessly applying rigorous science to problems that changed the way we think about profound questions of self and identity, and the limits of current scientific knowledge and methodologies. Her theory of endosymbiosis, controversial at first and now enshrined in biology textbooks, showed that in evolution, radical cooperation is just as potent a force as deathly competition. One great example involves mitochondria, the tiny micron-size power plants inside our cells. According to endosymbiotic theory, these used to be freely living bacteria that joined our ancestral cells in a mutually beneficial, symbiotic, relationship. The association became so tight that eventually

Lynn Margulis, 2005.

the partners joined together to form a new kind of organism. In fact, Lynn taught us that many of life's most important evolutionary innovations resulted from assimilation, fusion of formerly separate individuals, the formation of cooperatives, and the sharing of genes and traits. Survival of the fittest still applies, but often the fittest are those assemblages of organisms that creatively merge into new kinds of individuals.

I had the pleasure of knowing Lynn for five decades. Her son Dorion Sagan, an accomplished science writer (also, I would say, a natural philosopher), became her longtime collaborator on several clever and important books that dazzle with literary and scientific brilliance. In the mid-1960s, Dorion and I, with our parents involved in the Bostonian scientific community, became fast friends at an early age. So I also got to experience some of Lynn's style of mind nurturing. She was supportive, or at least tolerant, of our various ad hoc creative projects, experiments, and stunts, including some involving pyrotechnics that today would probably have the neighbors calling Homeland Security. She was awfully busy, and not always around, but occasional outings with her to places like science museums were thrilling. She spoke to kids as if they could understand adult concepts, which they generally can. I still remember puzzling over something she said about metabolism when I was, oh, probably ten.

It is remarkable how many astrobiologists today remember Lynn as an essential mentor.[3] Many also describe her as an old and valued friend. She had an outsize influence on the development of exobiology, and then astrobiology, at NASA, lending her biological wisdom and perspective to an agency heavily biased toward physical science.

Many people have heard of the Gaia hypothesis, which Margulis formulated, along with Jim Lovelock, in the 1970s. As a concept, it has often been abused as a New Age catchall, or mistakenly derided as pseudoscience, but at its core is an insight about the relationship between planets and life that has changed our understanding of both. In his work toward finding life on Mars, Lovelock realized that life must be, inherently, a planet-altering phenomenon. So he and Margulis, chemist and biologist, applied their collective symbiotic mind to the question of what life does to its environments. In exploring this question, they realized that the distinction between the "living" and "nonliving" parts of Earth was not as clear-cut as we thought. Lynn added her focus on the role of cooperative evolutionary networks in biological systems. Studying Earth's global biosphere together, they realized that it has some of the properties of a life form. It seems to display "homeostasis," or self-regulation. Your own internal homeostasis maintains your temperature, the pH of your blood, and a wide range of other properties within an optimum, healthy range that keeps you alive. Many of Earth's life-sustaining qualities exhibit remarkable stability. The temperature range of the climate; the oxygen content of the atmosphere; the pH, chemistry, and salinity of the ocean—all these are biologically mediated. All have, for hundreds of millions of years, stayed within a range where life can thrive. Lovelock and Margulis surmised that the totality of life is interacting with its environments in ways that regulate these global qualities. They recognized that Earth is, in a sense, a living organism. Lovelock named this creature Gaia.

Margulis and Lovelock showed that the Darwinian picture of biological evolution is incomplete. Darwin identified the mechanism by which life adapts due to changes in the environment, and thus allowed us to see that all life on Earth is a continuum, a proliferation, a genetic diaspora from a common root. In the Darwinian view, Earth was essentially a stage with a series of changing backdrops to which life had to adjust. Yet, what or who was changing the sets? Margulis and Lovelock proposed that the drama of life does not unfold on the stage of a dead Earth, but that, rather, the stage itself is animated, part of a larger living entity, Gaia, composed of the biosphere together with the "nonliving" components that shape, respond to, and cycle through the biota of Earth. Yes, life adapts to environmental change, shaping itself through natural selection. Yet life also pushes back and changes the environment, alters the planet. This is now as obvious as the air you are breathing, which has been oxygenated by life. So evolution is not a series of adaptations to inanimate events, but a system of feedbacks, an exchange. Life has not simply molded itself to the shifting contours of a dynamic Earth. Rather, life and Earth have shaped each other as they've coevolved. When you start looking at the planet in this way, then you see coral reefs, limestone cliffs, deltas, bogs, and islands of bat guano as parts of this larger animated entity. You realize that the entire skin of Earth, and its depths as well, are indeed alive.

Lynn Margulis's influence on science will continue to emerge and dazzle gradually. When she left this earth, our global mind, our noösphere, lost one blazing neuron, which, during its time, sparked great thoughts that linger. She challenged us to rethink our cozy established ideas, definitions, narratives, and categories of living things, and she changed the way we see life, evolution, our planet, our cells, ourselves. Recently the science news is full of stories of the "microbiota," detailing how you and I are mostly microbes, with the bulk of our biomass and metabolism, in sickness and in health, dominated by the complex ecologies within us. Lynn was always telling anyone who would listen, "You are not an individual. You are a community." The discovery of the human microbiota is redefining what it means to be human and how we relate to the rest of life. When I read about this unfolding revolution, I think, "Yes, this is where Lynn was pointing us all along."

What Does Earth Want?

The acceptance of the Gaia hypothesis was, and remains, slow, halting, and incomplete. There are several reasons for this. One is just the usual inertia, the standard conservative reluctance to accept new ways of thinking. Yet Gaia was also accused of being vague and shifting. Some complained that the "Gaians" had failed to present an original, well-defined, testable scientific proposition. How can you evaluate, oppose, or embrace an idea that is not clearly stated, or that seems to mean different things to different people? There was certainly some truth to this. Gaia has been stated many different ways. Also, it didn't help that Margulis and Lovelock were more than willing to mix science with philosophy and poetry, and they didn't mind controversy; in fact, I'd say they enjoyed and courted it.

Gaia, subversively, blurs the boundaries between the scientific and the nonscientific. This may be one of its most valuable aspects, but is also a big reason that the scientific establishment has had so much trouble with it. Saying that "Earth is alive" is, of course, asking for it. The statement is both true and not true, profoundly insightful yet subject to infinite reinterpretation, and not a scientific statement that can be tested.

Yet Earth does have previously unrecognized qualities that make it more like a living organism than we once knew. It seems to exhibit self-regulation arising from the collective interactions of numerous component systems. But you would not confuse it with any other entity we considered to be an organism. It has never reproduced.* It's been "alive" for billions of years. It also happens to be a planet. So why even ask if Earth is an organism? Well, are *you* an organism? You probably like to think so,

---

* Not yet. A truly self-sufficient, sustainable colony in orbit or on Mars would represent an act of biospheric reproduction.

but like Earth, you're also essentially a community of interacting parts, many of which are themselves organisms and some of which are not made of living cells. Whether or not you have reproduced, your DNA molecules are billions of years old and will be quasi-immortal for a long while, at least as long as Earth life lasts.* Now think of parasites living their whole life in the body of their host, or bacteria living in your gut, unaware that their "world" is a living organism and that their survival depends on the maintenance of an environment that is synonymous with the health of that organism. Are we not like such dependent bacteria living within our host, Gaia? If it induces such thoughts, then the hypothesis is not only controversial, but useful. It gets us to consider what we really mean by "living" and "organism."

A more science-y phrasing of the hypothesis would be something like "The sum total of life on Earth interacts with other planetary subsystems to form a complex, self-regulating entity that maintains conditions within which life can continue to thrive, and that is observable in terms of the stable, nonequilibrium composition of the atmosphere and other environmental variables." This, however, is rather ponderous and unpoetic compared to "This planet is alive and we'll call it Gaia."

Does Gaia have a sense of purpose? A big stumbling block to widespread acceptance of the Gaia hypothesis came over the question of *teleology*. A teleological explanation is one that justifies the way something is by invoking its functional purpose, as in "These boots have heels because they were made for walking." The notion that Gaia, the living planet, has some sense of purpose was seeded in the early papers of Margulis and Lovelock: their first published paper on the topic was called "Atmospheric

---

* Say what? Think about it. DNA is the molecule that copies itself and divides, copies and divides. It's all the same molecule, in all organisms, going back billions of years to the common ancestor.

Homeostasis by and for the Biosphere." The "by and for" is what caused all the hubbub—not so much the *by*, which merely suggests that it is the biota that do the self-regulation, but the *for*, which seems to imply that Gaia is not only living but also serving, and perhaps also seeking, that Gaia is *trying* to optimize conditions on Earth for life.

Many of the arguments about Gaia became stuck on this aspect. Lovelock and Margulis probably lost a lot of potential supporters because of the (to many) apparent fallacy of teleology. In later statements they backed off on this aspect, claiming that they had been misunderstood and had never meant to imply conscious control by the biosphere. I think there was more to it, though.

At first I thought of the teleological language as somewhat annoying overreach on their part, language that muddied the waters. Now I'm not so sure. Anyone who knows and has followed Lynn and Jim knows that both relished their status as iconoclasts who shook up the world of science with their disturbing ideas. Lovelock in particular has cultivated his position as misunderstood scientist on the fringe. So, I also had the sense that there was deliberate obfuscation here, and that really, maybe a little, they were both toying with us.

There's a trickster element to this line of argument—that Gaia is *seeking* something, not merely exhibiting stabilizing behavior that arises out of the complex interactions of a global geobiosystem, but acting on its own behalf, with life taking care of life because it wants to—because the more you think about it, if you really think about it, this starts to mess with our notions, our assumptions, about consciousness and intentionality.

Just as we should question how Gaia could be considered to want anything, to possess any kind of desire or direction, we should ask the same thing about a human brain. Of Earth, we can ask: how can this system of interacting parts, each clearly without conscious intent, following physical laws and evolutionary processes, be said to manifest conscious intentionality? Then, with the next thought, using all the concentrated neuronal intensity we can muster, we can ponder the exact same mystery about our own thoughts and intentions. Here in our heads, the physical parts are not evolving populations of organisms and landscapes, volcanoes, chemical cycles, rivers, forests, and reefs. Rather, in the brain, the interacting components are populations of neurons with ions migrating in and out of their dendritic branches, chemical flows of neurotransmitters, and shifting patterns of networks and connections. So here, too, we can ask: how can this system of interacting parts, each clearly without conscious intent, following physical laws and evolutionary processes, manifest conscious intentionality?

Innumerable conferences have been held and volumes written about this. The results can be neatly summarized in four words: we do not know.[4]

Life Goes Deep

Owing partly to rhetorical overreaches and shifting stances on teleology, Gaia is sometimes wrongly dismissed as a discredited idea. In fact, the essential insights of Margulis and Lovelock have become deeply ingrained in our views of biology, of Earth, and of the deep and subtle interplay between them and in our growing awareness of the nebulous distinction between the living and nonliving parts of our planet.

The truth is, despite its widespread moniker, Gaia is not really a hypothesis. It's a perspective, an approach from within which to pursue the science of a life on a planet, a living planet, which is not the same as a planet with life on it—that's really the point, simple but profound. Because life is not a minor afterthought on an already functioning Earth, but an integral part of the planet's evolution and behavior. Over the last few decades, the Gaians have pretty much won the battle. The opposition never actually surrendered or admitted defeat, but mainstream earth science has dropped its disciplinary shields and joined forces with chemistry, climatology, theoretical biology, and several other "-ologies" and renamed itself "earth system science."

The Gaia approach, prompted by the space age comparison of Earth with its apparently lifeless neighbors, has led to a deepening realization of how thoroughly altered our planet is by its inhabitants. When we compare the life story of Earth to that of its siblings, we see that very early on in its development, as soon as the sterilizing impact rain subsided so that life could get a toehold, Earth started down a different path. Ever since that juncture, life and Earth have been coevolving in a continuing dance.

As we've studied Earth with space age tools, seen her whole from a distance, drilled the depths of the ocean floor, and, with the magic glasses of multispectral imaging, mapped the global biogeochemical cycles of elements, nutrients, and energy, we've learned that life's influence is more profound and pervasive than we ever suspected.

I've already discussed the flagrant disequilibrium in the chemical makeup of Earth's atmosphere. All this oxygen we take for granted is the by-product of life intervening in our planet's geochemical cycles: harvesting solar energy to split water molecules, keeping the hydrogen atoms and reacting them with CO_2 to make organic food and body parts, but spitting the oxygen back out. In Earth's upper atmosphere some of this oxygen, under the influence of ultraviolet light, is transformed into ozone, O_3, which shields Earth's surface from deadly ultraviolet, making the land surface habitable. When it appeared, this shield allowed life to leave the ocean and the continents to become green with forests. That's right: it was life that rendered the once deadly continents habitable for life.

The more we look through a Gaian lens, the more we see that nearly every aspect of our planet has been biologically distorted beyond recognition. Earth's rocks contain more than four thousand different minerals (the crystalline molecules that make up rocks). This is a much more varied smorgasbord of mineral types than we have seen on any other world. Geochemists studying the mineral history of Earth have concluded that by far the majority of these would not exist without the presence of life on our planet.[5] So, on Earth's life-altered surface, the very rocks themselves are biological by-products. A big leap in this mineral diversity occurred after life oxygenated Earth's atmosphere, leading to a plethora of new oxidized minerals that sprinkled colorful rocks throughout Earth's sediments. Observed on a distant planet, such vast and varied mineral diversity could be a sign of a living world, so this is a potential biosignature (or Gaiasignature) we can add to the more commonly cited Lovelock criterion of searching for atmospheric gases that have been knocked out of equilibrium by life. In fact, minerals and life seem to have fed off each other going all the way back to the beginning. Evidence has increased that minerals were vital catalysts and physical substrates for the origin of life on Earth. Is it really a huge leap, then, to regard the mineral surface of Earth as part of a global living system, part of the body of Gaia?

What about plate tectonics and the dynamics of Earth's deep interior? At first glance this seems like a giant mechanical system—I used the heat engine metaphor earlier—that does not depend upon biology, but rather (lucky for life), supports it. Also, although we're probably still largely ignorant about the deeply buried parts of Earth's biosphere, it's unlikely there are any living organisms deeper than a couple of miles down in the crust, where it gets too hot for organic molecules. Yet, just as we've found that life's sway has extended into the upper atmosphere, creating the ozone layer that allowed the biosphere to envelop the continents, more and more we see that life has also influenced these deeper subterranean realms. Over its long life, Gaia has altered not just the skin but also the guts of Earth, pulling carbon from the mantle and piling it on the surface in sedimentary rocks, and sequestering massive amounts of nitrogen from the air into ammonia stored inside the crystals of mantle rocks.

By controlling the chemical state of the atmosphere, life has also altered the rocks it comes into contact with, and so oxygenated the crust and mantle of Earth. This changes the material properties of the rocks, how they bend and break, squish, fold, and melt under various forces and conditions. All the clay minerals produced by Earth's biosphere soften Earth's crust— the crust of a lifeless planet is harder—helping to lubricate the plate tectonic engine. As I described in the previous chapter, the wetness of Earth seems to explain why plate tectonics has persisted on Earth and not on its dry twin, Venus. One of the more extreme claims of the Gaia camp, at present neither proven nor refuted, is that the influence of life over the eons has helped Earth hold on to her life-giving water, while Venus and Mars, lifeless through most of their existence, lost theirs. If so, then life may indeed be responsible for Earth's plate tectonics. One of the original architects of plate tectonic theory, Norm Sleep from Stanford, has become thoroughly convinced that life is deeply implicated in the overall physical dynamics of Earth, including the "nonliving" interior domain. In describing the cumulative, long-term influence of life on geology, continent building, and plate tectonics, he wrote, "The net effect is Gaian. That is, life has modified Earth to its advantage."[6] The more we study Earth, the more we see this. Life has got Earth in its clutches. Earth is a biologically modulated planet through and through. In a nontrivial way, it is a living planet.

Gaiasignatures

Now, forty years after *Viking* landed on Mars, we've learned that planets are common, including those similar in size to Earth and at the right distance from their stars to allow oceans of liquid water. Also, Lovelock's radical idea—pay attention to the atmosphere and look for drastic departures from the expected mixture of gases—now forms the cornerstone of our life-detection strategies. Gaian thinking has crept into our ideas about evolution and the habitability of exoplanets, revising notions of the "habitable zone." We're realizing that it is not enough to determine basic physical properties of a planet, its size and distance from a star, in order to determine its habitability. Life itself, once it gets started, can make or keep a planet habitable. Perhaps, in some instances, life can also destroy the habitability of a planet, as it almost did on Earth during the Great Oxygenation Event (sometimes called the oxygen catastrophe) of 2.1 billion years ago, which I'll discuss in chapter 3. As my colleague Colin Goldblatt, a sharp young climate modeler from the University of Victoria, once said, "The defining characteristic of Earth is planetary scale life. Earth teaches us that habitability and inhabitance are inseparable."

In my book *Lonely Planets* (2003), I describe what I call the "Living Worlds hypothesis," which is Gaian thinking applied to astrobiology. Perhaps life everywhere is intrinsically a planetary-scale phenomenon with a cosmological life span— that is, a life expectancy measured in billions of years, the timescale that defines the lives of planets, stars, and the universe.

Organisms and species do not have cosmological life spans. Gaia does, and this is perhaps a general property of living worlds. Influenced greatly by Lovelock and Margulis, I've argued that we are unlikely to find surface life on a planet that has not severely and flagrantly altered its own atmosphere. According to this idea, a planet cannot be "slightly alive" any more than a person can (at least not for long), and an aged planet such as Mars, if it is not obviously, conspicuously alive like Earth, is probably completely dead.[7] A living world may require more than temporary little pockets of water and energy as surely exist underground on Mars. It may require continuous and vigorous internally driven geological activity. I believe that only a planet that is "alive" in the geological sense is likely to be "alive" in the biological sense. Without plate tectonics, without deep, robust global biogeochemical cycles which life could feed off and, eventually, entrain itself within, life may never have been able to establish itself as a permanent feature of Mars, as it did on Earth.

As far as we can tell, around the time when life was starting on Earth, both Venus and Mars shared the same characteristics that enabled life to get going here: they were wet, they were rocky, they had thick atmospheres and vigorous geologic activity. Comparative planetology seems to be telling us that the conditions needed for the origin of life might be the norm for rocky worlds. One real possibility is that Mars and/or Venus also had an origin of life, but that life did not stick, couldn't persist, on either of these worlds. It was not able to take root and become embedded as a permanent planetary feature, as it did on Earth. This may be a common outcome: planets that have an origin of life, perhaps even several, but that never develop a robust and self-sustaining global biosphere. What is really rare and unusual about Earth is that beneficial conditions for life have persisted over billions of years. This may have been more than luck.

When we stop thinking of planets as merely objects or places where living beings may or may not be present, but rather as themselves living or nonliving entities, it can color the way we think about the origin of life. Perhaps life is something that happens not *on* a planet but *to* a planet: it is something that a planet becomes.

Compared with these neighbors, Earth became something completely different: a biosphere, a living world. Yet when did this divergence happen? In thinking about this, I realized that, on Earth, the origin of life and the birth of Gaia must have been separate events. Life started out as some very simple organisms on a world not too different from these others, and then at some point it became integrated into the depths and cycles of the planet, transforming it into a biosphere. The origin of life was presumably a local event, one that started at one particular location on the planet, around four billion years ago, and then spread. Certainly by two billion years ago—after photosynthesis evolved and life was oxygenating the atmosphere, redirecting the further evolution of both life and solid planet—Earth had a global biosphere, or perhaps we should say that, by then, Earth had become a global biosphere. The oxygen catastrophe may have been the final violent spasm of the birth of Gaia.

Think of life as analogous to a fire.[8] If you've ever tried to start a campfire, you know it's easy to ignite some sparks and a little flicker of flame, but then it's hard to keep these initial flames going. At first you have to tend to the fire, blowing until you're faint, to supply more oxygen, or it will just die out. That's always the tricky part: keeping it burning before it has really caught on. Then it reaches a critical point, where the fire is really roaring. It's got a bed of hot coals and its heat is generating its own circulation pattern, sucking in oxygen, fanning its own flames. At that point it becomes self-sustaining, and you can go grab a beer and watch for shooting stars.

I wonder if the first life on a planet isn't like those first sparks and those unsteady little flames. The earliest stages of life may be extremely vulnerable, and there may be a point where, once life becomes a planetary phenomenon, enmeshed in the global flows that support and fuel it, it feeds back on itself and becomes more like a self-sustaining fire, one that not only draws in its own air supply, but turns itself over and replenishes its own fuel. A mature biosphere seems to create the conditions for life to continue and flourish.

A "living worlds" perspective implies that after billions of years, life will either be absent from a planet or, as on Earth, have thoroughly taken over and become an integral part of all global processes. Signs of life will be everywhere. Once life has taken hold of a planet, once it has become a planetary-scale entity (a global organism, if you will), it may be very hard to kill. Certainly life has seen Earth through many huge changes, some quite traumatic. Life here is remarkably robust and persistent. It seems to have a kind of immortality. Call it quasi-immortality, because the planet won't be around forever, and it may not be habitable for its entire lifetime. Individuals are here for but an instant. Whole species come and go, usually in timescales barely long enough to get the planet's attention. Yet life as a whole persists. This gives us a different way to think about ourselves. The scientific revolution has revealed us, as individuals, to be incredibly tiny and ephemeral, and our entire existence, not just as individuals but even as a species, to be brief and insubstantial against the larger temporal backdrop of cosmic evolution. If, however, we choose to identify with the biosphere, then we, Gaia, have been here for quite some time, for perhaps three billion years in a universe that seems to be about thirteen billion years old. We've been alive for a quarter of all time. That's something.

Uniquely Attributable

The origin of life on Earth was not just the beginning of the evolution of species, the fount of diversity that eventually begat algae blooms, aspen groves, barrier reefs, walrus huddles, and gorilla troops. From a planetary evolution perspective, this development was a major branching point that opened up a gateway to a fundamentally different future. Then, when life went global, and went deep, planet Earth headed irreversibly down the path not taken by its siblings.

Now, very recently, out of this biologically altered Earth, another kind of change has suddenly emerged and is rewriting the rules of planetary evolution. On the nightside of Earth, the lights are switching on, indicating that something new is happening and someone new is home. Has another gateway opened? Could the planet be at a new branching point?

The view from space sheds light on the multitude of rapid changes inscribed on our planet by our industrial society. The orbital technology enabling this observation is itself one of the strange and striking aspects of the transition now gripping Earth. If up to now the defining characteristic of Earth has been planetary-scale life, then what about these planetary-scale lights? Might this spreading, luminous net be part of a new defining characteristic?

Even in the very brief time we've been watching ourselves from afar, the picture has noticeably shifted. Since the start of the space age, the atmospheric composition has changed considerably, with carbon dioxide rising by 30 percent. The Amazon Basin has become deforested to an extent visible from the Moon. Our blue oceans are increasingly crisscrossed by linear clouds emanating from airplanes and ships. At night the radiance of offshore oil platforms and expanding fishing fleets pierces the dark seas. The lights have continued to come on, defining coastlines and national borders more starkly, most noticeably in India, where population has nearly tripled during the space age, and in China, where it has more than doubled. Had you been studying Earth over the eons you would have noticed all this, along with one other recent and dramatic change seen in the radio part of the electromagnetic spectrum.

In December 1990, the *Galileo* spacecraft, using our home planet as a gravitational boomerang, zoomed rapidly toward Earth, swung six hundred miles over Antarctica and then rapidly receded toward its final destination of Jupiter. This afforded an opportunity for a close encounter with Earth as it might appear to an alien spacecraft, and Carl Sagan proposed using this as a "control experiment for the search for extraterrestrial life by modern interplanetary spacecraft." The instruments detected the spectral signature of the chlorophyll from green plants, and signs of an obviously life-altered atmosphere. As Sagan and colleagues wrote in their paper "A search for life on Earth from the Galileo spacecraft" published in *Nature,* they

> found evidence of abundant gaseous oxygen, a widely distributed surface pigment with a sharp absorption edge in the red part of the visible spectrum, and atmospheric methane in extreme thermodynamic disequilibrium; together, these are strongly suggestive of life on Earth.

There was one additional, very noticeable phenomenon observed during that encounter. As Sagan et al. describe it,

> the presence of narrow-band, pulsed, amplitude-modulated radio transmission seems uniquely attributable to intelligence... Most of the evidence uncovered by Galileo would have been discovered by a similar fly-by spacecraft as long ago as about 2

billion years. In contrast, modulated radio transmissions could not have been detected before this century.

Galileo was able to detect the noisy radio chatter of the civilization that made it. Presumably any aliens studying our planet would notice the same thing, though whether this cacophony is actually attributable to "intelligence" depends on what we think we mean by that term (a subject I'll return to in chapter 6).

Life on Earth has both withstood and itself induced many catastrophic global changes. Now an entirely new catastrophe is unfolding that is, in some key respects, unlike anything that has ever happened to our planet. In the next two chapters we'll look at how this new transformation compares with the other major changes Earth has experienced, in an effort to see our time, and our role, more clearly.

MONKEY WITH THE WORLD

Who can stop what must arrive now? Something new is
waiting to be born.

—Robert Hunter

Something New

Have you noticed that something strange is happening to Earth?
Take a good long look at this world: a dazzling blue orb festooned
with spiraling clouds, spinning through star-flecked darkness;
dayside glinting in slowly brightening sun; winter-white pulsing
between north and south as Earth ambles through its orbit.

Now imagine you are a very patient alien regarding Earth
over the eons. If you've been watching carefully for, say, the last
several billion years, you've seen a lot happen: the brown continents drifting around the oceanic globe, coalescing and breaking apart, animated pieces in a morphing spherical puzzle; the
polar caps growing and shrinking, advancing and retreating,

as climate rocks between ice age and hothouse. Throughout all these changes, the nightside remains a nearly unbroken black, and the dayside continents are the stark, dull gray of bare rock. After four billion lonely years, a green fringe first edges over the land, and the night starts to sparkle with occasional forest fires. Still, for the longest time, the unlit hemisphere remains as black as the starry space surrounding it, the dark interrupted only by these fleeting fires, and by occasional flash of lightning or splash of aurora—until, very recently, in the last few hundred years, just a twitch in the life of the planet: whoa—what is this?

Something new! Suddenly the planet lights up, in a peculiar, spidering pattern that seems to reflect an organic process but something else as well. Something... cognitive? Starting in a few isolated river valleys and coastal areas, glowing points appear, abruptly dotting the night, then stitching together and spreading along widening and brightening webs, hugging the shores and eventually growing in loose nodal patterns across the interiors of the lands. [See page 1 of the photo insert.] On the dayside, a mesh of dark lines becomes visible, winding between the locations of those night lights, each swiftly surrounded by a growing verdant grid of novel angular geometry. Soon regular movements of small wave-generating objects start crossing the oceans, and bright linear clouds start streaking the skies. At the same time a host of other accelerating changes are observable in the atmosphere, the land, the oceans, and the ice.

Finally, just sixty years ago, a blink-and-you-missed-it interval in this fast-forward view, began the curious anti-accretion, with small bits of Earth stuff jumping into space. Little insect-like constructions of refined metal, bristling with sensors, thrusters, and radio antennae, started leaping off-planet, sallying first to the nearest worlds and then to those farther afield, sending pictures and other information home to their inquisitive builders, signaling the arrival on Earth of curiosity and gravity-defying technology. Yes, after billions of years of geology as usual, something new and strange is definitely happening here.

What is the meaning of these new changes?

The Evolving Past

One of the things I do for fun is read out-of-date popular science books, paying attention to what, from our perspective, is right, and what is obviously wrong. It animates the shifting of scientific verities and provokes thoughts about which of our current sureties will soon become antiquated. George Gamow's 1941 *Biography of the Earth* provides a terrific snapshot of scientific thought after "modern physics" but before the space age. By then, atomic physics and radioactivity (which Gamow, one of the twentieth century's great physicists, had helped explain) had allowed us to use Earth's rocks as clocks, recording the timing of ancient geological events in patterns of isotopes. So Gamow knows that our planet is billions of years old. He has the order of magnitude right, but he states confidently that Earth is two billion years old, less than half its true age.*

Some ideas in the book seem surprisingly current and sophisticated. Gamow accurately describes the structure of Earth's deep interior as derived from seismic data. He describes the requirements for life on a planet the way a modern astrobiologist would, in terms of the need for liquid water, energy, and nutrients.

This all seems prescient and accurate. But then he describes how Earth, along with the other planets of the solar system, was

---

* Gamow did the math correctly, but the decay rates of radioactive elements were not yet well known when his book was published.

born when the gravity from a passing star yanked a huge stream of gas out of the Sun. This, he says, explains why the interior of Earth is so hot. Because such close stellar encounters are extremely unlikely, it means that our solar system is a freak of nature and that stars with planets must be incredibly rare. He explains how the Moon split off from Earth, leaving a giant gaping wound: the Pacific "ring of fire." On the idea of continental drift, he is dismissive. Although he acknowledges the attractiveness of the concept for explaining the puzzle fit of map shapes and fossil findings, he contends that it must be wrong, because Earth's rigid outer shell is just too stiff and unbreakable to permit such motion. He suggests instead that it is the thermal contraction of Earth's crust that causes the cracking and shifting that makes mountains.*

Less than eighty years later, we know all these ideas about origins (of Earth, the planets, the Moon, and mountain ranges) to be completely wrong. We're certainly not any smarter today than they were then, and almost nobody is as smart as George Gamow was. We just have better data, much of it from space exploration. As we navigate this age of extrasolar planet discovery, supercomputer climate predictions, personalized gene sequencing, and in-depth planetary exploration, I wonder how our confident statements about what we know now will seem to anybody reading them in another fifty years. Scientific knowledge is always a work in progress. Keep this in mind when I (in this chapter) describe Earth history and (in future chapters) discuss human prehistory. Our best geology books are always only rough drafts, though they continue to get better as we get more data against which to check our account. The story has changed significantly in just the last few years, and will undoubtedly continue to do so.

---

* Gamow had the wrong planet. It turns out that it is on Mercury where much of global tectonics has been dominated by the cooling and shrinking of the planet.

On the wall in my office at the Library of Congress, where I have been writing much of this book, there's a poster showing the history of Earth with a multicolored, annotated cartoon of the layers in the geologic timescale. I love the fact that, even over the last year, some parts of this poster have become obsolete. It says that the oldest-known rock is 4 billion years old, but in February 2014 it was confirmed that there are some rocks in Australia that are much older than this, dating to 4.5 billion years, back almost to the origin of Earth itself, a time when, we thought, there were no rocks, just a hellish sea of molten magma. The poster shows a giant question mark for the cause and timing of "the Great Dying" of 250 million years ago. Yet, as I'll explain shortly, that question mark has recently become somewhat smaller.

This is our story, and we're not sticking to it. We'll keep changing our account as the evidence comes in, usually in a slow trickle of discovery, but sometimes in dramatic revelations or revolutions, like occasional asteroid strikes to our accumulating edifice of knowledge.

Living Dangerously

Life on Earth has had a tumultuous history. Extinction is a fact of life, as certain as death. More than 99 percent of species ever to have graced Earth have gone the way of all flesh. The rate of extinction has never been smooth and steady. It tends to come in pulses, when the slate gets wiped clean, making room for new evolutionary creation.

The geologic record shows that, repeatedly throughout Earth history, the environment has changed suddenly and lethally. At such times, the tree of life has been severely trimmed, beaten back, burned by frost, half-drowned, choked with poison gas, or singed by lava. Yet it always grows back dense with new species that would never have appeared without these brutal prunings. Earth's biosphere is robust, or as Lynn Margulis memorably put it, "Gaia is a tough bitch." Nearly wiped out many times (at least as measured by diversity), life has always bounced back quickly. What has happened to Earth to cause these repeated and disastrous die-offs? The causes are many.

In the 1980s, in the aftermath of the Alvarez revelation about the asteroid that brought an end to life in the Cretaceous, University of Chicago paleontologists Jack Sepkoski and David Raup surveyed the data of extinction rates over time and mapped out the history of "mass extinctions," those times in Earth history when the majority of species vanished suddenly. Their landmark analysis focused attention on the five largest die-offs, and these became known as the Big Five extinctions. In reality, the history of life shows a complex, variable, and always changing pattern of diversity, with a wide spectrum of extinction events of varying severity, usually showing a precipitous decline followed by a more gradual increase. The Big Five extinctions are simply slightly more intense than the sixth- or seventh-largest declines, and you could just as easily discuss the Big Seven or Big Ten, depending on how you sliced up the data. This tally refers only to extinctions since the "Cambrian explosion," the sudden proliferation of complex animal life 542 million years ago, and it neglects extinction events that happened earlier, during several billion years of evolution dominated by simpler organisms. So, though the changes occurring right now are often referred to as the beginning of a possible "sixth extinction," take this with a grain of salt. It's good to focus attention on the dramatic loss of species currently under way and how this fits into the history of extinction events on Earth, but this also reinforces an incomplete picture of Earth's dynamic history.

Post-Alvarez, some scientists, swept along in the new outer

space catastrophism, wanted to blame all mass extinctions on asteroids and comets. Yet further evidence has not supported this. Some episodes of mass death are clearly associated with more earthly causes, such as massive volcanic floods or changes in sea level, and some have been triggered by life itself, with biological evolution feeding back on the biosphere, as successful new life-forms have altered the world in ways that doomed established life. The causes of many mass extinctions are still being uncovered and debated, but the intellectual tumult prompted by the Alvarez hypothesis has left us with an increased appreciation for the role played by catastrophe in the history of life.

Here, I am using the word *catastrophe* in its scientific sense, to mean a sudden and dramatic change in the state of a planet. This doesn't necessarily imply a value judgment. Catastrophic planetary change is not in itself a bad thing. Without repeated catastrophe we wouldn't be here; nor would most any kind of life you value, except possibly some microbes. For the species experiencing them, such changes are indeed catastrophes in the more common sense of the word, meaning a terrible and disastrous turn of events. Yet, for the biosphere, catastrophes are also opportunities. The end-Cretaceous extinction event was obviously an awful disaster for the majority of species that did not survive, such as the dinosaurs, and was probably no picnic for the 25 percent or so that did squeak through. All that death and destruction because the planet was simply in the wrong place at the wrong time, in the path of a wandering space rock—but it opened up a multitude of niches for further evolution, allowing, for example, mammals to get a furry little leg up. Certainly human beings would not be here without this random collision. So if you are a fan of the cinema, the symphony, or the space program, you might consider this to have been a good day for Earth. Or if you value any form of art, music, literature, science, architecture, dance, philosophy, or anything else humans produce, or if there is anyone whom you love dearly, you should be glad this happened. I am, but when I say so, I have to acknowledge I'm expressing appreciation for a mass extinction, which is kind of a strange thought to have.

The changes occurring on Earth now are clearly another catastrophe, in the "rapid change of state" sense, and they certainly seem like a catastrophe in the other sense, too, for all the other species affected, for people in places starting to feel the bite of global change, and perhaps for our immediate descendants. Whatever happens, no matter what path we take from this point on, we have already left our mark and changed the course of evolution. Yet millions of years from now, how will this seem? Tragic? Or like the dinosaur deaths and other extinctions seem today, a necessary prelude to the precious and unique evolutionary creations enabled in their aftermath? I suppose the answer will depend on who, if anyone, is here to judge. Evolutionary history can be written only by survivors, and not just any survivors. To be able to study and write planetary history requires some special skills. Some of those same unique powers that allow us to change the world and unleash a new kind of mass extinction also have given us at least limited ability to decipher and understand our past, to see what's coming, and to mourn what is being lost. Whether we can gain the power over ourselves required to avert the worst of it is another question—but I'm getting ahead of myself.

Catastrophe is not only a mixed bag, but also unavoidable, especially on a world such as this one. Although no place in the universe is safe from nature's weapons of mass destruction, not all planets are equally vulnerable. Earth is a minefield of lethal hazards. Is this simply a universal fact of planetary existence? No. Other solid planets have more stable tectonic plates and relatively quiescent surfaces. So why are we cursed to inhabit a planet with the strange combination of factors that makes life here so risky? Actually, this is the only kind of planet we could come from. A good planet like ours, one that is suitable for life, will always be especially accident-prone.

Other than the external threat of asteroid and comet strikes (a common hazard throughout the solar system), all Earth's natural hazards go hand in hand with its habitability. Indeed, Earth seems to be constructed almost perfectly to create calamities, with an outer shell thick enough to be rigid but thin enough to be breakable. Earthquakes, tsunamis, and extreme volcanic transformations are the price we pay for the life-sustaining gift of plate tectonics.

Likewise, Earth's atmosphere is primed to do damage by just the same qualities that make it so nurturing. Sun-animated, water-saturated, and seasonally shifting, our dynamic atmosphere maintains our comfortable climate. Its motions feed and water us, but can also swirl into violent storms that flood villages and splinter cities. Tornadoes, floods, blizzards, and climate shifts are collateral damage from life-enabling flows of energy, water, and chemical elements. All are also symptoms of Earth's destructive/creative energy flows.*

On a planet like ours, lively and deadly conditions are two sides of the same coin. For planetary life, dangerously is the only way to live. Space exploration and comparative planetology have shown us that other planets, and especially those most similar to Earth, have histories of catastrophic environmental changes. Yet the majority of other solid planets are more stable, and consequently less fertile. On Mars you would not have to worry about Marsquakes or eruptions. The surface is wind-carved and frost-heaved, but it overlies a dead crust that is far too cold and

---

* We may even want to add the right combination of natural disasters to our list of "bioindicators" on distant planets. Only dangerous places will ultimately reward our searches for alien biology. As we explore the universe, we should seek other places that are comfortable for life—but not too comfortable.

thick to break into competing plates. As a result, there are no seismic hazards or active volcanoes on Mars, and it's no coincidence that there are (apparently) no Martians who can worry about such things.

Inside and out, Earth is a confluence of chaotic subsystems, a perfect recipe for unpredictable, fluctuating conditions and behavior. The most extreme of these vacillations cause severe global changes that lead to mass extinctions. Calamity is built into Earth's DNA, an unavoidable feature of a living world.

A Taxonomy of Catastrophe

Each December, San Francisco hosts a gargantuan gathering of the nerds. The American Geophysical Union (AGU) meeting is the largest annual congregation of international earth and planetary scientists anywhere in the world. An invited AGU talk carries some geek cred, so in 2012, just as I was embarking on this book project, I was psyched to be invited to talk about catastrophic planetary changes at a special session on "Planetary Evolution and the Fate of Planetary Habitability." The session organizer, Nathalie Cabrol, an eclectic and adventurous planetary geologist at the SETI Institute, was (I think) expecting me to speak about runaway greenhouses and other natural climate catastrophes on Earth-like planets. I told her I had been taking some time away from my usual research to look at Earth's current catastrophe from a planetary perspective, and I asked her if I could speak on this topic at her session. Fortunately for me, Nathalie is a broad-minded and good-natured scientist, and she agreed, so I got my first chance to try out these ideas on an international scientific audience.

For AGU, more than any other conference I've attended, you've got to gear yourself up for the onslaught, because over the years it's grown to an overwhelming size. All week long the great glass escalators of the Moscone Center, like nonstop conveyer belts at a brainiac factory, carry an endless stream of nametag-wearing geoscientists up through the vast, airy mezzanine and disgorge them into the intellectual and social scrum. Open spigots of beer or coffee, depending on the time of day, can be found at the densest parts of the nerd swarm. These days, it's become impossible for anyone, PhD or average Joe, to keep track of all the breaking science about our changing planet and the myriad ways we humans are implicated in disturbing it. Reflecting this insane proliferation of planetary information, there is no way to follow everything going on at AGU. Still, it's a great place to take the pulse of the geoscience community, to see what everyone is talking about. That year the new word buzzing through the hive of twenty-four thousand researchers gathered for science, networking, and microbrews was *Anthropocene*, meaning "age of humanity." Though its status as an official geological time period was the subject of an ongoing debate, it was starting to become common as an informal moniker for our present-day epoch, acknowledging humanity's dominant influence and placing us in the landscape of deep time.

At the 2012 AGU meeting, I was one of several speakers addressing the Anthropocene from different angles. In quickly scanning through the unwieldy, phone book–size program, I was amused to find, among the jargon-filled titles, one presentation called simply "Is Earth Fucked?"[1]

It is—let's face it—a good question. Although we don't usually, at least in this forum, put it so crudely, this anxiety is largely what drives the new Anthropocene buzz. How can we wrap our heads around the totality of the human influence on Earth? Do we really hold the fate of this planet in our unschooled and unsteady hands? Can we learn to handle that responsibility? My talk, with the comparatively tame title "Global Changes of the Fourth Kind: Assessing the Anthropocene in the Context of Comparative Planetology," was an attempt at framing an answer.

Today our planet seems to be in crisis, but let's calm down and look at this in perspective. It's normal for a planet to be rocked by occasional calamity. Earth wasn't born yesterday. Clearly the human presence is stressing it out in new ways, but this old world has been through a lot, including a long series of catastrophic changes, several of which might have nearly wiped out all life. Given all that Earth has gone through, what really is so new and so different about now, about us? If the world has suffered dramatic and deadly fluctuations before, if climate has been so hot that the polar caps completely melted off and so cold that the whole Earth was frozen over, if (as I'll explain) other species have also thoroughly polluted the atmosphere and threatened the future of the entire biosphere, then how can we claim that what is happening here now is so unprecedented as to mark a new phase in geological history? We have the sense that there is some essential difference about the current change, but what exactly? Is that just a self-centered impulse? What is really so astoundingly new and different about the transformation of Earth under the influence of humanity?

In my research, I've studied the major transformations that planets can suffer, and lately I've been thinking about how the change occurring now on Earth looks from a planetary perspective that emphasizes deep time and whole planet transformations. In order to assess how we rate as change agents, I've tried to distill all the major types of global catastrophes into a simplified taxonomy of planetary traumas. In looking, from an astrobiology perspective, at the revolutions that can rock worlds, I've concluded that all planetary catastrophes can be grouped into four major categories, four kinds of planetary change, classified with respect to the role of life and the agency of mind. These four kinds of planetary change are:

1. *random change,*
2. *biological change,*
3. *inadvertent change, and*
4. *intentional change.*

The first two have profoundly affected life on Earth throughout the history of the biosphere. The third describes a catastrophe that is occurring for the first time now and is rapidly remaking our world. The fourth is currently occurring in very limited ways, but it suggests a potential, a kind of change that may be key to survival of intelligent technological life here and on planets elsewhere in the universe. Indeed, I will argue that the fourth kind of planetary change may suggest the best way to *define* a certain kind of intelligent life. Planetary changes of the fourth kind will, I suspect, be the key to our survival.

Planetary Changes of the First Kind: Random Catastrophes

The universe can be a dangerous place. Some catastrophes arise from the random acts of nature, those drastic events that "just happen" to a planet, when massive physical change arrives suddenly from within or beyond.

The canonical example is the big comet crash that ended the Cretaceous period. This was the most recent of the Big Five mass extinctions,* one of the largest ever, and is probably the

---

* There were also substantial extinctions at 34, 37, and 57 million years ago that did not make the Big Five.

most famous, as it terminated the reign of those paleontological rock stars the dinosaurs. The giant lizards ruled Earth for 135 million years, but their time in the sun ended instantly one day 66 million years ago when a comet crashed down in what is now the Yucatan Peninsula of Mexico. It left a smoking crater one hundred miles across, and a planet shrouded in dust and suddenly wiped clean of its top predators and most other species. Among the few survivors were some tiny early mammals who, in their meekness, inherited the earth and evolved, eventually, into us. [An artist's conception of one planetary change of the first kind is shown on page 2 of the photo insert.]

Planets have bad days, and there's a lot of stray stuff out there in the solar system. Such collisions will occur, with varying rates depending on the local populations and orbits of asteroids and comets, on every planet in the universe.

There are other possible sources of mass death from the sky that must sometimes cause mass extinctions on planets with life, and may have on Earth, although none has yet been definitely implicated. The most fearsome of these are gamma ray bursts, unbelievably monstrous explosions caused (we think) by the violent collapse of a star into a black hole that then glows briefly with the light of a million trillion suns. We've observed these occurring only in distant galaxies, which is a good thing. We're glad they don't happen very often in our own galaxy, as they would definitely leave a swath of mass extinction on nearby planets. These and the more frequent supernova explosions, when massive stars blow their guts into space, are possible causes of mass extinctions that would also go in the category of planetary changes of the first kind, or random acts of nature.

Several other mass extinctions were brought about by random changes that came not from deep space but from deep within Earth. As the continents slowly drift around the globe, riding the churning mantle, their shifting arrangement deflects and rearranges prevailing winds and ocean currents, sometimes altering global circulation patterns and triggering big climate changes. A few times, all the world's land surface has clumped together into one giant supercontinent, surrounded by one global ocean. Even supercontinents drift, and 450 million years ago the planet was put out of sorts when the supercontinent Gondwana drifted down over the South Pole. As this one and only huge continent completely froze over, the ice and snow piled up, miles thick. So much of the world's water became locked into these colossal drifts that sea level plummeted. During several million years in this strange configuration, there were multiple glaciations punctuated with more mild interglacials. Both sea level and ocean chemistry fluctuated sharply, dooming many species dependent upon the disappearing and changing coastal environments. This caused the Ordovician–Silurian extinction, probably the second-most extreme mass extinction in Earth history. It seems that the perpetrator was just a "series of unfortunate events," a random convergence between global cycles of plate tectonics, hydrology, atmospheric circulation, and orbital oscillations to produce a rare climate condition that threw Gaia's homeostasis out of whack long enough to give the tree of life a severe pruning.*

Another kind of threat from Earth's insides is the occasional rare but intense pulse of volcanic activity capable of causing major climate change, poisoning the oceans, and radically disrupting the biosphere. Earth's internal heat engine generally runs smoothly, with subducting slabs balancing rising plumes, but occasionally it falters, stutters, or revs up and lurches forward. Sometimes a superheated plume of material rises to the surface from deep within Earth's mantle and unleashes a mighty gusher

---

* A minority (but not crazy) opinion among scientists is that this extinction was actually caused by an extraterrestrial gamma ray burst.

of volcanic magma that floods vast areas with smooth basaltic magma and pumps the air full of greenhouse gases. Earth's surface contains several huge areas of volcanic rock, extending across millions of square miles, created in such outpourings. These include the Ontong Java Plateau, spread along the ocean floor in the South Pacific; the Siberian Traps; the Deccan Traps, in India; and the Columbia River Flood Basalts in the northwestern United States. Several of these outsize outbursts of lava and volcanic gas have been precisely dated to times in Earth's history when the climate surged and species died en masse.

Among the largest is the Central Atlantic Magmatic Province. Its remains today are found in vast deposits of basaltic rock spanning the Atlantic from Brazil through eastern North America to Africa to northwest France. Its formation 201 million years ago, one of the most massive floods of volcanic rock in the history of the planet, caused the Triassic–Jurassic mass extinction, one of the Big Five. The volcanism produced a great surge of CO_2 emission, and as a result Earth was wracked by climate change, sea level fluctuations, and ocean acidification— sound familiar?—causing a sudden decrease in biodiversity. As with all mass extinctions, there were ultimately winners as well as losers. The Triassic–Jurassic extinction cleared the way for dinosaurs to dominate the continents for the next 136 million years, until the day that asteroid came along.

A similar outpouring of magma, forming another of Earth's largest volcanic areas, has recently been convincingly implicated in the most severe mass extinction Earth has ever experienced.

The Worst Thing That Ever Happened

Two hundred fifty-two million years ago, the Permian geological period, a fertile time of vast conifer forests crawling with diverse reptiles and amphibians, was brought to an abrupt and brutal end. Suddenly animals were going extinct at horrifying rates. More than 90 percent of all species living in the seas and more than 70 percent of land animals vanished completely. Among the multitude of victims were all the trilobites, those classic, collectible horseshoe crab-like fossils; almost all the spiraling, segmented ammonites; most fish and reptiles; and the largest insects ever to live on Earth, the *Meganeuroposis permiana*, or griffinflies, dragonfly-like creatures with wingspans up to twenty-eight inches. At least judging from the number and variety of creatures suddenly wiped away forever, it was the worst thing that has ever happened on Earth. This event, the greatest loss of biodiversity in the history of animal life, is officially called the end-Permian, Permo–Triassic, or just P–T extinction, as it marks the boundary between the Permian and Triassic geological periods. Unofficially, but widely, it is called the Great Dying.

What the hell happened? The geological record shows that many Earth systems went haywire at this time. Ocean temperatures shot up in a way that implies rapid global warming of 8 to 10 degrees Celsius (15 to 18 degrees Fahrenheit). There was some sort of sudden and mysterious disruption of the global carbon cycle, and ocean chemistry went through sweeping changes, with drastic acidification and a big decrease in dissolved oxygen content.

Two theories have long competed. One is (you guessed it) a huge and deadly extraterrestrial impact, much larger even than the dino-killing one at the end of the Cretaceous. Many scientists have advocated for an impact, and over the past few decades several teams have even announced with great fanfare that they had finally found the "smoking gun," the remnants of the giant crater, buried on the ocean floor or beneath Antarctic ice. In each case, the evidence has been contested and the community has remained unconvinced.* In the last several decades, the end-Cretaceous impact has taught us a lot. We've learned that, in addition to a crater, a giant impact leaves all kinds of other evidence. This collateral environmental damage includes ash layers, highly shocked mineral grains, and vast beds composed of tiny spheres of melted rock. If there was a colossal impact at the end of the Permian, we should have found some of these other signs by now. Rather, as the geological record is studied in more detail, an impact explanation for the Great Dying seems less and less likely.

The other leading theory has blamed the Great Dying on an enormous spurt of volcanic activity, with the obvious suspect being the Siberian Traps. We've known for years that this mammoth volcanic deposit in the vast northern Asian province of Siberia, consisting of seven hundred thousand cubic miles of basaltic rock, was formed at *around* the right time. Yet it has been hard to really pin down the timing of both eruptions and extinctions well enough to know if they could be convincingly linked. Without better data, it was like trying to solve a crime where you knew only roughly when it occurred and that your suspect was probably in the area.

It appears that the culprit has now been identified, thanks to recent breakthroughs in the accuracy of this geochronological detective work. Many efforts to decipher the history of the P–T extinction have focused on the Meishan region of China, where a section of rock, exposed in the walls of an old limestone quarry, abounds with fossils from the very end of the Permian and the beginning of the Triassic. These fossil-rich

---

* "The community" is appropriately stubborn, curmudgeonly, conservative, and not easily swayed.

rocks are interlaced with layers of volcanic ash that, in principle, can be dated very precisely. However, these rocks are a quarter of a billion years old, so an error of just 1 percent in your age determination would mean it was off by 2.5 million years. In order to nail down the chronology of extinction well enough to derive cause and effect, you need to say with confidence what happened over a matter of thousands of years. This requires outlandishly precise lab work.

In 2013, MIT geologist Sam Bowring, working with his graduate student Seth Burgess and with Shu-zhong Shen from the Nanjing Institute of Geology and Paleontology, used a new high-resolution dating technique, analyzing atoms of uranium and lead that, counted properly, transformed the tiny zircon crystals in these rocks into incredibly accurate timepieces. Their work showed that the extinctions in China happened quickly and simultaneously with the Siberian eruptions. The die-offs at Meishan were complete within about twenty thousand years, which is more than ten times faster than previous estimates. In a geologic time frame, this is nearly instantaneous—truly catastrophic. Improved dating of volcanic rocks from the Siberian Traps suggested that the rates of volcanism were extreme, among the highest ever in planetary history, with the bulk of the vast igneous province forming in less than a million years. The environmental effects of all that erupting lava and gas were likely peaking at just the right time to cause the Great Dying.

Yet climate models show that even the great volume of CO_2 released through this sudden volcanism should not have caused the amount of climate change seen, unless the effect was somehow amplified. Recent research has identified several ways the climate impact may have been magnified. The area where the Siberian Traps formed was rich in organic sediments, so heat from the rising magma may have cooked huge deposits of coal, carbonates, and other sedimentary rocks, causing them to release vast amounts of methane and sulfur dioxide (both strong greenhouse gases), and may have started coal fires that raged for millennia, releasing even more greenhouse gases. When these additional sources are added into climate models, the Siberian Traps volcanism can cause the observed climate change of 8 degrees Celsius. The magma percolating through organic sediments may also have released large quantities of mercury and other toxic metals, adding another lethal element to the environment, as well as organohalide gases (including CFCs) that could have destroyed the ozone layer, which protects land-based life from DNA-destroying ultraviolet radiation. All the volcanic sulfur would have caused intense acid rain, which, along with the pulse of CO_2, would have acidified the oceans. The fossil record is consistent with this: it shows that the dying was especially concentrated among shallow-water species with shells that would have been dissolved in acidic ocean conditions. In summary, a volcanic pulse that sudden and huge can cause all hell to break loose.

The exact kill mechanisms are still being worked out, but the chronology has now convincingly implicated the volcanic culprit. At just the precise geological moment when most species suddenly dropped dead, a hot plume rising from the mantle caused enormous floods of volcanic magma to pour forth from the Siberian ground, warming Earth, acidifying the oceans, and creating a host of other extreme environmental changes. Most of life just couldn't cope.

Whether death comes from the sky or from deep within Earth, planetary changes of the first kind are characterized by random forces of nature acting in ways where life is simply an innocent victim of circumstance. Not so with the next category of catastrophe.

Planetary Changes of the Second Kind—Biological Catastrophes

Then there are those mass extinctions where life is not so innocent. Biological evolution can also cause catastrophic global transformations. Sometimes life brings disaster upon itself.

Recently some scientists have taken another look at the cause of the Great Dying, the most extreme crisis Gaia has ever experienced. They've found that while there was undoubtedly a volcanic trigger, there may also have been a biological bomb. The improved time stamping I've just discussed has made it abundantly clear that the Siberian Traps volcanism is implicated. Yet the climate modeling of this event doesn't obviously add up. The predicted temperature does not clearly rise to the level indicated in the rocks. The question has lingered as to whether the impressive quantity of gases released would have been sufficient to have caused such an apocalypse. Could the physical consequences of the volcanism have been greatly amplified by a positive biological feedback? The idea arose when Dan Rothman, a geochemist at MIT, was studying the graphs showing changes in the carbon cycle at the time of the Great Dying. He noticed something about the shape of the curves. He saw a pattern of extremely rapid and accelerating atmospheric and oceanic change that did not look like the simple signature of a volcanic injection of gas into the environment. Rather, it looked to him like a "superexponential" pattern, the kind of increase you see in a gas being exhaled from a rapidly multiplying biological population. He and his colleagues did some detailed mathematical analysis of the geological record, and their results confirmed that the pattern of change was more consistent with biological growth.[2] They connected the Great Dying with the evolution and rapid growth of a new type of bacteria that produces methane and multiplies rapidly when it is supplied with the metal nickel, a trace element that was delivered in abundance by the Deccan Traps volcanism. They proposed that the climate change had been enhanced by the rapid multiplication of these "methanogenic" bacteria that convert organic food into methane and flood the air with this very strong greenhouse gas.[3] Methane is a much more powerful absorber of infrared radiation than CO_2, and its sudden rise could explain the precipitous global warming observed at the time of the extinctions.

A burst of atmospheric methane would disturb more than the climate. Methane also destroys oxygen and would have contributed to the anoxic (oxygen-depleted) conditions in the ocean that, along with acidification, contributed to the Great Dying. Less oxygen in surface waters would have meant more sulfide, possibly releasing toxic amounts of hydrogen sulfide (the poisonous gas that gives rotten eggs their awful smell), helping to doom land animals. Organics from dying organisms may have fed further growth of methanogenic bacteria, producing an accelerating feedback cycle of animal death, bacterial growth, methane emission, climate change, and extinction.

So, the Great Dying may well have been biologically induced, or at least amplified, with massive volcanism and microscopic bacteria conspiring to cause the greatest calamity in the history of life. It's a crazy story, and it just might be true. I don't know if this wild new idea will stand the test of time, but it serves to illustrate the kinds of feedbacks that can plausibly occur between self-multiplying biological and random geological elements within the complex Earth system. Many of the biggest extinctions in Earth history probably resulted from intertwined changes of the first and second kind.

Now, traveling about twice as far back again in Earth time, back to 542 million years ago, we encounter another sudden massive wave of extinction instigated by runaway biology: The Cambrian substrate revolution, when the texture and chemistry of the seafloor (at that time, the entire habitable surface of the planet) was rapidly remade by a spurt of biological innovation that caused both mass death and fantastic new evolutionary opportunity.

This happened nearly simultaneously with the onset of the Cambrian explosion, one of the most profound and portentous transitions in the history of Earth, when life suddenly became complex. Prior to 542 million years ago, almost all life consisted of microbes in the ocean. Cells had already experimented with many colonial, communal living arrangements, and a few multicellular forms that can be considered early animals had appeared. Yet, by and large up to this point, life was simple and single-celled. Then a catastrophe occurred (in the sense of sudden change), a dramatic increase in biodiversity, an anti-extinction, if you will. In a flash, there were not just *some* animal forms, but *all* of them. In the fossil record, all modern body types appear together at this moment. We don't know why this happened when it did. It's likely that the explosion had to wait until oxygen levels rose high enough to support the greater energy needs of larger bodies, and it came right on the heels of a "snowball Earth" episode, when Earth's climate temporarily lost its balance and the whole planet froze over. It's almost as if Earth's biosphere was so relieved finally to melt its way out from under all that ice that it celebrated with an exuberant burst of animal evolution. Then, almost as abruptly as they appeared, many of these earliest animals suddenly went extinct. The exact cause is not known, but one suspect is the Cambrian substrate revolution.

Up to this point the seafloor was covered with microbial mats, soft, slimy, layered carpets of bacteria that lay upon the ground like a thick coating of living Jell-O. Many of the earliest animals were well adapted to resting upon this mat-covered sea-floor, and grazing on its surface. The soil beneath the mats was devoid of oxygen and inhabited by bacteria that emitted hydrogen sulfide, poisonous to animal life. Yet among the multiple animal inventions at the dawn of the Cambrian were new forms of mobility. A number of critters began burrowing into these mats and braving the toxic soils beneath them. These digging, tunneling creatures are known as "bioturbators," basically life that stirs up the ground. In fact, the fossils that officially define the beginning of the Cambrian in the geological record, *Trichophycus pedum*, are identified by the looping, burrowing patterns they left in the seafloor. Once this bioturbating disturbance of the Cambrian seafloor got started, it began changing the texture and chemistry of the dirt in ways that made it more inviting for other burrowing creatures. Oxygen began penetrating the loosened soil beneath the mats, neutralizing the lethal hydrogen sulfide. This started a positive feedback. The more oxygenated the soils were, the less dangerous they were for animal life, allowing more animals to evolve to burrow underneath. This additionally broke up and oxygenated the ground, inviting subsequent waves of animal invasion. The era of a seafloor covered in unbroken microbial mats, which had lasted for hundreds of millions of years, was soon over. This may explain the sudden extinction of many of the early Cambrian animal forms, as the environment into which they had evolved, a world of endless soft microbial mats upon which to sit and feed, rapidly disappeared, shrinking to a few isolated rocky coastal areas. The evolution of the new burrowing lifestyle greatly changed the world in which early animal life had appeared, determining the winners and losers during that time of rapid evolutionary change.

Now if we travel still farther back in time, about four times again as far, to over 2 billion years ago (halfway back from now to the beginning of the world), we come to the most dramatic example of a biologically induced catastrophe: the radical and abrupt chemical change that swept Earth's surface and atmosphere after life perfected the use of solar energy. Around 2.5 billion years ago the oceans of the world were suddenly filled with efficient, self-multiplying solar cells. These cyanobacteria, sometimes called blue-green algae, proliferated around the planet and collectively transformed it in the most promising and devastating way. Life had flirted with solar energy earlier, but the cyanobacteria evolved much more efficient chemical pathways to exploit photons of sunlight, using their energy to build useful organic chemicals. This development was both wonderful and horrible.

Wonderful, because life gained the ability to harvest sunlight for energy. What a fantastic breakthrough! Earth's biosphere learned to plug into the best power source the universe has to offer, tapping a nearby star to manufacture food from light. This liberated huge amounts of free energy for the biosphere, enabling the profuse diversification of life that has come in its wake, turning the planet green with phytoplankton, ferns, forests, and flowers, and the fish, flamingos, ferrets, and frogs that feed on them, furthering the journey from light to flesh.

But first it wrecked the world. It filled the atmosphere with a dangerous, corrosive, poisonous gas that brought about horrible worldwide death and destruction. What was that terrible pollutant? It was O_2, oxygen!

With oxygenic photosynthesis, life harnessed the Sun, using the energy to split up H_2O molecules, keeping the H to react with CO_2 and make organic biomolecules (i.e., food) and discarding the O. Even after this started, it took a huge amount of time for oxygen to build up in the atmosphere, because it reacts with nearly everything. For hundreds of millions of years, the cyanobacteria kept spitting out oxygen, but all the excess was hungrily snapped up by the abundant iron in Earth's crust and interior. Yet eventually, around 2.4 billion years ago, the crust was thoroughly oxidized and there was no more available iron lying around. The oxygen was finally free to accumulate, and the amount of O_2 in the air shot up—which sounds great to us, but believe it or not, oxygen is an awful poison that reacts violently with and destroys the molecules of life. When life made oxygen, it was literally playing with fire. When we burn wood, coal, or oil, we are taking advantage of this tendency for oxygen to combust organics. Burning is the reversal of photosynthesis. In a flash of flame, energy once gained from the Sun is released again. When you bask in the glow of a wood fire, you are enjoying, years later, the heat and warmth from all the sunny days when that tree was growing. For all those years, the energy was stored in the organic matter of wood. Similarly, when we burn fossil fuels, we are releasing solar energy extracted and stored by plants millions of years ago. Burning snaps up oxygen, returns organic carbon to CO_2, and throws the energy back out.

Oxygen and organics do not play well together unsupervised. Exploiting that imbalance, our animal ancestors evolved respiration. Respiration is controlled combustion that uses enzymes to carefully marshal that powerful energy release, harvesting it so that instead of dissipating in a wasteful flame, it is instead converted into phosphate bonds. These are the little chemical batteries that store energy within each of our cells and release it only when we need it. Before we evolved that ability, however, the buildup of corrosive oxygen in the atmosphere was massively fatal for most of the species that existed on Earth at the time. For much of the biosphere, the rapid oxygenation of Earth 2.1 billion years ago meant game over.

This Great Oxygenation Event was the most extreme chemical transformation Earth ever experienced—and it gets worse. The consequences for life were even more dire because not only was oxygen poisonous to organic life, but to add grave insult to fatal injury, its sudden rise seems also to have caused one of the worst climate disasters Earth has suffered.

The Great Oxygenation Event was contemporaneous with one of the most severe ice ages this world has ever known, an event known to geo-nerds as the Paleoproterozoic Snowball Earth episode.* This was probably no coincidence. At the time, Earth's climate was likely being kept above freezing by a methane greenhouse. Methane is such a powerful infrared absorber that a very small amount of it can significantly warm a planet. It is also, however, an organic molecule that is easily and eagerly consumed by oxygen. So when all that oxygen released by the cyanobacteria built up in the atmosphere, it quickly destroyed the methane greenhouse, the atmosphere suddenly became more transparent to infrared radiation, and the temperature plummeted, plunging our planet into a complete global freeze. Such a deeply frozen condition could even potentially become a permanent dead-end state for a planet like Earth. Under what circumstances such a perma-freeze planet might emerge is a classic problem I had to solve in grad school, and later made my students solve. Simple climate models show that a frozen-over world is stable because ice has a very high albedo, meaning it reflects most of the sunlight striking it, which keeps the planet cold and maintains the reflective ice. Fortunately, the real world, more complex than a simple climate model, is full of competing feedbacks and holes in the ice. Ocean currents thin the ice in places, creating patches of open water that, along with areas of dirty ice, absorb more sunlight. Then what really springs the trap is that volcanoes keep pumping out CO_2 and other greenhouse gases, oblivious to the freeze-out. Along with the slowly warming Sun, that was enough

---

* The Paleoproterozoic distinguishes it from a few other "snowball Earth" episodes that happened much later in Earth history.

to melt the world. Life persevered, and eventually thrived in the newly oxygenated world. Evolution, the ultimate opportunist, turned destruction into creation. Some organisms developed the chemical machinery to derive energy from the extreme, deadly reactivity of oxygen with organic matter and to funnel it into useful cellular work. That's called respiration, and it allowed the evolution of new animal life. This led, among many other things, to us.

For a planet like ours, it turns out, it's not so easy being blue-green. The wild global success of the cyanobacteria, enabled by their use of solar energy, caused one of the closest calls to total extinction our planet has ever faced.

The Anthropocene Dilemma

Those irresponsible cyanobacteria. They not only caused a mass extinction, but they nearly wiped out all life with their careless climate interference. They discovered a powerful new energy source and recklessly exploited it, bringing ruin to all species that couldn't get with their program, taking the world to the brink of ecocide. Talk about unintended consequences! Unwittingly, they changed the world and made it unlivable for many, perhaps most, of their fellow travelers.

Of course we don't really think of them as irresponsible. They're just bacteria. Yet today we see ourselves behaving in a similar way, and we look on with horror* because our actions do seem deeply irresponsible. So what have we got that the cyanobacteria didn't have?

Just as the collective action of billions of cyanobacteria, each responding to its own survival needs, once led to a catastrophic

---

* Those who are paying attention.

transformation in the chemical state of Earth's atmosphere and oceans, so today the collective action of a new species is causing sudden global transformation. Yet there are novel aspects of this transition that signal a new kind of global change, neither random nor biological.

In a sense, of course, this new catastrophe is biological. It is brought about by living organisms: human beings. Yet to call it merely another biological catastrophe would be to ignore mechanisms at work that are unprecedented in the history of the planet. Sure, you could make a good argument that this current change, like the oxygen catastrophe, is brought to you by bacteria. Our multicellular ancestors evolved from collectives of single-celled organisms. Our individual cells still resemble them to the point where you can reasonably view yourself as a kibbutz of bacterial cells that act out the coherent pattern of activity that defines your self. That's true even for the parts of you (the minority, it turns out) that are actually "you" in the sense of being coded in the chromosomes you got from your parents. Now we've discovered that this parentally inherited part is but a portion of you that forms a habitat for the rest, for your microbiota, the shifting, porous microbial community that makes up the rest of you. So, certainly it can be said that whatever humans do is being done by bacteria, but not just by bacteria. We are also obviously something completely new that other bacteria, not organized in this way, are not; and we are something that other animals are clearly not, either. Something that can make art; invent, reinvent, and remember new tools; create, remember, and pass down stories, songs, dances, and hypotheses; record the past; imagine the future; plan and reflect; figure out laws of nature; walk on the Moon; and wreck the world.

From the planet's point of view, what is happening now is as much a departure from the biological change it has experienced for most of its lifetime as that biological change was a departure from the brief earlier time of naked physical changes. This new form of global change comes from biology, but it is not described by biology any more than biology is fully described by physics and chemistry.

Yes, I realize that such an assertion reeks of human exceptionalism, the hubristic claim that humans are not just another species but that we represent an entirely new phase in evolution. Yes, this is problematical, but also unavoidable. From a planetary evolution perspective, our significance as a new kind of change agent is at least as obvious as it is disturbing, and it's interesting to try to pinpoint what exactly is so new and different about us. My approach to this question is an astrobiological one. Imagine for the sake of argument that the evolutionary developments that led to this change represent a phenomenon that may occur throughout the universe, on certain lucky (or cursed?) planets. Whether, given what we know, this is a good assumption, and what implications that has for life and intelligence throughout the universe is a subject I'll return to. Just as we can speculate about the universal qualities of global biospheres, we can ask what emerging global technospheres should have in common. What might their essential characteristics be?

It would be wrong to hang it simply on the evolution of intelligence, or the development of language or technology. On Earth, none of these is unique to humanity, and other species possessing these qualities are not using them in ways that cause catastrophic global changes. Pods of whales have sophisticated language and engage in clever group planning and coordination when hunting. Many nonhuman primates creatively use sticks and stones to catch termites, reach honey, and open shells. Elephants modify branches into back scratchers and fly swatters. Numerous species of birds have been observed using twigs as simple tools, or adapting human artifacts for nest building or various other clever purposes. A popular online video shows a playful daredevil crow using a plastic lid repeatedly to sled down a steep snowy roof. After watching this a few times, it is not too hard to imagine, if human technical civilization does not persist on Earth, that some other species will come along in a few tens of million years and give it a go.

Yet, in other species, these activities have not led to an acceleration of technological innovation with widening planetary consequences. Something else is going on here. One species, through some combination of language, cognitive skills, social intelligence, and technical inventiveness, has become successful and powerful enough to start rapidly changing the world while, at first, having absolutely no awareness that it was doing so.

Such inadvertent global change, at least as a phase, seems inevitable for a young, expanding technological species. If members of a species successfully develop technology to effectively solve their local survival problems (such as how to grow and transport more crops or survive in the cold), but are operating under the reasonable assumption that the world is infinite, and still behaving according to the biological imperative to go forth and multiply, they'll increase their numbers and impact to the point where they start to have a global influence. It seems likely that this activity would usually precede any knowledge of its world-changing nature. Why should they imagine they could change the world?

There was always a seemingly infinite, immutable world beyond our sphere of influence, and we never perceived of our own mortal actions as altering the very world itself. The World— that is, all of nature, all that there is. Who could tinker with existence? Maybe gods, but not us. This illusion would be shared by any young species first wielding the power of world-changing technology. They would not know their own strength.

Such clever creatures will have global influence but likely will lack both awareness of this and self-control (or even a self?).

Even if or when they did realize what they were doing, could they act on this knowledge, even in the interest of their own survival, if they were not really set up for globally coherent action? In order to have self-control, one needs a sense of self. Do we have this, on the global level at which intentional action is now required? Recall the Asimov essay I describe in the first chapter. We are at a stage between the individual and the multiorganismic. Our planetary-scale challenges require us to perceive ourselves and act as the latter.

You cannot become aware of global problems, of your own global role, or think about responding globally until you know that you live on a planet and are able to observe and study its changing properties. For us these realizations required very sophisticated scientific knowledge, including space-based sensors, which could not have been built without the bootstrapped explosions in knowledge and innovation of the scientific and industrial revolutions. We could not really see our world, wholly and clearly, until our transformation of it was already well under way.

In planetary changes of both the third and fourth kinds, the world is changing through cognitive processes. Both involve intention, enabled by its great force multiplier, technology. Intentional actions require forethought, internal mental pictures of a future state that one's actions may bring about and, often, the ability to communicate this vision to a group and plan collectively. These are capabilities that have quite recently appeared in Earth's biosphere. Later I will parse this a little more carefully, taking up the question of what it means for a species to evolve these abilities, but for now, suffice it to say that this is a kind of activity that bacteria, fleas, and sea slugs do not undertake but that human beings do, and one that is quite clearly changing the world. So this qualifies as a new kind of planetary change, one that Earth has not encountered before.

What do I mean by technology? The invention and use of

tools to enhance one's ability to affect one's environment. With technology we transcend the physical and temporal limitations of our individual bodies, changing our environment by building "extrasomatic" (beyond the body) structures, including information storage that allows us to accumulate knowledge, recording and passing on stories, lessons, and plans. From stone chips to silicon chips, we've extended our reach, moving and shaping our surroundings with greater distance, force, and persistence than can be achieved with teeth, skin, muscle, and bone.

Key here is not simply the existence of intentionality, but the physical scale at which the consequences of intentional actions play out. Earlier I said that the cyanobacteria changed the world "unwittingly." Neither the third nor the fourth kind of change can be called "unwitting," as each arises from awareness and intention, manifesting at different levels and scales.

Consider the traffic at rush hour on Interstate Highway 10 through Los Angeles. You have many individuals acting to solve their own problems, such as how to get from Pasadena to Venice Beach without, at least until recently, much awareness of the cumulative, global effect of their actions. Each driver is at the wheel of their own car, capable of applying feedbacks, making course corrections to avoid problems. All are participating in a massive system of cooperative collective activity that, despite our obligatory bitching and complaining, actually works fantastically well. Collisions and acts of road rage are rare, even in California. Yet who is driving the global transportation system, applying feedbacks and making course corrections? Nobody? Or perhaps "the market," that metaphysical beast whose emotions, thoughts, and decisions we hear about daily in meta-metaphorical attempts to describe our collective actions to ourselves. To the extent that this system can be said to be under conscious control, it is an entirely different sort of consciousness than we each possess individually.

We have individual agency, and we have limited collective agency—that is, we have the ability to plan and act in groups. Indeed, this is perhaps the hallmark of our species. Yet we also act on scales beyond which we have any obvious agency. On a global scale, we watch, discuss, and describe our actions as if from a distance, seemingly unable to stop, change course, or control ourselves. And as we start to become aware of the multigenerational consequences of our actions but act largely without any long-term plan or sense that we can enact one, it is almost as if we are observing someone else doing these things. It's like a disconcerting dream where you see yourself running toward a cliff or committing a crime, but you can't do anything to stop yourself. There is certainly intentionality in the human actions that are causing planetary-scale changes, but it exists almost entirely at much smaller spatial and temporal scales than those at which these effects play out.

It is this combination of local technical success with global obliviousness that defines planetary changes of the third kind, or inadvertent catastrophe: that is, planetary-scale changes that result from intentional applications of technology but are not themselves intended.

As we become aware of the planetary consequences of our activities, but still lack any sense of direct control over them, we find ourselves in a dilemma. We knowingly use our technology on a scale that is changing the world, yet we have no global sense of intentionality. This is what I call the Anthropocene dilemma.

Planetary Changes of the Third Kind: Inadvertent Technological Catastrophe

And now the world is ours—or at least our responsibility. As those department store signs say, "If You Break It, You've Bought

It." Whether we like it or not, whether we deserve or can handle this responsibility—these are separate questions—somehow, without realizing it, we have assumed it. Human influence has taken firm hold of Earth, and letting go is not an option. There is no doubt that we have entered a new geological time period with a new type of force simultaneously upsetting multiple Earth systems. The case for anthropogenic global warming has been made. I don't feel that I need to waste a whole lot of ink laying out the evidence. It's been done well elsewhere, and it's not my purpose here to convince the holdouts, but rather to try to move the conversation along. Global warming is just one of many disturbances. Discussions framed around the Anthropocene epoch are broadening to include a wide array of anthropogenic changes. I'll offer a summary here, but from a planetary history point of view, there is no question that the Anthropocene reworking and reshuffling of Earth is rapid and extreme enough to be called a catastrophe.

We have all heard by now that we've pushed atmospheric CO_2 to dangerously high levels. You've seen the now-iconic graph of carbon dioxide ascending, oscillating through the seasons, but inexorably rising over the years. This, perhaps the most recognizable, frightening, and convincing diagram in all of earth science, is called the Keeling curve, after Charles David Keeling, who in 1958 somewhat quixotically began monitoring airborne carbon dioxide from Mauna Kea, in Hawaii. Over the intervening decades, up to his death in 2005, he staunchly maintained a continuous set of observations that allowed us to see what we are doing with stark clarity. After just a few years the emerging pattern in Keeling's curve was key to getting scientists to realize that human industrial intervention really was changing our atmosphere at an accelerating rate. Now it has become key for our efforts to communicate this reality to everyone else.

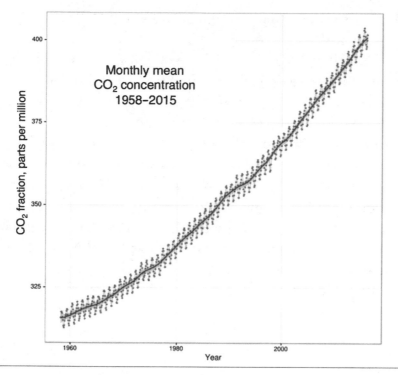

The Keeling curve shows CO_2 rising steadily over recent decades.

Keeling's measurements started a year before I was born (in 1959). Since then, the amount of CO_2 has risen by almost 30 percent. That's not a tweak; it's a jolt.

Our planet has not experienced four hundred parts per million of CO_2 for almost two million years. The details of our ancestry are still being worked out, but it seems that the last time there was this much greenhouse warming, some of the first humans (that is, members of the genus *Homo*) had recently appeared on Earth. These *Homo habilis* were using early stone tools along the drying rivers and spreading savannahs of East Africa as the climate warmed. Anatomically modern *Homo sapiens* first appeared in East Africa around three hundred thousand years ago. Nobody from our species has ever breathed air as thick with CO_2 as we do today.

Certainly if you look back in geological history you can find plenty of times when our planet had more carbon in the atmosphere. Yet it has rarely, if ever, in its entire 4.5-billion-year history, experienced the *rate* of atmospheric change we are inducing, in confluence with such a wide suite of other provocations, to the land surface, to the oceans, to the hydrological and biogeochemical cycles. When you study Earth's deep history, this convergence of changes at accelerating rates is what truly stands out about our time. Together, they represent an unprecedented challenge to the Earth system. The fact that the planet has never experienced anything like this limits the utility of the geological record for making detailed predictions. Still, we can safely predict some general consequences. Already we observe, in addition to the warming of the planet, a rapid loss of sea ice, the retreat of glaciers, rising sea level, wholesale migrations of species to higher latitudes, and large-scale deviations in precipitation patterns and ocean chemistry.

Despite popular assertions to the contrary, none of this climate change is permanent. Not on Earth time. Human time, you might say, is a different story, but our new awareness of the Anthropocene is forcing a reckoning between these two timescales. Some changes may seem "permanent for all practical purposes," yet when we make this distinction, we are removing ourselves from geological time. We need to start doing the opposite: seeing ourselves within the spatial and temporal landscape of the planet we inhabit, especially as, increasingly, we do more than just inhabit Earth.

Earth does, in fact, have natural buffers and correction mechanisms that, eventually, will push back and fix atmospheric imbalances. These systems are now being temporarily overwhelmed by our relentless industrial emissions. Over many millennia, these insults will gradually be corrected by slow but inexorable planetary cycles. The environmental perturbations we are causing will not be permanent, but the extinctions we are causing will be. Other signs of our having been here will also persist for as long as Earth lasts. We are now a part of Earth's geological record.[4] No matter what we do, no matter where we go from here, we have left our mark. One of those indelible signatures will be the sudden disappearance of certain fossils. Barring some determined intervention, CO_2 will remain at elevated levels for about one hundred thousand years. In addition to the direct climate effect of its infrared absorption, it dissolves in ocean water and creates carbonic acid, which corrodes shells and reefs. This acidification of Earth's oceans seems, by now, inevitable.[5] We can change course, save our civilization, limit the harm to vulnerable peoples, and rescue many species that are currently threatened, but much damage has already been done. I'm afraid we may lose our coral reefs.

Slightly comforting, perhaps, is the knowledge that over the ages our planet has lost its reefs several times. During ancient episodes of high CO_2 and ocean acidification, the magnificent coral conurbations have disappeared entirely—and then somehow have returned. So it may be that, even if they are soon dissolved by our short-sightedness, in a million years Earth will once again have great barrier reefs. Yet our great-great-grandchildren may know them only as curiosities in natural history museums.

Other greenhouse gases are also being forced to unnatural levels, the most notable being methane from intensive agriculture: the flatulence of cattle and the bacterial burps of rice paddies. Yet atmospheric change and global warming are only the most well-known examples in an array of accelerating planetary provocations. They're the ones that have gotten our attention, but there are many others. The carbon cycle has grabbed the headlines, but we've also seized hold of many other major geochemical cycles. Through production of fertilizers, we've radically altered Earth's nitrogen cycle. The sulfur cycle has become dominated by industrial emissions. We've dammed rivers so thoroughly that there is now more than five times as much fresh water captured in reservoirs as there is remaining in all the wild rivers and streams of Earth! That is not a minor change. It's fair to say that we've domesticated a major part of the water cycle of this planet. Earth's vibrant hydrosphere, arguably our planet's most distinctive feature, has to some degree become an artifact of human civilization.

Every year now, humans constructing roads, buildings, and farms displace ten times more dirt than the combined erosive forces of wind, rain, earthquakes, and tides. Simply measured by the amount of stuff we move around, we have become the undisputed world heavyweight champions of change.

The sky itself has been altered, muted. Those same spreading lights that, seen from space, mark our age as different, also, here on the ground, fill our nights with scattered artificial illumination. Ironically, in this age when we've finally figured out what the stars are and where we stand in relation to them, we've also distanced ourselves from the direct experience of them. Yet we and the stars go way back together. We're made of stellar remnants and ashes, and as long as we've been human they've guided and inspired us, served as compass, calendar, and clock, oriented us in our wanderings as we peopled the continents and crossed the seas. Even now our interplanetary spacecraft use star sensors to find their way. The stars have humbled us with their beauty and filled us with curiosity and wonder. Yet as we haphazardly spill light from our mushrooming cities and roads, we carelessly push back the night. We've so intensively urbanized our populations that, for most of us, the stars have largely receded from view. I've heard it said, and it may be true, that for the first time ever the majority of children being born today will never in their lives directly see the Milky Way galaxy. The cosmic connection has never been closer or more remote.

Age of Plastic

We've created major new geographic features of this planet, including the Northern Pacific plastic gyre, otherwise known as the Great Pacific Garbage Patch. This semipermanent vortex of debris comes from partially decayed plastic toys, trash, and manufacturing products washed out of storm drains and into the coastal waters of Japan, the United States, and farther afield, gathered and trapped by the ocean currents and winds, extending over an area that is larger than the United States. There's a nearly mythical quality to this structure. For so long we imagined that the world, and in particular the ocean, was infinite. Even when we knew it wasn't, literally, we still fancied it was so large that we could just throw stuff away and it would disappear.* We never worried about where "away" was. There was a time when our numbers were sufficiently few and our construction materials so impermanent that the things we threw away really did disappear into the cyclic reclaiming and repurposing of Earth and its biota. This is no longer the case. We've found out where "away" is. It is, of course, many places, but some of what we throw there ends up in a giant gyre in the Northern Pacific.

In the previous chapter, I recount how the takeover of Earth by life introduced a plethora of new materials here not found on dead worlds, including the majority of minerals in the crust. Now this new global force, the influence of mind and technology, has again flooded the planet with novel substances.[6] We've

---

* Remember, "dilution is the solution to pollution"—until it's not anymore.

created newfangled materials that are rapidly becoming integrated into Earth's oceans, landscapes, ecologies, and rock cycle. My friend Odile Madden, a materials scientist, is fascinated by plastic. Not like the guy in *The Graduate* (whom she is so sick of hearing jokes about), but as principal investigator for the Age of Plastic research program at the Smithsonian's Museum Conservation Institute. She is also part of our informal "Washington Anthropocene Group," a semiregular gathering of like-minded folks I first convened at the Library of Congress, where she regales us with tales from the age of plastic. She's convinced me that plastics make a great proxy for tracking the human influence on Earth in the Anthropocene age.

Plastics are synthetic polymers that were created early in the Industrial Revolution and began to abet and replace naturally occurring polymers such as cotton and rubber. Polymers are long-chain molecules, usually based on carbon. If that sounds familiar it's because that is basically what we terrestrial creatures are, inasmuch as proteins and DNA are themselves long-chain carbon molecules. Yet our proteins work by being modular, flexible, and loose. They fold into intricate 3-D sculptures whose complex shapes run our internal chemical factories by controlling their interactions with the other molecules in our cells. Plastic polymers have rigid, repetitive structures that make them, as molecules, too inflexible for biology but, as materials, stronger and more versatile than living tissues. So we, watery cells of folding, mutable long-chain carbon molecules, have discovered, invented, and refined these materials made of less flexible carbon chains, and have used them to extend our reach into the world.

At first we made them from modified natural materials, but later primarily as synthetic petroleum products. When we started using them we didn't have a long-term plan for what would happen to them. Why should we have? We never had a plan for rock, wood, or steel, which we just left to erode, rot, or rust in place. Who ever thought that the world would start to fill up with our stuff? After World War II, manufacturing techniques perfected for wartime production were used to produce a profusion of new industrial and consumer products, and as with so many other Anthropocene signatures, their use exploded around 1950, and our plastic things started to pile up by roadsides and in landfills and to drift across the seas.

We often use the word *plastic* as a synonym for cheap, fake, or insubstantial, but Odile, in her talks, reminds us of the many ways it has made our lives better. One small but hugely significant example she cites is in the area of medical technology, where plastic products such as blood bags for IVs and inexpensive syringes have helped to spread affordable and sanitary medicine throughout the world. As Odile puts it, we have a love/hate relationship with plastics. We're repulsed by the way they are accumulating in wild places and animal guts, yet we would never give up all the benefits they provide.

Odile thinks we are still in the early part of the plastic age. She reminds us that other ages defined by materials (the Bronze Age, the Iron Age) lasted a lot longer than this, and that we are still on a learning curve, figuring out how to use them well. She and her colleagues have gone on research expeditions to study the plastic debris washed up on the remote Alaskan coastline, analyzing the composition and figuring out the sources. There they find massive amounts of derelict fishing gear, packing and construction materials, and random consumer items (little orange Nerf balls and fake flowers for hummingbird feeders), largely from container ship spills. On one beach they found hundreds of NFL team fly swatters, made in Asia, destined for American markets. Though this sounds dismal, Odile adds, "What we didn't see was Alaska clogged with plastic. Nature was much bigger than the garbage, which felt very good."

Still, in some places nature is being overwhelmed. Midway Island is named for being a lone, tiny waypoint between Asia and North America, the kind of place that used to be considered the middle of nowhere. Yet it is situated in an ideal location for collecting plastic flotsam from the Great Pacific Garbage Patch. No people live on Midway Island, but our stuff is there in abundance, brought with the currents of the gyre, and birds see our colorful plastic bits, mistake them for food, and bring them to their nesting chicks. You've probably seen those heartbreaking images and videos of the sick and dying baby albatrosses. About one-third of the young birds die, many choking on or starving from the plastic diet. Odile and her team have examined plastic retrieved from inside those dead birds and found mostly beverage caps, but also pieces of plastic cutlery, cigarette lighters, and fishing gear.

Odile studies what becomes of discarded plastic, how it breaks down and interacts with its environs. Contrary to the widespread myth that plastic lasts forever, it breaks down chemically and physically, reacting to radiation and agitation. It crumbles into smaller and smaller pieces, becoming minuscule and then microscopic, and then only molecular fragments of the larger polymers. These are ingested by the tiniest of marine organisms and make their way through the food chain. Perhaps appropriately, some of our plastics and their chemical products end up in our own tissue, blood, and guts.

Plastic does take one form, recently discovered, that may in fact last if not forever then for a significant amount of geological time. Some washes up on beaches and ends up in bonfires. In 2006, a strange new kind of rock was discovered on some Hawaiian beaches containing melted plastic binding together bits of rock, sand, shell, and other materials. Some of this tough material, now called *plastiglomerate*, will become buried and last for many millions of years, a new rock type that has suddenly appeared in Earth's strata, marking the time when people built things of plastic, scattered them widely, and sometimes burned them on beaches for celebration or warmth.

The plastic flamingos are coming home to roost. Our response is characteristically human: slowly, haphazardly, and unevenly we are moving to incorporate awareness of the new reality, and taking baby steps toward mitigation. Policies and habits are shifting toward less wanton discarding of plastics, and plastics themselves are evolving to incorporate what we've learned about their larger role in the world. Early on in the age of plastic, scientists and engineers were concerned only with how to make materials that were useful and stable for as long as possible. Now materials scientists such as Odile Madden are learning more about the long-term global consequences of our choices of materials and manufacturing processes. They are thinking about the entire life cycle and how to design plastics with "engineered instability" so that products may disappear gracefully, becoming food instead of hazards or poisons.

Studies are even under way to create technologies to clean up the Garbage Patch. In 2012, a Dutch teenager named Boyan Slat proposed, as part of a high school science project, an idea for a machine that could start to remove plastic from the gyre. Equipped with giant arms to pull in debris, and designed with filters to protect sea life, the solar-array-powered machine would be anchored to the seafloor and would collect plastic for recycling. The sale of the recyclable material it collected would help offset its cost. Many experts have examined Boyan's proposal and found it potentially feasible. His machine would not clean up the microscopic fragments, but by getting rid of the larger pieces of plastic, it would cut off the source of those fragments and dissipate the gyre. I don't know if this or some other innovative cleanup technology will be the way we rid our oceans of this unintended monstrosity, but I bet in fifty years the Garbage

Patch will be smaller than today, and in one hundred years it will be gone entirely. That's too late for generations of poor albatrosses on Midway Island, and other victims of our lack of foresight, but in the life of Gaia, the Garbage Patch will have been a momentary annoyance, here just long enough to leave some odd deposits, pressed like plastic flowers between pages in the long book of geological stratum.

The Homogenocene

Long before the Industrial Revolution or even the Agricultural Revolution, our species had already established a pattern of global ecological disruption. As much as we've altered the course of any river, we have also diverted the stream of biological evolution. It's normal for successful species to rework their environments, forcing other species to adapt and squeezing some out entirely. We've been doing this for a long time, but at some point we were no longer just another species messing with all the others. We became a major new kind of evolutionary force, determining through our actions which species around the globe would flourish and which would flop.

Our success as hunter-gatherers allowed us to successfully inhabit every continent. The toll was particularly heavy on the largest animals because they were either food, threat, or competition. When human beings first spread around the planet, we quickly wiped them out everywhere we went, creating a wake of extinction behind our waves of global migration. As we spread, the "charismatic megafauna" fell away from every continent we colonized: the giant kangaroos and marsupial lions of Australia; the mammoths, mastodons, and giant ground sloths of North America; the elephant birds and giant lemurs of Madagascar;

and so on: when *Homo sapiens* showed up, the big animals soon died off.

Even before the Agricultural Revolution, we were already remaking landscapes, setting fires to drive game from the forest. Then we started clearing fields for planting, and reengineering plants and animals to suit our needs for food, materials, labor, companionship, intoxication, and beauty.

For millennia we've coevolved with our flora and fauna, manipulating some species and being manipulated by others. While we've driven many to extinction, to others we've brought absurd levels of global success. We think of eucalyptus trees as native to Australia, but they were rare there when humans showed up forty-five thousand years ago. Our use of fire to clear other species helped these fire-resistant trees eventually thrive and spread throughout that continent. When I first visited Northern California, I assumed that eucalyptuses were native there, as they so dominate the forested hills. Yet they were introduced only during the gold rush, in the 1850s, when they were thought to be a promising source of timber for railroad ties. They turned out to be lousy for that purpose, but they took well to the dry, sunny hills, where they displaced many native trees. They are now regarded as an invasive species and, ironically, given their history in Australia, have become a major fire hazard in some areas. Eucalyptuses are blamed for fueling the 1991 firestorm that destroyed thousands of homes in the Oakland Hills. If trees could dream, the eucalyptus would never have dreamed of occupying such a large territory without our help.

Domestic cattle are now among the dominant species living on Earth—if you call that living. The combined biomass of all domesticated animals is now some twenty-five times that of all remaining wild terrestrial mammals. Some of our favored plants have become among the most widely propagated on the planet. Wheat, rice, coffee, and cannabis, to name a few, have gone worldwide by giving us what we want. You have to wonder who has been using whom, because in terms of evolutionary success, these plants have done well by us[7]—others, not so much. Through our reworking of landscapes, especially for agriculture, we have destroyed and altered habitats and, often without realizing it, created new ones.

We've remixed the very mechanisms of evolution and created an embarrassment of vectors for creatures to spread between distant and once-isolated locales. Life is opportunistic, and the global success of humans has presented many irresistible openings for world travel and adventure. In the cargo holds and bilgewater of ships; in the bulging microbial manifest of passenger jets; in the cool, leafy folds of produce trucks; and in the fur and guts of livestock and pets, we've offered transport for innumerable species. In this way, we've changed the geometry of the planet. Before we came along, the world was discontinuous. Oceans, deserts, and mountain ranges formed impenetrable barriers, breaking Earth into separate regions where populations could evolve independently, and then be isolated or merged by continental drift and climate change. Now we've created pathways around all those borders, and to some degree the planet is one continuous habitat.

Some species are primed to take advantage of this new human-altered geography. Many of these we consider pests, but we've invited them in, or at least given them a lift, a place to stay, and a way to make a living. Some have followed in our footsteps and gone global. *Linepithema humile* is a species of ant native to Northern Argentina that has, with our help, become a new kind of global superorganism. They first found their way from Buenos Aires to New Orleans in the 1890s, stowed away on ships, perhaps in dirt used for ballast or in bags of coffee or sugar. There they prospered, easily outcompeting all local ant species and spreading throughout the Southeastern U.S. In the early twentieth century, these eager immigrants made their way West, apparently by riding the rails, hopping trains to California. In midcentury, with the new highway system, they spread up and down the West Coast. While ruthlessly aggressive toward other species, they are also unusually cooperative with their own kind. Ants from neighboring nests don't fight, as with other species, but welcome one another as family, forming vast, continuous networks connected by underground tunnels. Thus, in effect, they form supercolonies that can spread and grow over hundreds or even thousands of miles. By now one vast supercolony extends from Oregon down to Mexico, and there are others thriving in most southern states.

Unfortunately, the *Linepithema humile* don't play well with others. They have severely disrupted many ecosystems, decimating animals and birds and trees that depend on the local ants for food or seed dispersal. For example, the horned lizard of coastal California has been dying off, as its normal diet of local ants has succumbed to the aggressive colonists. The result is a dramatic decrease in biodiversity in many of the occupied landscapes. The invasive Argentine ants are also farmers. They breed aphids—milking them for a sugary secretion. These ant farmers sometimes successfully compete against human farmers when their aphids destroy our crops.

What will stop them? Perhaps nothing we do, at least not on purpose. In parts of the southern United States they seems to have met their match in another invasive species we've brought to our shores, the Asian needle ant, or *Pachycondyla chinensis*, who may be even better adapted for invading "our" territory.[8] Also, in California, as of this writing, the drought seems to be holding the Argentine supercolony in check for now, as it thrives only around moisture.

The combination of extreme aggression toward outsiders

and extreme cooperation with their own kind, combined with their facility at co-opting human transport and industry to spread around the planet, has made them a globalized success story. These adaptable stowaways have spread to every continent except Antarctica, forming supercolonies in Europe, South Africa, Japan, Australia, and many Pacific islands. The largest of these extends from southern Italy through western Spain, a distance of more than twelve hundred miles. When placed together in close quarters, ants from the European and Californian supercolonies, or any of the others dispersed around the globe, recognize each other as family. Thus, with our unsuspecting help, they have perhaps formed the world's first global megacolony, a planet-enshrouding superorganism enabled by this newfangled human-connected evolutionary topology.

The Argentinian ant is just one dramatic example of how human transport, travel, and migration have provided opportunities for some species to thrive in new ways, usually at the expense of others. Historian Charles Mann has suggested that this mashed-up new biological phase we've induced be called the Homogenocene, as we have blended so many evolving populations, once geographically dispersed and isolated, into one homogenized genetic broth. In the 1970s Carl Sagan used to zip around Ithaca, New York, with a bumper sticker on his orange Porsche reading, "Reunite Gondwanaland!" referring to the time when all Earth's continents were merged into one supercontinent. I guess, biologically, this is what we've now done. We've seriously rearranged the evolutionary geometry of the world—and all this biological meddling I've described doesn't even include the intentional manipulation of genomes, the engineering of new organisms, or even the potential resurrection of old ones.

New World

This is not your grandmother's Earth. In 2011, a collection of scientific papers was published in a specially themed issue of the *Philosophical Transactions of the Royal Society* entitled *The Anthropocene: A New Epoch of Geological Time?* Several groups of scientists and other scholars laid out the evidence for, and discussed the implications of, the new era we have entered, one characterized by human activity as a geological force. One of these papers, entitled "The Anthropocene: Conceptual and Historical Perspectives," by Will Steffen, Jacques Grinevald, Paul Crutzen, and

Socio-economic trends

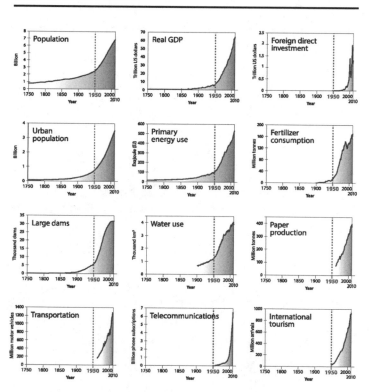

John McNeill, included a two-part figure that masterfully made the case in graphic form. The two parts could almost be labeled "Cause" and "Effect." The first set of graphs on page 131 shows the changes, since the start of the Industrial Revolution in the mid-eighteenth century, in multifarious measures of human activities that have caused changes to Earth systems.

Whatever you choose to measure, be it global population, the damming of rivers, increases in communication or transport technology, or the relentless spread of McDonald's restaurants, the pattern is similar: a gradual but accelerating influence until about 1950. After that point, everything starts shooting exponentially off the charts, in the phenomenon known to scholars of the Anthropocene as "the Great Acceleration."

Earth system trends

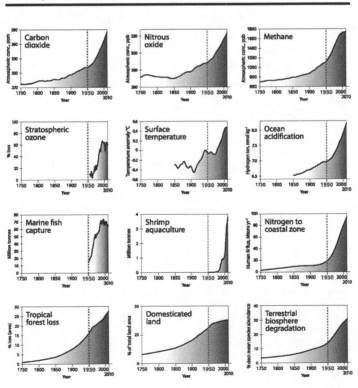

The next set of graphs on page 132, from the same paper, shows different measures of the global-scale *effect* of all that increased human activity on various natural systems.

Many of the modifications graphed here, such as changes in atmospheric gases, losses in biodiversity, exploitation of fisheries, and loss of rainforests, also show the Great Acceleration in human impacts over the last sixty-five years.

Environmental historian John McNeill, another member of our local Washington Anthropocene Group, is a coauthor of the just-mentioned study compiling measures demonstrating the existence of the Great Acceleration. He has been a prime mover in the collaboration between history and geology that is needed to study the Anthropocene. He's also in possession of a dry, self-deprecating wit. After one of our Anthropocene gatherings at the Library of Congress, he joked to me, "Well, of course you know we cherry-picked that data. We only included things that fit this pattern." This led to some enjoyable speculating on possible measures of human activity they could have included that would not have fit the pattern so well: the number of dirigibles made per year has not increased according to this pattern. The average number of cats per household has not increased exponentially. This made, briefly, for a fun game. It's not too hard to come up with metrics that have not kept pace with the Great Acceleration. Yet, in the end, this exercise in contrarian thinking only reinforces the point: any set of *meaningful* measures will reveal the same pattern.

Venus and the Ozone

An instructive example of inadvertent global change is the near-destruction of Earth's protective ozone layer. Had it unfolded slightly differently, this could very easily have been a frightening story of epidemic cancer deaths and crop failures. It is now a story of a close call, a tragedy narrowly averted. Everyone's heard of it, but most people don't know that the problem was discovered in time because we were exploring the planet Venus.

Pages of our missing planetary operation manual are scattered around the solar system. Sometimes in our wanderings, poking around out of pure curiosity, we stumble upon fragments of priceless practical knowledge directly applicable to understanding our influence on the home world. Many such insights have come from studying "Earth's twin." Venus sometimes seems uncannily constructed as a convenient foil for our discovery of the Anthropocene Earth. Our nearest neighbor is a place where aspects of many of our self-made problems, our planetary changes of the third kind, exist naturally, in exaggerated form. There is, of course, greenhouse warming gone off the rails, which I describe in chapter 1. We also find in the clouds of Venus the most extreme case of acid rain you could imagine, with sulfuric clouds so acidic that their pH is less than zero.

In the 1970s we also discovered that something very strange, involving chlorine and oxygen compounds, was going on in Venus's upper atmosphere. Early spacecraft investigations revealed that ultraviolet light is not affecting the gases there the way we thought it should. According to seemingly obvious chemistry, the ubiquitous CO_2 drifting up from the lower atmosphere wouldn't stand a chance at higher altitudes. It should be continually broken up by energetic ultraviolet rays from the Sun, which should be splitting off oxygen atoms and reassembling them into other compounds, such as ozone (O_3). Yet, for some reason, that's not happening. The level of CO_2 we measured in the high atmosphere was much higher than in our models, and all the predicted ozone and other oxygen compounds were nearly absent. Some unknown process was, it seemed, protecting CO_2 from the anticipated destruction. This unexpected and strange stability presented a puzzle, which was solved by Michael McElroy at Harvard and Ron Prinn[9] at MIT, two atmospheric scientists whose careers have straddled earth and planetary science. The answer, they found, lay in the highly reactive element chlorine. Even minuscule amounts of chlorine in such an environment wreak outsize havoc on oxygen compounds, catalyzing their destruction and reconstituting CO_2. Modeling Venus in the early 1970s, McElroy and Prinn showed that you would not expect ozone to survive in an environment where stray chlorine atoms were running wild. Right around the same time, Jim Lovelock was making the first observations showing that chemicals called chlorofluorocarbons (CFCs) leaking from old fridges and sprayed from aerosol cans were accumulating in Earth's atmosphere, and likely seeping up into the stratosphere.

CFCs were once hailed as an environmentally responsible alternative. Developed as a more benign kind of refrigerant than the toxic and stinky molecules they supplanted, such as ammonia and sulfur dioxide, they seemed like an important breakthrough for health and safety. Their use was quite clever— but not quite clever enough. The road to planetary hell is paved with ingenious solutions and good intentions. Scientists trumpeted the fact that these gases were completely safe because they are chemically inert, meaning they don't react with anything. This is true—at least in the troposphere, the turbulent bottom part of the atmosphere where we live*—but nobody had

---

* The troposphere is roughly the lowest ten miles of the atmosphere, although in the polar regions it extends only four or five miles up, depending on the season.

thought through every possible consequence of introducing them into the environment in large quantities. We never can. When these gases eventually leaked up into the stratosphere, twenty miles up, they went through a transformation driven by the intense solar ultraviolet light up there. Energetic photons readily split off chlorine atoms (the first *C* in CFC). Once liberated, this chlorine started doing what Venus taught us chlorine does: attacking and destroying ozone.

Nobody had worried about the effects of ultraviolet light on CFCs. When we introduce a new product, we don't usually ask what will happen to it under intense ultraviolet irradiation. Why ask such a hypothetical question when we don't live on a planet with such a dangerous flux? We don't have to worry too much about the effect of UV radiation on our household and industrial products (not to mention our crops and our skin) *because* we have an ozone layer that shields the lower atmosphere and surface from solar ultraviolet rays. Yet what if that protection were suddenly being eroded?

Frank Sherwood (Sherry) Rowland and Mario Molina from the University of California put all this together and showed that the chlorine we were venting high into the sky could indeed be destroying Earth's ozone shield (the work for which, along with Paul Crutzen, they won the Nobel Prize in 1995). Subsequent ground-based, airborne, and satellite observations confirmed that Earth's ozone was indeed in decline, most notably in the seasonal opening of a dangerous "ozone hole" over the southern high latitudes. This was a frightening discovery because this powerful radiation not only splits up CO_2 and CFCs, it also damages the fragile organic molecules that compose our cells, and in particular wreaks havoc with DNA, causing mutations and cancer.

Had we not started exploring Venus in the 1960s and '70s, we would certainly have noticed what was happening to the ozone, eventually. Still, it's a good thing we figured this out when we did. If you want to see what a planet with no ozone layer is like, take a look at Mars, where the surface is bathed in deadly ultraviolet radiation. Billions of years ago, back before Earth had enough oxygen to make ozone, all life was confined to the ocean, because the land was sterilized by Mars-like solar radiation. Earth's stratospheric ozone screen, a by-product of the Great Oxygenation Event more than two billion years ago, made the land surface habitable. Without knowing it, we had recently started reversing this. We had been inadvertently tinkering with one of Earth's basic life-support mechanisms.

This is a classic example of unintended consequences. What gets you is the "unknown unknowns," and that's why we need as much out-of-the-box, off-the-planet perspective as we can gain. As we increasingly perturb our world, we need to educate ourselves broadly about all aspects of planetary systems. This often means following our curiosity beyond what seems immediately practical, filling in gaps in our knowledge of planetary behavior. The more we know, the less likely we are to make dumb mistakes. In this case, the insights gained from our efforts simply to understand weird chemistry in the uppermost air of Venus contributed directly to our discovery of something dangerous happening to the ozone of Earth.

We did not intend to destroy the ozone layer, any more than the cyanobacteria intended to create it in the first place. Unlike them, however, we had the ability to see what we had begun, and to choose not to finish it. This ozone story has a happy ending. The wide recognition of the problem in the 1980s prompted global discussions. Solutions were identified, replacement products that could serve the same purposes without the dangerous side-effects. Agreements were made. Treaties were signed. The Vienna Convention for the Protection of the Ozone Layer entered into force in 1988, and was followed by the stronger Montreal Protocol on Substances That Deplete the Ozone Layer in 1989, which included legally binding reduction targets for the use of CFCs. And guess what? It's working. Despite some defiance, lawbreaking, and profiteering, by and large the protocol is holding, the emissions of CFCs have leveled off, and the ozone hole is on its way to recovery. There is no quick fix because the inherent timescale of the problem is long. It takes fifty to one hundred years for a CFC molecule to diffuse into the stratosphere and get broken down by sunlight. It takes a roughly equivalent time for the natural chemical cycles in the stratosphere to repair the damage. Vigilance is required. Yet, so far, so good. The ozone hole is still there, but it is no longer increasing in size. Projections show it on schedule to being fully repaired by mid-twenty-first century.

We should feel good about this, and what it implies about us. We recognized the danger, worked the problem, and are well on our way to fixing it. So while the ozone story began as an example of the third kind of planetary change, inadvertent catastrophe, it became an example of something else: an intentional planetary change—what I call a planetary change of the fourth kind.

Self-Control

Planet Earth is being rapidly remade by forces it has never previously encountered. The explosive expansion of human numbers and technological imprint is transforming the environment of every species, including that of *Homo sapiens*. Are we active participants in this transformation? It depends on what we mean by "we" and "active." Certainly we're not just spectators, but considered as a global entity, are we, humanity, even smart and in control enough to be considered a perpetrator? Or are we more like a partially awake automaton, slave to our biological imperatives to grow and compete and grow some more, even when we're overfilling our container like a root-bound plant? Is there even any kind of a coherent "we" who is responsible for our behavior on a global level? Not obviously, not in the same clear-cut sense as when you can ask, "Whose idea was this?" or "Who made a mess out of this kitchen?" or "Who shot J. R.?" and identify a culprit. Certainly there is no individual who decided to do all this, and it's not as if we all got together and decided to make these alterations to our world. Rather, we have now attained global impact without any sense of global control: the Anthropocene dilemma.

Here, I find it useful to point out the distinction between cleverness and wisdom. Cleverness is the ability to solve problems through invention and innovation. We've got that in spades. Wisdom is the ability to apply experience, awareness of context, and prior knowledge of consequences, and fold all that into action. There, we seem to be more challenged. Even to the extent (which is rapidly growing) that we are aware of our global influence, our capability to apply this knowledge and change course, even in the interest of self-preservation, is not clear. We're finding that, for now, our great cleverness has outstripped our wisdom.

We could also look at this in terms of systems theory. As I describe earlier in this book when discussing planetary climate evolution, the behavior of complex systems with many interacting, changing parts is often controlled by feedbacks, positive ones acting as destabilizing forces and negative ones acting to increase stability. An example I use is the positive feedback that long ago led to a runaway greenhouse on Venus, dooming the oceans and the once more moderate climate of our sister planet. As organisms, we can sense changes in our surroundings and in ourselves as we interact with our environment, and modify our actions accordingly. When we do so, we generally apply negative feedbacks to maintain equilibrium or stay on course in some way. In our daily lives we do this literally all the time. Any action that involves any kind of balance or finesse usually involves a lot of this, much of it unconscious, on autopilot. When you walk across a room, you are constantly sensing how you stand with respect to your center of gravity, and adjusting accordingly without needing to realize what you're doing. Driving down a road, you sense if you are getting too close to the center line or the curb, or to the car in front, and without even being aware of it, you nudge yourself toward safety. The autopilot comparison is apt, by the way, because when we design control systems for aircraft or spacecraft, we're mimicking an aspect of our own cognitive capacities. Machines with sensors, controls, circuits, and algorithms do what we do: sense their environment and make adjustments to maintain some desired condition. Negative feedback also describes conscious self-control. When you realize you are getting a little tipsy, you can (I hope) think to yourself, "I guess I've had enough," and moderate your intake.

One thing we humans can also do is intentionally choose to put functions on autopilot that were once under conscious control. We do this when we learn to operate machinery or software or to master a musical instrument. I've been playing guitar since the fifth grade, and while I'm no Hendrix, I'm not a beginner. When I was just getting started and wanted to jam with someone, I would have to ask what key they were playing in, then think, "B minor? Okay, move hand to seventh fret and play within this pattern for a minor key." Now I can hear the song and watch my fingers move to the right place and start doing the right thing. When you are first learning to drive, you actually need to think to yourself, "Uh-oh, I am going too fast. I'd better step on the brake"—which is why student drivers should be given wide berth. When you're more experienced, you just sense you are speeding, and your foot applies pressure. You develop reflexes to act more quickly, circumventing the thought process.

Enacting such control consciously or unconsciously merely requires an ability to sense your surroundings, including the effects of your own activity, and to develop mechanisms to apply this information as input to modify your continued activity. This is a fundamental property of a cognitive creature, but for the most part in our global activities, we do not have this basic ability to temper our actions with compensating reactions. In this way, collectively, we exhibit less of a certain hallmark of intelligence than does an infant learning to walk. As our ability to change the world increases, so does our potential to create dangerous positive feedbacks, which can lead to runaway changes in the environment. What about the potential for cultivating habits and mechanisms for negative feedbacks that would counter these? Can we, global humanity, sense our environment and act on that input?

There are certainly some negative feedbacks operating today in our economy and our global technosphere. The "law" of supply and demand is one example. This resembles the unconscious feedbacks undertaken by an individual, but it is weak at best, prone to oscillation and instability. To control our capacity for runaway global behaviors, an important first step is wide awareness of global consequences. Knowing what we are doing is necessary, but is it sufficient? Without strong mechanisms to apply such knowledge as input to future decisions, negative, stabilizing feedbacks cannot compete with the positive, destabilizing ones. Many of our activities are currently creating a pattern, seen in the graphs just shown, of exponential, or runaway, change. We're sending all the indicators off the charts. Such a pattern is characteristic of a complex system dominated by positive feedbacks. This is a precarious situation. It can't last.

Yet the existence and application of world-changing technology need not lead to runaway states or wild swings in environment. The solution is to increase our capacity for plane-tary changes of the fourth kind: intentional changes.

What do I mean by intentional planetary changes? Global-scale changes that result from actions undertaken with forethought by an intelligent technological species aware of its actions and their consequences. Yes, these are loaded terms. Here I actually mean to suggest a reconsideration of what we mean by "intelligent life," one that I find useful for discussing the role of life and mind in planetary evolution (and also for the *I* in SETI, as I will discuss in chapter 6). Because when some form of life gets to the point where it has this capacity to change the world on purpose, it can take on a powerful new role in the life of the planet. It's fine and customary to define intelligence as a kind of neuronal activity or capacity for logic or abstract thought or in terms of the relationship between a creature with a nervous system and its environment. Yet here we're trying to understand a stage in the development of a biosphere, when an unprecedented kind of phenomenon, a new force in planetary evolution, starts to manifest itself and may or may not become a stable factor in the long-term evolution of a planet. We could then consider classifying intelligence by the effect it has on planetary evolution. There is a kind of cognitive activity that results in rampant, unchecked, unplanned global change of the kind we're seeing today, and there is another level, what we might call true intelligence, or planetary intelligence, of more globally coordinated cognitive activity, that can result in more stabilizing behavior.

As I will discuss further, when we think of looking for the signs of intelligent civilizations elsewhere in the universe, this is the only kind of intelligence that it makes sense to search for. Such a defini-tion also leaves it undetermined whether intelligent life has arrived on Earth. That's part of why I like it. It is not so self-congratulatory and does not imply that we are some kind of standard for intelli-gence in the universe. It gives us something to try for.

4

PLANETARY CHANGES OF THE FOURTH KIND

What is the use of a house if you haven't got a tolerable planet to put it on?

—Henry David Thoreau

Changing the World on Purpose

Some will reflexively feel that any attempt to purposefully change Earth is a bad idea, that the very notion is hubristic and foolish. People intuitively love the natural and suspect the artificial. It's not nice to fool Mother Nature, and it's painfully easy to come up with examples of vain human attempts to control or outsmart nature that have backfired badly. From this you might conclude that the best course for us is to take a "hands off" approach to our planet, but at this point even agreeing to take our hands off would in itself be a huge intentional planetary change. We no longer have the option to avoid choices about how, collectively, globally, to apply Earth-changing technology.

We have to choose, and "None of the above" is not a possibility. Even if you were the most radical green who ever lived, and you insisted that we must instantly shut down every car and factory, you would be advocating for conscious intervention in planetary functioning, and therefore for a massive planetary change of the fourth kind.

This includes any collective corrective measures we take to fix environmental problems of our own making. A global agreement to *stop* a change of the third kind is a change of the fourth kind. This is true whether it comes about through a "top-down" global covenant or a successful grass-roots movement to change mass behavior. So, if purposeful global change is ill-advised, then any appeal for global action on climate issues is wrong-headed. Cynical voices will object that a vision of world-changing technology applied collectively, thoughtfully, and carefully, with wisdom, is utopian, futile, or impossible. Yet we've already seen that it is not. I'm describing something that obviously does not require a perfect, unified global society, because it is already occurring even on our fractious, troubled orb. The ozone recovery is an excellent and hopeful example. Broad-based curiosity-driven science allowed us to recognize the problem. We saw what we were doing, and this incited a global conversation that, within a few years, led to effective action. Now the ozone hole is on track to being just a memory by about 2050.

Yes, there are uglier wrinkles in the ozone story. As a scientific consensus emerged, it was met with a great deal of resistance. The manufacturers of CFCs mounted a concerted disinformation campaign, which included hiring scientific "experts" to cloud the issue and lobbyists to insist that more research was needed before any action could be taken. The scientific evidence was called a hoax, and many smart people who fancied themselves skeptics were sucked in by an orchestrated public backlash.

A frightening turning point came in 1985, when the Antarctic ozone hole was discovered to be already well developed and growing. The situation was worse than had been predicted. Soon the DuPont Corporation, the biggest manufacturer of CFCs, which had led the well-funded denial campaign, did an about-face and joined in calls for a regulated global phase-out. Why did they do this? They didn't suddenly abandon the profit motive in favor of global responsibility. They realized that a phase-out was inevitable and hoped to influence the terms of the treaty to make it gradual enough to preserve profits. It helped that they had by then patented the important replacement gases and stood to profit further. But, so what? The science ultimately gained widespread acceptance. Forward-looking corporations saw the writing on the wall and realized that the smart money would follow the quest for environmentally safe alternatives.[1] A further wrinkle is that some of those replacement gases, substituted for CFCs, now turn out to be very strong greenhouse gases, so introducing them fixes one problem but exacerbates another. This demonstrates that, at our current level of knowledge (or lack thereof), even considered and constructive global actions are still not immune from unintended consequences. So these gases will in turn be phased out in favor of others. Ozone recovery is not a done deal—we have to stay on track, but there is no real reason at this point to doubt that we will.

In some ways the ozone problem is a relatively easy one to solve compared to others that loom. In the larger scheme of things, the solution required relatively minor economic disruption. Nonetheless, it serves as an important "proof of concept." Clearly we have the capacity for global changes of the fourth kind. There is nothing impossible or magical about the transition from inadvertent to intentional planetary change.

There are other success stories. The fight against infectious diseases is notable here, an ongoing struggle that has in many ways been made more difficult by the modern vectors of world-wide trade and travel. Yet the global public health community has accomplished miracles and wonders. One Anthropocene extinction that will not provoke too much mourning is the elimination of the smallpox virus from the planet. The near accomplishment of this merciful task has been possible only through a huge planetary-scale coordinated effort.*

The Twisted Gift of Global Warming

I'll give a few more illustrations of planetary changes of the fourth kind. They range from fully or nearly achieved global goals, to urgent near-term priorities, to far-future possibilities sometimes explored in science fiction. What these examples have in common is that they are all purposefully chosen planetary-scale changes.

Among these, the most pressing goal is the relinquishment of fossil fuels and the diversification of energy sources required to keep global warming in check. This is a change, a global project, that is currently under discussion. When I've said this at lectures for savvy audiences in Washington, DC, that the problem is "under discussion," it has sometimes elicited ripples of mordant laughter from the cognoscenti. Chalk that up to my killer comic timing, but I don't mean it as a joke. Yes, of course the actions mustered thus far seem like too little too late. At this writing it appears that there is virtually no way we will be able to avoid some level of global catastrophic change as a result of our carbon emissions. The ultimate magnitude of these changes, and whether we can act to avert a massive twenty-first-century human

---

* Remaining populations exist only in two research facilities (we hope), one in Russia and one in the United States. The decision to finalize this extinction would be a political one.

disaster, one on the scale of the tragic global conflicts and famines of the twentieth century, is still an open question, one of the great dramas of our age. Its outcome hinges on the uncertain interactions between two complex systems, both impossible to predict in detail: mass human behavior and the response of the global climate system. Yet things have moved hugely in that there is now widespread global awareness of this situation, with much fevered debate and gnashing of teeth. The existence of this agitated global conversation is something new and positive.

We scientists have been discussing the problem for half a century, at first as a somewhat abstract concern and then with an increasing sense of reality and urgency. It started to emerge as a visible public policy question in the 1980s. A key milestone was the testimony of climatologist Jim Hansen before the U.S. Congress in June 1988, during which he declared that "It is time to stop waffling so much and say that the evidence is pretty strong that the greenhouse effect is here."[2]

The more widespread debate really only started up, however, in the mid-2000s. The film *An Inconvenient Truth*, which helped launch a more visible public discourse, first in the United States and then worldwide, was released in 2006. Consensus on answers may still be sorely lacking, but the question is now undeniably on the world's radar screen. Whether or not any solutions enacted this year, or next year, are to your liking, or seem equal to the problem at hand, the emergence of this global awareness and deliberation is a hugely significant change, and a necessary prelude to any substantive action.

Keep in mind that we have never dealt with anything like this before. How will we power our global civilization without wrecking the natural systems upon which we, and the rest of the biosphere, depend? With this question there is a sense in which global humanity is trying to make up its mind about something important for the very first time.

To put it mildly, not everyone agrees on the answer. Yet now the question looms large, even for those who put all their energy into trying to wish or talk it away. It is not going away. Around the world, everyone—at least those who are not completely consumed with warfare or daily survival—is talking about it. To those who are filled with urgent concern this seems to be happening very slowly. However, considering how hard it is for this problem to compete for people's attention with concerns that seem more immediate and pressing, and how long it takes for an idea to diffuse around the globe and sink in to wide public consciousness, this is actually a rapid change in the right direction. Maybe this comes from my perspective as a planetary scientist, where anything changing on the scale of a decade seems lightning fast. To some of us who model planetary radiation balance, who were following this for many decades before anyone else was paying attention, it's a huge relief that now everyone is.

This is only one of many issues that will force us to confront ourselves as global actors in order to maintain a thriving civilization, but it is the first one to get our attention in this way. Perhaps it is at first shaking us too gently, but it is awakening us to our planetary nature. This is the twisted gift of global warming.

Planetary Defense

What else goes in this category of purposeful interference with the fate of the Earth? Active defense against dangerous asteroids and comets. As I describe in chapter 1, a big insight from planetary exploration has been the important role played by small bodies crashing into larger ones. Earth has suffered catastrophic collisions, and will do so in the future, unless we (or others) intervene. Studies of cratered surfaces throughout the solar system and telescopic surveys of the population of small orbiting objects give us a rough idea of the level of danger, the "expectation frequency," of impacts of a given size. Some historical events remind us that the threat is real. On June 30, 1908, a small comet exploded over the remote forest near the Tunguska River, in Russia, flattening some eighty million trees and generating a powerful shock wave that knocked people off their feet hundreds of miles away. This was the largest impact explosion in recorded history, with an estimated energy equivalent of some ten to thirty megatons of TNT, or roughly the yield of the largest nuclear bombs ever built. Had this occurred close to a modern city, it could have caused millions of casualties.

Impacts from space are an ongoing phenomenon, with the smaller objects arriving much more frequently. In February 2014 a tiny* asteroid entered the atmosphere over Chelyabinsk, Russia, formed a meteor that appeared brighter than the Sun and then exploded, causing thousands of injuries, many to people who had been drawn to windows by the initial glow, only to have the blast wave splatter them with broken glass. It's just a matter of time before our luck runs out and an inhabited area suffers a more catastrophic impact. These historically recent impacts, including potential city destroyers such as Tunguska, are small potatoes in planetary history, not world-changing events of evolutionary significance.

There are some real monsters lurking out there. Every one hundred million years or so, on average, Earth gets clobbered by an object greater than six miles across, the size of the end-Cretaceous impactor, the one that did in the dinosaurs. Smaller (but still sizable) bodies fall more frequently. A half-mile-diameter object could cause a short-lived but intense climate disaster, raising enough dust to seriously disrupt global agriculture for long enough to cause mass starvation. There is

---

* Roughly twenty meters in diameter.

about a one-in-five-thousand chance of something this size hitting Earth in the next century. Depending on where it fell, it could cause other horrible effects, such as gargantuan, historically unprecedented tsunami waves sweeping over densely populated continents. However, we are no longer helpless against such outrageous misfortunes. Science to the rescue. We now can detect, and likely intercept, such an intruder.

Planetary defense is being taken seriously in the planetary science and space engineering communities, and several defense mechanisms are currently being studied. Probably the worst idea, although it makes for flashy Hollywood movie plots, is to send up a nuke and blow the offending object to bits. Instead of one giant asteroid heading for Earth, you might now have an unstoppable radioactive swarm of smaller objects. More promising is the idea of affixing small thrusters to an asteroid and gently pushing it off its collision course. If the interloper were identified many years out from its projected impact date, then a small change in trajectory would be enough. Or, if we used a "gravity tractor," we wouldn't even have to touch the wrong-way asteroid: park a massive spacecraft alongside it, and the gravitational force acting over time will be sufficient to pull it into a nonmenacing path.

We don't yet have a planetary defense system in place, but there are many feasible ideas. No matter what, the first step is simply identifying the objects out there that might pose a threat. Several observational programs are under way to do just that. Unless we discover a dangerous object that is going to hit in the next decade or two, a possibility that diminishes as our telescopic surveys improve, we should have plenty of time to stop a monster asteroid.

I am slightly more concerned about the possibility of a dangerous comet heading our way. Unlike the asteroids (rocky and metallic remnants of planet formation that mostly wander the inner solar system and a vast belt between Mars and Jupiter), the icy comets lurk, largely undetected, in the cold, dark outer fringes of the solar system. We know that comets can also impact Earth, and one might be harder to thwart than an asteroid.

In 1994, the eyes of the world's astronomers were glued to telescopes trained on Jupiter to watch as pieces of Comet Shoemaker–Levy 9, gravitationally disrupted by a previous near miss with Jupiter, and strung out into a line of icy fragments, smashed one by one into the swirling atmosphere of the giant planet. With some friends, I watched it happen through a telescope at the University of Colorado in Boulder. It was quite the spectacle, the giant, dark impact scars appearing and then rotating around with the planet—breathtaking, and both comforting and disturbing. The fact that the predicted timing of the impacts was exactly correct was reassuring. We really are pretty good at detecting orbiting objects and predicting their motions with incredible accuracy. That part we have down. When a respected dynamical astronomer declares that a certain object will or will not hit Earth in such-and-such a year, you should believe her. Still, it was a little disorienting, strange, and surreal to see these large black flaws suddenly appear: enormous new features abruptly altering the age-old familiar geography of the bright, banded giant. It was also surprising to realize the extent to which nobody knew how to predict the consequences of these impacts on Jupiter itself. The range of predicted effects was huge, from massive visible flashes to barely noticeable changes. Observing the outcome in real time, with our own eyes affixed to telescope eyepieces as well as a battery of instruments measuring the flash and the aftereffects, was awe-inspiring and instructive. The experience reminded us vividly that violent planetary collisions are a fact of life in the modern solar system, not only something that happened in the ancient past.

In January 2013, sharp-eyed Australian telescopic observer

Rob McNaught found a new comet. This in itself was not surprising. He's discovered more than eighty of them. Once, when I was but a wee postdoc, I spent an evening observing with him through his telescope at Siding Spring Observatory, in eastern Australia, after a day spent exploring the adjacent Warrumbungle National Park, where I first saw tree-nested koalas and galloping emus. When night fell, I had my first really good look at the southern sky, another wild new landscape populated with exotic beasts. I could not have had a better guide, and with McNaught's deep-sky familiarity and encyclopedic knowledge, it was clear that he would instantly notice any faint intruders wandering into this territory. Yet what was surprising about this particular new comet was that it seemed to be on a direct path to collide with the planet Mars on October 19, 2014. Initial estimates suggested that the icy nucleus of the comet was possibly up to thirty miles across, much larger than the doomsday object that struck Earth sixty-five million years ago, leaving fire, darkness, and mass extinction in its wake.

Such a colossal collision might spell big trouble for our spacecraft currently at Mars, but, man, would it ever provide us with front-row seats to a phenomenal planet-shaking collision. And you thought the Shoemaker–Levy 9 smash-up with Jupiter in 1994 was cool! Adding to the strange sense of celestial coincidence, this comet was fated to arrive at Mars less than a month after two new spacecraft. NASA's Mars Atmosphere and Volatile Evolution (MAVEN) entered orbit on September 22, followed by India's Mars Orbiter Mission on September 27.

As it turned out (unfortunately?), upon further observation and calculation, this comet was much smaller, less than half a mile across, and would miss Mars by about eighty thousand miles. This was still close enough to splatter the atmosphere with hydrogen and dust, producing atmospheric effects that were observable with telescopes and spacecraft. When the close approach occurred, it provided a great opportunity to learn more about the infrequent (on human timescales) but inevitable interactions of planets with comets. For me there was also an unsettling aspect. We hear more about the impact threat from stray near-Earth asteroids, but a comet like this one, plunging at a frightening pace from the dark periphery of the solar system, would be a more formidable threat. In contrast to the many years or even decades of warning we'd likely have for a menacing asteroid, a comet can appear with little notice. This Martian near miss occurred less than two years after McNaught's discovery. If the wrong comet appeared, we might have only a similar warning interval between detection and Earth impact. The chances of this happening in any year are minuscule, but recent solar system history teaches us that, if we watch long enough, seemingly unlikely objects and events will eventually materialize. Such incidents remind us that the apparent isolation of planetary distance can be abruptly shattered and that, given enough time, the constancy and the safety of our world are illusory.

Most likely, assuming that our civilization sticks around on this planet for another few centuries, by the time another truly scary comet comes our way we'll have systems in place to detect and deflect it. In the meantime, I'm glad (for many reasons) that we are continuing to send spacecraft to comets and asteroids. In addition to providing pretty pictures and new information about planetary origins, such missions are also valuable for the longer-term project of threat mitigation. Deepened understanding of cometary structure and evolution will sharpen our ideas about how to redirect or disrupt one, should that be necessary one day. And I'm glad that the Rob McNaughts of this world are keeping watch for anything new coming in our direction.

The comet threat, while smaller, is also less predictable, and one could strike with little warning. So, once we ascertain that there are no threatening asteroids currently out there, we still shouldn't wait too long to design some kind of deflection capacity. We'll need it someday.* The creation of a planetary defense system will be an important event not just for humanity, but in the life of the planet: it will be that moment in planetary evolution at which life first develops the capacity to avert a threat that has been hanging over it since the world was born.

The Big Payback

If we hope to be an enduring entity on this planet, we need to start thinking like one. Asteroid defense need not currently be our most pressing concern, but it is in a category that we need to get much better at dealing with: those threats that lurk and linger, that are not imminent but are also not going anywhere, that will eventually be our undoing if we don't address them.

Just as we have unwittingly become global actors, we have also become long-term actors. We've set in motion processes that will play out over centuries and millennia. We *are* acting on multigenerational timescales, but there's nobody in charge of long-range planning, nobody remaining at the helm long enough to see obstacles coming at long distances down the time stream, and steer around them rather than into them. Unless we want to be driving blind, we have to have some awareness of where we're going on the larger scales, both spatial and temporal, on which we're now acting.

Once, our biggest worries came and went with seasonal cycles. Biological evolution had no reason to equip us for

---

* The joke is often made, at least among those in my line of work, that the difference between us and the dinosaurs is that they did not have a space program.

problems spanning longer than a human lifetime.* There's no reason our cognitive tool kit, well evolved to deal with Paleolithic survival challenges, should include the ability to think ten thousand years into the future. But we are also creatures of culture. When we began to develop oral histories and make plans together, probably first for hunting game, we transcended our genetic capacities. By telling stories, we found new sources of resilience, new ways to incorporate past experiences and meet future challenges. Now it is up to us to adapt again. We're awakening to several threats, some of our own making, that will require us to maintain constant attention, and action, over centuries and millennia. We need to continue to expand our time horizons. Cultural evolution has equipped us with institutions that endure for decades and centuries, but not too many that have maintained continuity over millennia. Religions are exceptions, and whatever institutions we develop to deal with asteroid defense or other very-long-term endeavors may have something to learn from those that have persevered.

We have a deficit of sustained attention, but this disorder seems especially acute today. We've never needed a long-term outlook more than we do now. Yet, at the same time, as the pace of technological and social change quickens, both the past and the future seem to recede in a fog of electronic distraction. Sometimes it seems we can barely hope to master the elusive present. However, technology has also gifted us with greater access to our past than we have ever enjoyed, in burgeoning scientific knowledge of the history of our species, biosphere, planet, and universe, in unprecedented instantaneous access to vast treasures of knowledge, and in the ability to share stories, old and new, with a global community. Likewise, our capacities for transmitting knowledge to our descendants, and

---

* A duration that itself has greatly increased since we were hunter-gatherers.

for forecasting and modeling future challenges, are growing in seemingly boundless ways. An expanding connection to the past and future is at our fingertips.

Planetary defense requires the kind of thinking we will need if our global civilization is going to survive the next couple of centuries. It's a good arena for us to practice the art of paying attention for longer than any of us is alive. By aiding our transition to this longer-term mode of awareness and action, the challenge of planetary defense, even absent an approaching asteroid, can help us to survive the next century or two with our civilization intact and thriving.

We've already begun the first phase of this: the astronomical surveys to know what Earth-crossing objects are out there. The next phase, building and operating a planetary defense system, should serve as one of many long-term goals of the world's space programs in the twenty-first century. There are many important reasons, both pragmatic and lofty, to maintain a space program, but this project surely provides a good motivation. However, it's not enough merely to solve the technical aspects, to learn to build the hardware for an asteroid defense system. We also need to plan for long-term, multigenerational continuity of our operational space capacities. Currently we have almost the opposite approach. Every Mars rover we build must be an entirely new design, even if building five more of an earlier design would make for good and cheaper exploration. We are so focused on innovation that we don't think in terms of stability and continuity. Yet it doesn't make sense to build an asteroid defense system meant to operate for only ten or twenty years. The technical puzzles occupy our attention now, but they will be the easy part. As with so many other global problems, the biggest challenges of planetary defense ultimately will not be technological but organizational, political, and cultural. We'll need to call upon them someday, for an urgent mission to save our civilization and possibly even our biosphere, so our space systems and the organizations that operate them must be built to last.

Space defense is ethically simpler than most other proposed or actual large-scale technological interventions with the planet. Here, human interests and the interests of the rest of the biosphere have obvious and large overlap.* An asteroid of sufficient size will ruin everyone's day. Doesn't it seem, then, that we have a clear moral obligation to the biosphere to pursue such a project?

Why? Well, how do you feel about mass extinction? When I've described past catastrophic die-offs, the disappearance of millions of species due to planetary changes of the first and second kind, do these stories seem sad to you? Are these tragedies? Or simply "nature taking its course" like a fuzzy bunny being devoured by a fleet fox in a nature documentary? Random, or "natural," catastrophes, even huge ones, are much easier to deal with than those which are caused, and therefore could be averted, by us. Because we have at least the possibility of foresight and avoidance, we see the mass extinction that we are currently starting as particularly tragic, and feel horrible responsibility for it.

Now, in this context, think about the morality of not just past and current extinctions but *future* ones. If we get through this next century with a stabilized population and a sustainable global energy system, then it may be time for payback. What could we ever do for Earth that would possibly help atone for the damage we are doing now?

Maybe humanity can build a planetary defense system to start to make amends for the extinction we are currently causing. As long as we, or our descendants, are on the case, Earth's

---

* This is true on most other issues, but often less obvious. In the long run, what's good for the biosphere is good for us.

biosphere never has to be decimated by another giant impact. Maybe this could be some kind of long-term compensation to the biosphere. First we need to get a handle on the killer asteroid that is our own reckless behavior. Then, in the future, we may well be able to prevent the next mass extinction.

Climate Control

If we're going to have a long future on this world, there is more that we'll have to do in the way of active interference with previously natural processes. Once we get over our current inadvertent climate vandalism, we could thoughtfully intervene against harmful climate swings. If we want our future here to be as long as our history has been, at some point we'll need to prevent future ice ages and episodes of natural global warming.

As city builders and history keepers, we've been around long enough for the climate to change, but it really hasn't—not much. We've been lucky. The current Holocene interglacial period has lasted about twelve thousand years, since the end of the Pleistocene. All of human civilization has developed in this stretch of relatively warm and stable climate. A few times, as a result of volcanic eruptions or changes in the Sun, we've experienced some small climate fluctuations, resulting in a few years of extreme cold or regional drought. These have caused rivers to freeze over in what we thought were temperate zones, and once-fertile valleys to turn to dust, hinting at what climate change can do. Yet, going back to before the Agricultural Revolution, ten thousand or so years ago, and certainly over recorded history, no human has ever experienced the larger swings in climate that are routine on Earth over millions of years. We've been lulled into an illusion of stasis by unusual climate stability during our short time here.

If you include the rest of our history as a species (most of it), before we started keeping continuous track of ourselves, you'll find the story is different. Over longer timescales, Earth's climate has gone through large swings and, left to its own devices, will continue to do so. Blame it on Jupiter.

Climate cycles on Earth are, in large part, a consequence of its existence as one planet in a solar system of many. Graphed over large stretches of Earth time, the complex warming and cooling oscillation of climate reveals polyrhythmic patterns. The major beats occur at intervals of 23,000, 41,000, and 100,000 years. We call these the Milanković cycles, after Milutin Milanković, the Serbian astronomer and mathematician who is considered one of the founders of planetary climatology. In 1916, Milanković published climate calculations of the surface temperatures on Mars, Venus, Mercury, and the Moon, some of which he had worked out while a prisoner of war during World War I. Yet he is mostly remembered for figuring out the relationship between planetary motions and climate change on Earth. Over hundreds of thousands of years, as first determined by Milanković, little gravitational nudges resulting from the motions of Jupiter, Saturn, and the Moon cause subtle but important wobbles in the orbit and spin of Earth. These induce slight changes in the seasonal distribution of sunlight, causing our planet to sway rhythmically between ice age and hothouse conditions. When Milanković died in 1958, his theory was not taken seriously. It took a while for our data to become good enough to prove him right.

Milanković's theory of the astronomical forcing of Earth's ice ages was resurrected and vindicated in the late 1970s, when ice cores first allowed us to reconstruct, in detail, the climate history of the last 450,000 years. A pattern jumped out, showing superimposed climate fluctuations, with pulses of 23,000, 41,000, and 100,000 years, neatly confirming Milanković's predictions (or retrodictions, that is, predictions of the as-yet-unknown past). More recently we have learned that some other planets in the solar system experience similar cycles—not surprising, as they travel the same spaceways of mutually interacting gravitational influences. Earth's rotational axis stays at a nearly constant twenty-three-and-a-half-degree tilt from the Sun, thanks to the steadying gravitational hand of our big moon. Compared to this, the Martian spin axis bobs up and down like a dreidel. Over a period of 120,000 years, the tilt of Mars varies from fifteen to thirty-five degrees.* Currently it sits about halfway between these extremes. The amount of solar energy at the poles is twice as high at the maximum tilt as at the minimum. As a result, the polar terrains of Mars are exquisitely marbled with interlaced layers of ice and dust recording complex climate fluctuations on the Red (and sometimes partially white) Planet. Titan undergoes a sixty-thousand-year climate cycle driven by its orbit around Saturn. The planets are all tugging on one another and rhythmically torqueing one another's climates.

We know that the timing of Earth's ice ages is determined by this interplanetary contra dance, but we're not exactly sure why. There is still plenty of mystery about how the very slight and subtle orbital changes in Earth's seasonal illumination translate into dramatic climate variations. Some amplifying feedback mechanisms seem to be responsible, but these are still being worked out and debated. Yet the pattern in the climate data bears an unmistakable orbital stamp.† Despite the claims of astrologers, there is no evidence that the position of the planets

---

* Occasionally, due to "chaotic obliquity variations," it may increase up to sixty degrees.

† You've heard that correlation is not causation, but that is not always the right assumption. A correlation can be so detailed and precise as to render the idea of lack of causation so unlikely as to be absurd. That is the case here.

at your time of birth affects your personality and fate. However, life on Earth and human evolution have been profoundly influenced by the motions of the planets over the ages as climate has danced to the music of the spheres.

Extreme climate change is like the impact danger in that it has not hit within the recorded memory of our civilization. We modern humans have had to infer its existence through the detailed scientific study of Earth and the rest of the solar system. Going back farther, however, we see that it has definitely shaped who we are. When we were hominids but not yet humans, we experienced extreme climate change. As I'll discuss in chapter 8, larger climate changes were extreme challenges to survival that altered human evolution. Perhaps some old stories and legends passed down by indigenous peoples refer to such cataclysmic events. In our prehistory as hunter-gatherers, we were better able to cope with minor climate change than we would be today. When we were a nomadic species, if a food or water source dried up, we could migrate to happier hunting grounds. Moving from place to place in response to shifting conditions, following the food and the water, was part of our modus operandi during ordinary times, so we were able to respond more adroitly when the world changed. Once we abandoned this lifestyle for stationary settlements with planted fields, we became much more vulnerable to changes in environment. As we built villages and city-states, and tied ourselves down to specific places, a blight or long-term drought could present a much more serious problem. Eventually we became even less resilient, dependent as we now are on global agriculture and a global economy that cannot as easily tolerate a change in conditions.

If we left Earth to its own devices for long enough, it would eventually enter another ice age. This would be much more extreme than the kinds of climate changes we are potentially facing now. During the last ice age, up until about thirteen thousand years ago, an unbroken sheet of ice two or three miles thick extended from the poles down to the latitude of Minneapolis. Sea level dropped by around three hundred feet, completely redrawing coastlines and the paths of rivers around the world. There is no way that a civilization of many billion humans, tied to cities and dependent upon a global system of agriculture and trade, could survive such a transition intact. We don't ever want to try to live through one, and if we get our act together, we will never have to. We have at least twenty thousand years to work the problem, probably much longer. Transitions between ice ages and interglacials are times of accelerated extinction. So, if we ever prevent an ice age, we will also save a lot of other species: perhaps, someday, another opportunity for payback.

Have we already injected enough CO_2 into the air to delay or prevent the next ice age? Quite possibly. Even if we completely stopped spewing carbon next Thursday, it would take about one hundred thousand years for the natural carbon cycle to draw it back down to preindustrial levels. That's probably long enough to delay the ending of the current interglacial and the onset of the next ice age. Yet this is not an easy thing to predict. Projecting climate on this hundred-thousand-year timescale is even harder than modeling the next hundred years (which, for obvious reasons is where most efforts have been focused).

Since, as I've indicated, we don't fully understand the mechanism by which the orbital Milanković forcing causes large climate swings, there is a lot of uncertainty here. The last warm interglacial before the one we're in now, the Eemian, about one hundred twenty-five thousand years ago, lasted for about ten thousand years. This length is typical. Most of these warm periods persist for roughly the length that ours has already lasted. Yet the one we're in now seems to be weird, and not just because of us. Some models predict that, without human interference, the ice would return within fifteen thousand or twenty thousand years. Others suggest that, due to Earth's orbit currently being in a phase of low eccentricity—meaning that it is now nearly circular, compared to other epochs, where it is slightly more egg-shaped—we inhabit an interglacial of exceptionally long duration. Models incorporating this fact suggest that, even without us, the ice would not return for another fifty thousand years. In other words, if we don't screw it up, sending climate careering beyond the safe zone, our luck might hold for quite some time. This warm, stable climate our civilization has enjoyed for ten millennia, and come to take for granted, might last for five times again as long. Yet what about looking farther into the future, beyond just the next ice age? Might we have initiated something more long term? Could we have seriously thrown Earth off its rhythm, perhaps even permanently halting the Milanković cycle of glaciations?

My young colleague Jacob Haqq-Misra has been studying this question. Early results from his modeling suggest we may be on our way to initiating such a change in Earth's behavior. Jacob is small and wiry, and a dynamo at pursuing his passions. He's a research scientist modeling planetary climate, but he also plays percussion in a touring jam band, helps run an internship program for students interested in astrobiology, and has followed his interest in environmental ethics with the same intensity as his scientific studies. My kind of guy. As a grad student at Penn State, Jacob studied with Jim Kasting, a pillar in the field. Kasting, who was himself a postdoc at Ames with Jim Pollack a decade before me, is now a sort of guru of planetary climate modeling, in the way Pollack once was. He has trained an influential cadre to be experts in big-picture climate modeling. Many of his former students and postdocs are now turning their attention to new climate modeling challenges such as exoplanets, the future of Earth's habitability, and (in Jacob's case) the long-term climate impact of current human industry. Jacob has started looking at how the amount of CO_2 we are projected to produce in both the most likely and worst-case scenarios will affect the long-term Milanković cycles of ice ages and interglacials.

His models are somewhat simple—intentionally so. As I've discussed, part of the art of planetary climate modeling is choosing a model that is appropriate for the task at hand. As the French say, *Pas besoin d'utiliser un marteau-pilon pour écraser une mouche,* or "You don't need a jackhammer to swat a fly." If you are simply trying to evaluate whether a given effect is significant, then you just need a model you know is good enough to get the magnitude approximately right. With this work, Jacob was not claiming to make a detailed and correct prediction of climate changes over the next several hundred thousand years, only to determine whether our influence could be strong enough to interrupt the Milanković oscillations over the coming millennia. His first model results suggest that the answer is yes. It is quite likely that we are extending the current interglacial, delaying the onset of the next full-on ice age that would otherwise start in perhaps fifty thousand years. Beyond that, though, Jacob's results suggest that no matter what we do from here, our climate influence will persist for much longer.

Even if our carbon emissions are modeled as one giant burp, as an instantaneous pulse that then ceases, the increased CO_2 abundance lingers in the air for hundreds of thousands of years due to the slow (to our sensibilities) operation of Earth's climate cycle, which will eventually resorb our excess carbon into the soils, sediments, rocks, and mantle. After several hundred thousand years, the climate would largely return to "normal." This is assuming we stop the Great Acceleration in our fossil fuel use pretty soon. That is actually a safe assumption on the long timescale of this exercise, because soon there won't be any fossil fuel left.[3] No matter what we do from here, Jacob's results show, we are likely to have interfered with the next few Milanković glaciations, stopping the cycle of ice ages that has been operating for hundreds of millions of years.

Would this be such a bad thing? As I've indicated, ice ages are not good for the biosphere. Okay, *good* is a loaded term. They cause extinctions. You might want to argue that extinctions are inevitable, like death and taxes, or even that some level of extinction is good, or at least normal, or that at the very least the biosphere has dealt with these deathly cold spells since time immemorial and come through just fine. From a certain vantage point, they are no more traumatic to the biosphere than are Earth's annual seasons. Yet that is a point of view that accepts extinction as a part of the way things are.

I'm not sure we will ever have that luxury again. It's a strange thought, but there may never again be such a thing as "natural" extinction.

In a way, we've been doing what all species do: altering our environment, forcing others to adapt, wiping out those who can't. Nothing new about that. For our entire history, human beings have been causing brutal extinctions of other species, but here's a hopeful thought: twenty-first-century humans could be the first ever to decide *not* to behave like this. From now on, or as long as humans (or thinking creatures descended from or created by us) are here on the planet, extinction will be a choice. Obviously the current carnage, the mass extinction that we are now starting to manifest, has to stop. Then what? What extinction rate would we prefer? Do we wish to eliminate it entirely? Then what of biological innovation? If we choose to eliminate all extinction, then this is equivalent to saying we are the gods in charge of all future species. We will either have to play god or allow species to go extinct (which, I suppose, may still be a form of playing god). Something to think about.

However you come down on extinction, if you regard mass human starvation as a tragedy, an ice age would be bad news indeed. It's not the biosphere that is fragile, it's our civilization. We've inhabited all the coastal regions and become dependent upon agricultural activities that span the globe. In our ten-thousand-year run we've come to depend on a climate stability that will not last.

So, then, isn't it a good thing if we've eliminated the threat of ice ages for the next several million years? Well, it might be, but there's a hitch. Unfortunately, what Jacob's results actually show is that Earth can be expected to be stabilized for all this time in an *ice-free* state. We've not only stopped the climate pendulum from swinging, but we've also got it pinned at one extreme end of its range. An ice-free Earth may in itself not sound horrible, unless you're attached to the kind of planet we've always known and you are fond of penguins, polar bears, and the countless other species dependent on icy high-latitude environs. Yet we have no idea what this implies for the planet as a whole. How hot will the equatorial regions be in this new normal? What will happen to ocean circulation, sea level, precipitation patterns, and the rest of the biosphere? Nobody knows, but it is clear that this would be a drastically altered Earth.

Yet, there is a potentially more hopeful aspect to these results. They show that the magnitude of the natural Milanković oscillations is not so large as to be beyond the reach of human tinkering. The amount of forcing that is required to change Earth from ice age to hothouse can be modified even with our current technology, which certainly pales in comparison to the technology of a civilization that has learned to survive for another ten thousand years. So, if and when we develop the capacity to apply our powers with a little more collective will and a lot more finesse, to apply them sensibly in the service of a healthy biosphere and a stable, healthy human civilization, it shouldn't be that hard to adjust the climate of Earth to a desired state—not hard, at least, from a technical point of view. It wouldn't be like trying to move Earth in its orbit or change the brightness of the Sun, projects that are perhaps not physically impossible but are completely unimaginable for us. By contrast, Earth's greenhouse climate balance is clearly malleable. Obviously in the near term we quickly need to limit our CO_2 emissions and reduce our impact on the climate. We don't want to try to live on the perpetually ice-free Earth that would result from our present course. In the longer run, though, if we make it that far, we will likely choose not to take our hands completely off the scale. We would want to dampen these severe natural oscillations, not by pinning the system at either extreme end of its natural range. We'd want to choose a more moderate climate, something like the one we have become accustomed to.

Is it possible that the most recent ice age may be the last one ever? Given the scale of inadvertent human climate interference, and the potential scale of thoughtful human climate intervention, and given at least the possibility of our long-term habitation on Earth, I'm going to say yes, this is quite possible. With us here, Earth may have no more glacial cycle. Would that be a bad thing? I'm going to vote no. No more glaciers? That would be bad. No more ice ages? Good riddance.

Ultimately, the only way we could have such a prolonged and moderating effect is if we became a much more mature, thoughtful presence on this planet. If we simply use up our fossil fuels in an orgy of consumption, then our influence on the atmosphere will be catastrophic but geologically short-lived. After a few million years at the very most, the Milanković climate rhythms would resume their complex beat. The only way our influence could produce a more sustained interruption, sparing the planet the horror of future ice ages, is if we planned it that way. We would have to become something new, and fully make the transition to the kind of species that enacts planetary changes of the fourth kind. Our current planetary changes of the third kind might cause the glacial cycle to skip a few beats at most. If the cycle is shut down, it will be by a species that knows what it is doing and why.

When it comes to very-long-term climate change, the range of what is natural goes far beyond what we are used to, and extends into some pretty scary territory. Famine and mass death are also natural, but probably not what reflexive nature lovers have in mind. On its own, Earth is capricious and cruel. If we want to keep our civilization for long enough, we're going to have to deal with that. On the longer timescales of the ice ages, in order to prevent these deathly oscillations, we will have to learn to provide negative feedback that is missing from the natural Earth system.

Planetary Engineering

If we can talk about applying smart engineering to moderate the glacial cycle in the distant future, why not use it right now to deal with our much more urgent problem of anthropogenic global warming? Do we really need to give up our precious fossil fuels, or can we come up with some clever new way to interfere in the Earth system to counteract their undesirable side effects? Can we invent our way out of this problem, substituting innovation for self-restraint?

Thus we come to the thorny subject of Earth climate intervention, or "geoengineering," the idea that we could undertake purposeful climate manipulation to fix some of the problems we've inadvertently created. It may seem like I'm coming into this topic ass backward, introducing the contentious subject of present-day climate engineering after the more abstract topic of far-future climate meddling. Yet this is how I first learned about the subject: as a subset of the larger, more general problem of planetary engineering.

Despite the rush of recent attention, it's not a new idea. For a long time, planetary scientists and science-fiction writers have been exploring the problem of how to purposefully interfere in climate. In the years during which they were doing their fundamental studies comparing climates on Venus, Earth, and Mars, Carl Sagan and Jim Pollack also studied the problem of terraforming. This is the idea that someday we might be able to change the environment of another planet to be more Earth-like, turning it into a friendly place for terrestrial life. As a part of their wider study of climate manipulation they also looked into what is now called geoengineering.

When I was at NASA Ames Research Center as a postdoc working with Pollack, he and Sagan were working on a paper entitled "Planetary Engineering" for an upcoming book, *Resources of Near-Earth Space*, edited by my PhD adviser, John Lewis. Lewis was a hands-on editor who had a lot of input into their chapter.* Since my three main scientific mentors were all collaborating on a project about purposeful engineering of climates around the solar system, I heard a lot about this topic.

The idea of finding Earth-like havens elsewhere in the universe is an old one. In the 1600s, after we figured out that the planets were other worlds but still knew nothing of their environments, many scientists imagined them as essentially other Earths, inhabited by familiar kinds of life. Then, once we started to learn, first with telescopes and then with spacecraft, how alien those other planets actually are, people started wondering if we might someday remake them more in Earth's image. What

---

* John also possesses the most extensive personal science-fiction library I have ever seen, and an encyclopedic knowledge of "hard," science-based science fiction, which clearly came in handy for this project.

if at some point we wanted to change a planet such as Mars or Venus into a place where Earth life could thrive? How would we do it? Could we? Should we? Many scientists have seriously studied the problem for decades, with increasing sophistication as we've gathered data on the other planets and learned how climates can change.

Terraforming has been the subject of several technical workshops and has generated a sizable literature of peer-reviewed papers. Since we've become aware that we are now inadvertently "Veneraforming" Earth (slowly making it more Venus-like), these models have new relevancy. While I was at Ames another group of planetary climate modelers, including Chris McKay, Brian Toon, and Jim Kasting, was also working on a paper about terraforming Mars, which was published in *Nature* in 1991. So this community of scientists who had collectively figured out the Venus greenhouse, Martian dust storms, the Titan anti-greenhouse, volcanic climate catastrophes, giant impacts, and nuclear winter was also involved in studying terraforming. It makes sense when you consider how these studies are all connected, variations on the same overlying problem. For scientists interested in how climates can change, and how we might accidentally change a climate, it is irresistible and instructive also to explore how we might purposefully adjust a climate. It's the same class of problem we solve when we predict long-term climate change on Earth, using the same physics, the same equations and similar models. You ask: How does this climate system function and change? How stable is it against different kinds of perturbations? What would be the best leverage points if we wanted to alter it? Where are the greatest sensitivities and what would be the cheapest way to influence them? How would these changes affect the rest of the system? Would the new state be stable or require continual inputs and meddling? On what timescales?

Terraforming gives planetary climate modelers additional scenarios to study. It's also undeniable that most of us space geeks have been influenced in some way by science fiction, most went through at least enough of a phase to gain familiarity with the canon, so we've all glimpsed the dream of a multiplanet human future even if some regard it as more of a fairy tale.

Pollack and Sagan began their planetary engineering paper by mentioning global warming, ozone depletion, and nuclear winter, declaring, "These three processes demonstrate the general proposition that humans can now alter environments on a planetary scale...It seems possible, therefore," they wrote,

> that within the not too distant future human technology should be capable of even more major alterations...An important issue is whether any can cause improvements rather than deterioration in the planetary environment—perhaps with high-precision negative feedbacks.

Then they discussed each of the other three worlds of our solar system with solid surfaces and sizable atmospheres—those that could potentially be terraformed: Mars, Titan, and Venus. On Mars and Titan, the challenge would be to enhance the greenhouse effect, causing deliberate global warming, raising the temperature to the point where liquid water would be stable on the surface and plants could grow and begin to work their self-reinforcing magic. Eventually, with large increases in temperature, pressure, and oxygen level, people might someday live on Mars without need of an enclosed habitat or suit. Sagan and Pollack went through the different gases that might be released into the air on these worlds, and thus trigger feedbacks that might push them into warmer states.

Venus presents the opposite problem. How do you get rid of enough CO_2, or block enough sunlight, to battle that Godzilla of a greenhouse? Science fiction has traveled this subject deeply, with varying degrees of scientific realism. The best of it is pretty damn good. The word *terraforming* comes from a 1942 sci-fi story by Jack Williamson, but the topic was explored earlier, by British philosopher Olaf Stapledon in his masterful 1930 novel, *Last and First Men*, in which Venus is made habitable by using electrolysis to release oxygen from its oceans. (It seemed like a good idea at the time.[4]) The most sophisticated fictional exploration of terraforming is in the Mars trilogy of Kim Stanley Robinson (*Red Mars*, 1993; *Green Mars*, 1994; *Blue Mars*, 1996), which follows several generations of Martian settlers, and then terraformers, during a time when Earth suffers from ecological devastation. Through the views and experiences of these Earthlings who become Martians, Robinson explores in detail the physical, ethical, cultural, and political challenges of transforming worlds.

Is Mars Ours?

Okay, sci-fi is one thing, but is this even a serious idea? It sounds far-fetched, but so would have a modern city in the Nevada desert or high above the Arctic Circle if you had tried to describe one to our hunter-gatherer ancestors on the East African savannah. Spreading to new environments, including places that once seemed uninhabitable, and modifying them to suit our needs is something humans have always done—and now we've learned the hard way that we have the power to change whole worlds. Perhaps we can eventually learn to do it on purpose.

What if it turns out there are already Martians? It seems pretty clear (I hope) that if we discover an indigenous biosphere thriving in current conditions, then it really changes the equation in a drastic way, and argues against any planetary engineering on Mars. In my view, if there are Martians, then Mars is theirs, and we should tread lightly there. We should explore but perhaps not exploit. Ironically, then, if we are too successful in our current mission to find possible signs of life, we may have to reconsider the next big mission of establishing human bases.

If we find that Mars has no indigenous life, then our choices about whether, and how, to go to Mars are much less complicated. In that case, I think, the ethical equation is reversed. The brightest prospects for Martian biology may lie in the future, because Earth life may one day choose to spread there. You might ask how we would ever determine, for sure, that Mars is lifeless. Fair question. Such a conclusion may be beyond our present-day science. Yet once we've explored Mars thoroughly, and explored the universe more widely, our knowledge of life and planets may become mature enough (e.g., light-years beyond our present understanding) for us to recognize a planet with absolutely no life. In that case, I think a pretty good argument can be made for the moral imperative to bring life to Mars and Mars to life. Why? Because life is the most precious thing of all. Shouldn't we try to spread it? If you lived in a thriving, luxuriant garden next door to a vacant dirt lot, wouldn't you want to toss some seeds over the fence, turn that emptiness into another garden?

Also, life on Earth is tough but not invulnerable. Things can happen to planets that would wipe out all life. Several times our planet has possibly come close. What if the largest mass extinction events had been a little larger? Knowing of another planetary biosphere, either because we discover or implant one, would mean knowing that the existence of all life in the universe does not ride on the fate of Earth. Think of an endangered species living only in one threatened habitat. Any reasonable environmental or conservation ethic would demand our trying to locate that species more widely. Well, as long as we are confined to one planet then all life on Earth is, at some level, like that threatened species. Once our biosphere becomes a multiplanet phenomenon, it is no longer vulnerable to a planet-destroying disaster. It will have come a huge step closer to immortality.

If at some point we have the opportunity to safeguard life against such a possibility, then, by all that we hold dear, wouldn't it be our highest moral duty to do so? Earlier I argue that once we gained knowledge of asteroid strikes and the potential to avert them, that became our responsibility. An extension of this same logic could imply a moral imperative to terraform. The greenest of the green philosophies, which treasure the intrinsic value of life above anything human made or influenced, should assign great importance to having life's domain spread across many worlds.

I've led public discussions on terraforming in classrooms and museums, and once in a church, and found that it gets people talking about our attitudes toward life, nature, and human responsibility and action. I sometimes start by showing a picture taken by one of our Mars rovers depicting its lonely tracks winding back into the distance across a vast, otherwise untouched Martian plain, and ask, "What do you see here? How does it make you feel?"

Mars rover *Opportunity* looks back over where it's been.

Dune buggy tracks on a beach of Cape Cod.

Do you see this as an act of vandalism, as the despoliation of an otherwise pristine landscape, equivalent to dune buggy tracks grinding through fragile dunes where seabirds are trying to nest?

Or do you see it as the mark of an evolutionary moment, equivalent to the tentative tracks left on a lonely beach by the first creature ever to venture haltingly from the sea and onto the land?

Three-hundred-ninety-five-million-year-old animal tracks in Poland, some of the oldest traces of land animals.

I understand the former view, but I personally see it more as the latter. This discussion of how we see our presence on an otherwise lifeless world then progresses to the question of our right future action on that world. There is the obvious reaction of "Hands off the cosmos!" What unbelievable arrogance to think that we can have the wisdom or the know-how to "fix" another world? Haven't we done enough damage on this one? As Pollack and Sagan put it in their chapter, "Can we who have made such a mess of this world be trusted with others?"

I find the conversation worthwhile, and as you might have guessed, I like to defend the possible justifications for terraforming against the reflexive reactionary opinions it sometimes evokes. Yet do I think we should terraform the other planets in our solar system? Hell, no. Not if "we" is the human race today, with our current scientific knowledge and institutions. We don't know what we're doing, and it would be foolish to pretend we do. It may indeed be our destiny someday to cultivate Mars, but we have to do our homework first. We won't be ready until we've done much more outer and inner exploration and become much wiser about life, on Earth and off.

These grandiose schemes don't worry me though, because even if someone decided they wanted to terraform Mars, it's not like they could just start tomorrow. The option is not on the table. Mars is safe for now. But in the meantime, terraforming dreams can help us think through our perplexing situation as reluctant engineers of our home world. When we imagine ourselves intentionally changing the climate of a planet, it flexes the right mind-muscles, those that will feed our ability to manage our own planet. This is the kind of thinking that humanity will need in order to survive in the long run. We'll leave Mars alone for now, but we don't have that option when it comes to Earth.

Earth Interventions

When Pollack and Sagan turned from icy Mars and Titan to red-hot Venus, they considered how we might shut down the massive greenhouse, or mute it, leaving just enough warming to make a planet friendly for life and possibly even humans.

The solutions fall into one of two categories: find a way to block enough sunlight or get rid of a lot of CO_2. The first could be done either through injecting some kind of absorbing dust in the atmosphere (perhaps by grinding up a captured asteroid) or putting giant mirrors or sunshades in space. The second could be done by engineering some kind of microbes that spread and multiply in the clouds, metabolizing CO_2 into organic carbon. Or you could remove CO_2 by liquefying and sequestering large amounts at or under the surface, or by circulating hot water through the ground (once you've made it cool enough, through one of these other mechanisms, for liquid water to exist) to induce "weathering" chemical reactions that would pull CO_2 out of the atmosphere and bind it into carbonate rocks, just as in Earth's carbon cycle. Yet with Venus you have the additional problem that there is basically no water, beyond a trace amount in the clouds. If you really wanted to create a surface biosphere on Venus someday, you would have to import a lot of water, compensating for the oceans the planet lost, in its wild youth, to the runaway greenhouse catastrophe. Where would we ever find enough water to return oceans to Venus? Our solar system has plenty of water, out in its fringes, frozen in billions of comets. Someday, when we've learned how to move comets and asteroids around, the rewatering of Venus could be feasible.[5] Pollack and Sagan pointed out that with the same technology we would need to build an asteroid defense system, modifying at will the orbits of potentially dangerous objects, we could also eventually move spare objects around the solar system if we wanted to use them for terraforming.

Finally, they turned their attention toward the home world, beginning by stating that "an important special case for planetary engineering is to reverse significant perturbations (e.g. global warming) in the environment of our own planet. This is surely easier than planetary engineering on other worlds."

It is not a coincidence that the two major strategies discussed for terraforming Venus resemble more extreme versions of the same two methods now under discussion for possible geoengineering of Earth: solar radiation management (SRM) and carbon dioxide removal (CDR). Venus stands, once again, as a caricature of our home planet's environmental processes and challenges. To return that planet to habitable conditions, we would need to block enough sunlight or remove enough CO_2 and other greenhouse gases to turn down the thermostat by at least 360 degrees Celsius (or 650 degrees Fahrenheit!). On Earth we have to mitigate a potential rise of only a few degrees. That sounds easier. Yet this is our home, and it feels different to speak in a cavalier way about making adjustments to it. In contrast to our schemes for other planets, with Earth we don't generally talk about using brute-force methods such as deliberately crashing comets, blocking the Sun out of the sky, or filling the stratosphere with dust from ground-up asteroids. Instead, we typically look at enhancing or simulating known processes that are already at work on our planet.

The basic physics of solar radiation management is familiar to climate modelers. The anti-greenhouse effect is a known entity, an observed phenomenon in the stratosphere of Titan, in the dust storms of Mars, and during large volcanic events on Earth. It's also hypothesized and modeled to explain asteroid extinctions and nuclear winter. So couldn't we just induce a little bit of a controlled anti-greenhouse to mitigate the warming we've been causing with our CO_2 emissions? This might be done, for example, by spreading enough sulfur dioxide into the stratosphere to mimic the stratospheric injection by volcanic eruptions.

Approached from a simplified modeling perspective, it's a straightforward problem: Create an idealized model planet. Then introduce some aerosols to induce an anti-greenhouse and cool off the lower atmosphere and surface. This is exactly what I did for my PhD dissertation when I simulated the climate influence of impact-raised dust on the young Earth. Compared to most of these other real-life and modeled anti-greenhouses, the adjustment needed to counter global warming here on Earth is tiny, requiring only a minor change in the energy budget of the planet. A difference of about 1 percent in the albedo (reflectivity) of the planet creates about a 3 degree (Celsius) change in temperature, which is roughly doubled when you include amplifying feedback involving water vapor. If we did this, the sky on Earth would look a little different, but not noticeably darker. Sounds easy, right? Yet the closer you look at an actual planet (this one), the messier the problem becomes. We really don't know how it would play out in detail, globally. It would surely change precipitation patterns, likely causing droughts in some places and floods in others, but our models aren't good enough to reliably predict this.

We've seen in the geological and historical records of large volcanic eruptions (the closest natural analog to a stratospheric geoengineering effort) that the effects are indeed geographically uneven. The largest volcanic eruption in the twentieth century was the great 1912 explosion of Katmai, on the Alaska Peninsula, in southern Alaska. Alan Robock, an atmospheric physicist who has studied (alongside nuclear winter, global warming, and geoengineering) the historical effects of volcanoes on climate, found that the Katmai eruption caused a drop in precipitation that led to a regional drought in the Nile River Delta. Records of Nile flow show that the year after the eruption had the lowest flow measured all century. A weakened Indian monsoon and a massive famine in Africa were also triggered in the two years following the eruption. Given what we know now and, more important, what we don't know, there is no reason to doubt that "fixing" climate with SRM would cause similar catastrophic regional problems.

Even if we thought we knew enough to implement such a solution safely, or reached a point where the regional disruptions seemed the lesser of two evils—and just who is going to make *that* call?—there are several other reasons to be highly skeptical of SRM as a solution for global warming. For one thing, it doesn't do anything to address the dangerous ocean acidification caused by rising levels of CO_2. Even if we were able to cool Earth with a controlled anti-greenhouse, the oceans would still become more acidic, the coral reefs would die, and many species would go extinct, ultimately threatening the marine ecosystems we depend on for food. Further, if cooling Earth in this way led to any slacking off on efforts to reduce our carbon inputs—and one can imagine it might—this would make acidification even worse.

It also raises thorny problems of governance and continuity. A stratospheric cloud will stay aloft for only a couple of years. Such a system would have to be continually replenished and maintained. Who is going to be in charge? What happens to the climate if this effort is not continued? Sometimes I am asked by screenwriters to speculate on sci-fi scenarios that might be both plausible and make for good drama. Next time I'm asked to think of a realistic possible future where Earth has become an apocalyptic nightmare, I'll suggest one where a stratospheric geoengineering project is started but not maintained.

Other forms of SRM (including various methods to increase cloudiness over the oceans, designs for giant structures in space, and arrays of satellites to block or reflect sunlight) suffer from the same problems of inadequate governance and woefully incomplete knowledge. And none of them does anything about ocean acidification. At best, SRM seems like an untested form of methadone for our carbon habit, transitioning us to a new dependence that might be slightly more manageable but that surely has unknown side effects and doesn't really address our core problem.

The other major category of geoengineering being discussed, carbon dioxide removal, is more promising. If we can pull large amounts of CO_2 out of the atmosphere, then we will be working closer to the root of the problem, not just compensating crudely and incompletely for warming. Earth's carbon cycle already has large CO_2 "sinks" that remove the gas from the air. Perhaps there is a way we could tweak the cycle a bit, enhancing or mimicking these natural sinks to pull out a little more carbon and sequester it into some of its other, climate-neutral, organic or inorganic forms.

What if we could use the energy of sunlight to bind atmospheric CO_2 into organic carbon, reducing the atmospheric greenhouse and simultaneously producing food or fuel? This sounds almost too good to be true, but it goes on every day. We call it photosynthesis. Photosynthesis is brilliant. No wonder the biosphere went a little bit crazy and temporarily lost its balance when it discovered this winning trick. Perhaps there is a way that we can get in touch with our inner cyanobacteria, once again transforming the world with sunlight, only this time with purpose, with a plan, with negative feedback and the concept of "enough."

In their planetary engineering study, after considering how we could play doctor to climates around the solar system, and reviewing ideas for mitigating global warming on Earth, Pollack and Sagan concluded that most of the schemes for Earth were too risky. They did, however, strongly endorse one method of geoengineering:

> We advocate instead of particle shields or sunscreens, a well-tried biological solution...We propose reforesting the world, especially in the tropics, in accordance with the ancient oriental wisdom "He who causes trees to be planted, lives long."[6]

Plants and trees grow by taking carbon dioxide out of the air and fixing it into solid organic molecules, so if you increase the mass of trees, reversing the deforestation that is currently taking place, there will be a consequent drawdown of carbon dioxide. Also, tropical forests have so many other intrinsic values, such as fostering and preserving enormous biodiversity. An expanded version of this concept involves changing land use practices, for agriculture in particular, to increase uptake of CO_2. There are some ideas as simple as changing when and how farmers till their fields and encouraging changes to less meat-intensive diets, which collectively can make a large difference.

You may not think of planting trees and changing farming practices as forms of geoengineering. Those who consider themselves opponents of geoengineering generally do not. Yet I would argue that, in an important sense, they are. When we do these on a large enough scale, in a conscious effort to affect the climate of the planet, we are deliberately interfering in Earth's climate, committing acts of planetary engineering. Nobody could reasonably object to these verdant interventions. Yet, given all the carbon we are dumping into the air, and will be for decades to come, they will not, on their own, be sufficient.

There are other, more intrusive, proposals to artificially enhance the rate of photosynthesis on Earth. One is to dump massive amounts of iron flakes into large areas of the ocean. The growth of photosynthetic marine life in many ocean environments is "iron limited," meaning that what stops much more plankton from growing is a lack of that one essential nutrient. This is why we sometimes see large algae blooms when storms blow desert sands out into the deep ocean and the iron-starved creatures instantly start to multiply like crazy. Simulating this natural process with an input of iron would cause a greening of the surface waters. Doesn't that sound nice?

As with the ideas for changing Earth's reflectivity with stratospheric hazes, this one sounds great from the point of view of very basic physics or chemistry, where you picture a simplified cartoon planet with boxes representing the ocean or the sky and arrows of different sizes representing inputs and outflows of radiation or carbon. However, again, as we go beyond the cartoon and look more closely at the details and mysteries of the real planet, our confidence in our ability to do this declines rapidly. Given our current level of ignorance, seeding the deep ocean with iron on a large enough scale to make a significant dent in atmospheric carbon is an obvious invitation for unintended consequences.

Like a panicked bull charging into a shop of living china, we'd be altering complex ecological systems, with their inherent biological feedbacks, in unknown and unpredictable ways. There's a long list of research questions that would need to be answered before we could confidently try this on a large scale. It is possible that many of the side effects would be beneficial for ocean life. Increasing the fertility and biomass of the deep, remote oceans might help restore stocks depleted by overfishing. So, it could be a win, win, win for us, the algae, and the fish.

Yet we don't know what else such a large intervention might also do. For example, it could release massive amounts of poisonous biotoxins from some of the algae species that would multiply in these altered waters. Then there are also fundamental questions about how effective such a strategy would really be. It would remove carbon from the atmospheric system in a lasting way only if a significant portion of the newly sprouted plankton died, fell to the bottom of the ocean, and was buried beneath sediments, taking its carbon with it to the grave. If, instead, this strategy mostly stimulated the growth of other species in deeper waters that then breathed the CO_2 back into the air, there would be no net benefit for the climate. We should certainly continue to study this problem with better models and carefully controlled and monitored small-scale trials. Yet, like most brute-force geoengineering schemes, seeding the oceans with iron is a very long way from being ready for prime time. Ocean ecosystems are not something we want to play around unless we really know what we're doing.

Other ideas for enhanced photosynthesis involve using algae in offshore "bioreactors" to convert sunlight and atmospheric carbon into food and fuel. This is somewhat similar in concept to the iron fertilization schemes for the deep ocean, but it would be done intensively, in controlled and confined near-shore locations. In some of these designs the bioreactors are even combined with wastewater treatment plants, using the carbon in our sewage as food for the algae, which removes CO_2 from the atmosphere and creates clean water, renewable fuel, and fertilizer. Sounds too good to be true, but the basic science is sound. The question is whether it could be scaled up sufficiently to make a meaningful contribution to climate mitigation. Genetic modifications could be used to enhance the photosynthetic capacity of algae in these settings. Care and cleverness would be needed to ensure that such beasties were not inadvertently released into the ocean at large.

The solutions that make the most sense are variations on those that Gaia found long ago: use solar energy to capture carbon from the air and put it to work in the service of organic life. It seems that enhanced or artificial photosynthesis ought to be a big part of the long-term plan. Can we find ways to greatly intensify the carbon-capturing role of sunlight without dangerously interfering in other natural processes?

I don't intend for this to be an exhaustive list of ideas for removing CO_2 from the air, and I urge you to read up on it elsewhere.[7] Some of the ideas for "direct air capture" (that is, machines that can simply pull CO_2 right out of the air) are physically feasible but seem prohibitively expensive. However, this technology (shockingly, considering the stakes) has not yet been extensively researched. This is an area ripe for game-changing innovation. When I speak with groups of eager schoolkids hungry for knowledge of science, it always fills me with hope, knowing that soon some of them will know a lot more than I do about how planets work. When I think about discoveries that may come along this century and completely change the trajectory of the future, I suspect that one of these kids will invent a reasonably priced or (better yet) profitable way to remove CO_2 from the air and lock it away or convert it into something useful.

Whatever mechanism we choose, we must make sure we know how to turn it off, lest we meet with an ironic end. A carbon-removal process with no end point or Off switch could create a complete planetary disaster. As Robert Frost wrote, "for destruction ice is also great and would suffice." If we removed too much CO_2, the world would freeze.

No Quick Fix

It is possible to love planetary engineering but remain skeptical and wary of intrusive, heavy-handed geoengineering schemes. Carl Sagan cherished the idea of terraforming. He wrote about it in *The Cosmic Connection*, his first popular book, and in *Pale Blue Dot*, his last on a space theme. He was always looking for ways to update terraforming schemes to take advantage of new discoveries and ideas in planetary science. In 1996, I was studying the stability of climate on Venus with my grad student Mark Bullock at the University of Colorado. We were exploring the climate feedbacks that arise because of chemical reactions between greenhouse gases and surface minerals, using techniques we learned from Jim Pollack, who worked closely with us until his death in 1994. We discovered a stable climate state that Venus might enter into that would be much cooler, though still a lot hotter than Earth. We published a paper about this in the *Journal of Geophysical Research*,[8] and a letter soon arrived from Carl, who was at the time being treated for what would prove to be fatal myelodysplasia:

> Dear David
>
> I read your most interesting paper in *JGR Planets*. When you get a moment, why don't you think about terraforming Venus?...How would you make a massive decrease in surface temperatures?

It seemed as though Carl fired this off the day after the paper came out. He didn't miss much. This reveals something that many people don't know about him. Even while he was engaging more and more visibly in public life, he was always still voraciously reading the literature, doing research, and carrying on vigorous scientific discussions with seemingly everyone in the planetary science community. He remained a card-carrying working scientist until the end. After I received that note, in early May, we had a fun exchange of letters about climate engineering. It was one of the last conversations we ever had. In December of that year, his weakened immune system could not fight off a sudden bout of pneumonia, and he died at the age of sixty-two.

Carl was way ahead of the curve on the consequences of anthropogenic climate change. He was into warning about global warming before it was cool. He was also a fan of terraforming, and saw becoming a multiplanet species as part of the long-term plan of humanity. Yet he was also wary of foolish human tinkering. At the end of their 1993 paper, he and Pollack concluded that:

> Clearly more work is needed, but comparatively inexpensive and environmentally prudent methods of mitigating greenhouse warming on Earth may be within reach in the next few decades.

Well, here we are now, after the next few decades. As I write this, nearly a quarter century has passed since their paper was written. When I look over the current literature on geoengineering, it remains clear that "more work is needed."

I believe that were he here today, Carl would urge extreme caution about geoengineering. I can close my eyes and hear his deep, resonant voice saying, "It would be an act of consummate recklessness and arrogance to try to jury-rig the only home we have, to attempt a quick-and-dirty repair job on the complex, deeply mysterious, and exquisitely balanced global mechanisms that all human life, and all other life we know of, depends upon, when we are still so ignorant about their functioning." Yes, I know Carl would have found a way to say it more eloquently, but I feel pretty confident I know where he would stand on the issue.

We are not ready to terraform Mars, or intrusively reengineer Earth, any more than Neolithic humans were equipped to design modern urban sewage systems or the inhabitants of King Arthur's Court were able to understand the technology of Mark Twain's Connecticut Yankee. Attempting such a fix now would be completely insane. Or, as my friend, climate scientist Ray Pierrehumbert, one of the coauthors on the recent National Research Council report on geoengineering by solar radiation management,[9] put it, trying to do this anytime soon would be "wildly, utterly, howlingly barking mad." Though there are a few loud voices advocating them, there is not really very much support for these risky quick-fix geoengineering schemes, and as people look into them more carefully, this support will continue to decline. The National Research Council reports released in 2015 endorsed further study but also stated strongly that none of the intrusive climate-intervention schemes should be implemented. The conversation is valuable if anything because it highlights the uncertainties, the incompleteness of our knowledge, and the fact that we really have no choice but to control our CO_2 habit. Given how much we have yet to learn, we really should not be let anywhere near the controls of planet Earth.

Only, it's a little late for that. Without realizing it, we have grabbed hold of those controls and we're...not exactly driving. We're not in control, but we're madly spinning the wheel, pushing buttons, and turning knobs. We have to try to understand how this planet works, ease up on the switches gradually without letting go entirely, and try to reengage some of the autopilot mechanisms we've disabled, while we work on a longer-term plan.

Sagan and Pollack ended their review of planetary engineering with a statement about Earth:

A short-term imperative for planetary engineering exists only for one world in the solar system, our own. Careless or reckless applications of human technological genius have put the global environment at risk in several different ways. The Earth is not a disposable planet. It is just conceivable, as we have discussed, that some of the techniques that in the long term might be applied to engineering other worlds might also be utilized to ameliorate the damage being done to this one. Perhaps a safe way to test our protocols is to implement them in carefully circumscribed ways on other worlds. But considering the relative urgencies, a useful indication of when the human species is ready to consider planetary engineering seriously is when we have put our own world right. We can consider it a test of the depth of our understanding and our commitment. The first step in engineering the solar system is to guarantee the habitability of the Earth.

Roger that, Houston. Yet how do we guarantee that Earth remains habitable, not just for life (which is not really threatened) but for the present biota and a society of (approaching) ten billion human beings? As I've discussed, there is no easy, quick fix. Or, rather, there may conceivably be one, but we are not in a position to know about it and safely apply it. Perhaps some wise, advanced technical civilization somewhere in the universe knows exactly how to cure what ails us (more on this in chapter 6). At some future date people may look back on our time and say, "If only they had known." Just as we can look back and say, "If only we had known about antibiotics in the Middle Ages, the tragedy of the Black Death in Europe could have been avoided," or "If we had known about the smallpox vaccine in the sixteenth century, the history of indigenous Americans could have been very different and much less tragic." The aversion of tragedy sometimes comes down not to what is physically possible or impossible, but what is known when.

Some have, understandably, called for a moratorium not just

on the deployment of these technologies, but even on studying them. The Australian environmental philosopher Clive Hamilton wrote an editorial in *Nature* entitled "No, We Should Not Just 'At Least Do the Research.'" Along with many others, he feels that even by studying how we might undertake these radical geoengineering proposals, we are making them seem like more acceptable options. He is right to voice concern about this. If we operate with the fantasy that a quick fix is near at hand, and it lessens motivation or pressure to reduce carbon emissions, then the existence of the research program itself would indeed be harmful. Certainly if we are studying these options, we should be extremely careful not to give the impression that—well, that they are options. Yet those who advocate a research prohibition are wrong for three reasons.

First, studying this makes us smarter about the climate system. Understanding human interactions with other planetary subsystems is now part of learning about Earth. We are already "engineering" it in various ways. Okay, *engineering* is not quite the right word, as it implies some larger degree of understanding than we have. We are perhaps engineering Earth only in the way that your infant is "engineering" your home media system when she sticks cookies in the DVD slot. Yet if you wanted to figure out why it wasn't working in the right way, you would not want to ignore this activity.*

We cannot model the climate system without including, in our equations, factors representing human actions—or we could, but these studies would be irrelevant representations of a world we do not actually inhabit. In fact, there is reason for us to pay particular attention to these terms in the equation: they are the ones that we ought to be able to change. Our collective

---

* Surely, in just a few years, there will be no more DVDs and your child, slightly older, will be the only one who understands your new technology.

behavior is now a component of the climate system. So when we talk about changing our own behavior in ways that will have a beneficial influence, we are in fact discussing a kind of climate engineering. We should focus first on our unintended contributions. Only when we get a handle on that will we possibly have graduated to the point of being able to experiment with other terms of the equation.

Second, we know we will need to intervene eventually, some millennia hence, because eventually climate will change dangerously if we let it. At that point our intervention could save a lot of species from extinction. In the meantime, learning to think on these longer timescales is a necessary survival skill. As I've discussed, over the long run, if we have a long run, we will be morally obligated to interfere with Earth's climate through some form of geoengineering to stop the dangerous climate change that would occur if we left it alone. In the short run, we are obligated to try to have a long run.

Third, we need to be ready with risky rescue options just in case the very worst-case climate scenarios do unfold and force us to try something desperate. Global warming could conceivably become rapidly more intense than our best models predict. Complex systems with positive feedbacks do contain "tipping points," where new modes of behavior can suddenly emerge. Some of the scarier scenarios involve the release of vast methane deposits trapped on the seafloor, or frozen in the Arctic tundra. There are projections where the release of these reserves starts to warm Earth, triggering further release, and causing a rapid runaway effect. There are other scary outcomes in which the Antarctic ice sheets could collapse rapidly and spectacularly, causing a sudden huge rise in sea level. Earth's climate record seems to show some episodes where the system jumped quickly to a significantly warmer state.

I don't believe these are the most likely paths for Earth's

climate in the next couple of centuries. My sense is that there are too many complex negative feedbacks built into the system to let one of these positive feedbacks completely take over, and even given the unprecedented provocation we're introducing, the system tends to move slowly on human time. The climate change I'm personally most worried about is the kind we *can* reliably predict will progress throughout the next century if we continue on our current course: the slow, inexorable warming and acidification that is already under way. But the fact that we can't rule out the scarier, rougher scenarios is deeply unnerving, considering what is at stake. Given this, it is our duty to research the more intrusive geoengineering options now. At the same time, we need to be doing everything we can so that we never need to use them.

Would you be comfortable about someone trying to reengineer, during flight, an airplane that you and your family were flying in? No, of course not. Unless you knew the plane was in trouble and going down. Then you would probably very much want them to try something. We don't want to be in that position with our planet, and with a little luck and a fast learning curve, we won't be. Still, just as NASA builds safety contingency plans into missions, and hopes never to use them, we need to have contingency plans for the possibility that planetary climate could take a sudden turn for the worse.

We Are Apprentice Planetary Engineers

We have no choice at this point but to engineer Earth, in the sense that we need to thoughtfully interact with it. Our actual choice is between methods: what kind of planetary engineering we should undertake, how and when we should intervene in the Earth system. We should regard ourselves now as apprentice planetary engineers, easing up on those behaviors that have been throwing the system out of balance, taking those steps we know are safe, and learning all that we can about how the system works so that by the time we need to call upon more intensive interventions, we will be ready to do so safely and wisely.

The most necessary and near-term geoengineering strategies are the least intrusive: moving to alternative energy, reducing our emissions of greenhouse gases, intensifying agriculture to take up a smaller land footprint, and halting and reversing deforestation. Then there are somewhat more intrusive schemes that must be employed only with caution, after further study, and starting on a small scale, such as ocean fertilization. On the other end of the continuum are those that, given our current level of knowledge, would be far too risky to attempt, such as artificially clouding the stratosphere.

On the most basic level of systems engineering, we can see that we have been unconsciously acting to promote positive feedbacks in the climate system, and these increase instability. Yet the mere fact that we are now studying this system, including our own role in it, potentiates a huge change to planetary dynamics. With awareness comes the possibility to alter, and reverse our role. First we can learn to stop our unconscious destabilization. Then we can choose to introduce more negative feedbacks. We can shift to being stabilizers.

The best way to change the equation right now is to enact a "sign change" on one quantity in particular: the rate of accelerating CO_2 input into the atmosphere. Our urgent task is to shift that from positive to negative. The more we learn about how the system works, and the role we are playing in it—and the more we share and spread this knowledge—the more impetus is created to enact this change. That's going to be tricky, but we are going to do it. There's no doubt about that. Fossil fuels won't be around forever and, over time, will only become more expensive, and their extraction more difficult and ruinous. Timing is everything, though. It takes a generation to transform a civilization's energy system, so we had better hurry up. Alternative energies are getting cheaper and more widespread. Fossil fuels are gaining more and more of a deserved stigma, but economic forces are slow to change. Also there is another dynamic at play that needs to be spelled out even in this very basic discussion of energy futures: the developing world needs and deserves a lot more energy. Hundreds of millions are in need of lights, sanitation, and so much that we in my country take for granted. If we want them to do this without burning coal—and believe me, we do—then we have to help them.

The Earth system can absorb a certain outflow of CO_2— it's been doing that for a lot longer than we've been here. It's easy to imagine a global technological civilization that dumps some amount of carbon into the atmosphere without causing long-term climate change. If we hadn't already pushed things so far out of whack, with an excess atmospheric carbon load that will last for thousands of centuries, this inevitable deceleration of our carbon effluence might ultimately prove to be enough. Yet given how far we are pushing the system, we are going to need to think about taking our level of engineering intervention up a notch and enacting a second "sign change" in the equations: transform the total human input of CO_2 into the system from positive to negative, actively removing some CO_2 to help restore balance. As I've just discussed, the most unobtrusive way to do that would be by planting lots of trees. The more intrusive ways include ocean fertilization, algae farms, genetic engineering, and artificial photosynthesis.

Some might object at all to calling the greener, less intrusive steps "geoengineering," but they are conscious choices, enacted at a global level, to change properties of our planet. The more hard-core, brute-force methods we usually associate with the term *geoengineering* are, appropriately, more controversial. We certainly don't want to make the cure worse than the disease. Still, they are all on a continuum of actions we might take purposefully to change the planet.

Returning to the medical analogy, consider the sometimes false dichotomy between "natural" medicine and more invasive and aggressive interventions. Any good doctor will tell you that you need both. It is always a good idea to do what you can to maintain and restore health with nutrition, diet, and lifestyle, fostering your body's own homeostatic tendency toward health and equilibrium. Yet sometimes you need medication or even surgery or chemotherapy. In these latter cases you wouldn't want to try a cure that had never been tried and tested, but if you found yourself or your loved one in rapidly declining health, you might gladly sign up for an experimental trial.

It's interesting to compare intrusive geoengineering with planetary defense against dangerous asteroids. Each of these is a capacity that, with luck, we won't need right away but we know we'll have to call on eventually. In both cases we can't afford not to study the problem to increase both our awareness of the dangers and our ability eventually to act. This means better and more complete Earth observations, better climate modeling, more thorough planetary exploration, and more complete surveying of the small-body population of the solar system, so we know just what we're dealing with. To gain both the knowledge and the wisdom to undertake such solutions, we'll need to keep exploring other worlds and integrating what we learn about how climate unfolds in different planetary scenarios.

On the longer timescales (tens of thousands of years) over which Earth will go through huge climate changes, and more likely than not enter into the path of some very dangerous space rocks, it will be essential to know how to meet these threats.

Certainly if there is still a human civilization on Earth

within one hundred thousand years (ten times longer than there has been one to this point), we will have employed both these defense mechanisms. Either one might be needed much sooner—in the case of climate, if worst-case scenarios and feedbacks are underestimated or unknown, and in the case of planetary defense, if we get unlucky or decide to mitigate against those smaller, more frequent impactors that would not decimate civilization but could destroy cities.

Especially as long as humanity, and all life, is confined to one planet, we are obligated to study these problems and learn how to intervene intelligently when the alternative is catastrophe.

Yet we are also morally forbidden from trying either of these technologies on any large scale at this point. Planetary defense, like climate engineering, is not without risk if attempted ignorantly. If we're changing the trajectory of giant space rocks that could cause calamitous impacts, we need to know what we're doing. Here also there will be difficult governance issues. In the case of long-term climate intervention, once we have moved beyond the emergency mitigation of dangerous warming, there will be the question of who gets to set the thermostat, and where it should be set. Governance challenges for planetary defense may arise, for example, if a city-destroying space rock is found to be heading for an impact point in a particular country. Altering its trajectory will mean, at first, that other countries are put in the crosshairs before it is safely deflected to miss Earth entirely. We'll need to temporarily increase the risk for some in order to save others. Those placed in harm's way will need reason to feel confident that the system will continue functioning long enough for us to finish the job. These problems can be surmounted, but we've got work to do on the technical, political, and cultural fronts.

In the case both of planetary defense and global warming, the larger challenge is the same: to become aware of ourselves as actors in a planetary system that is evolving over timescales much longer than our individual lives. We need a long-term plan for humanity's future. It needs to be flexible and to include a wide continuum of solutions, but we need to start enacting it now, knowing full well that the plan will change as events unfold and our knowledge increases. Part of dealing with uncertainty ought to be the realization that we need approaches that are sustained but flexible, subject to modification when we see how the planet actually responds to our poking and prodding. It is essential that we research much more extreme or risky maneuvers than we ever hope actually to use.

If we can ensure our own survival over (or even muddle through) the next few centuries, then we will eventually want to employ more obtrusive planetary engineering and learn to engineer our planet's climate, intentionally pulling it toward stability. So, if we have a long-term vision for ourselves, and I really don't think that is optional anymore, then it should include starting to learn what we'll eventually need to know to do that kind of geoengineering. That will include a great deal of self-knowledge as well as more sophisticated planetary science and engineering. When the technology is understood well enough to be deployed without unintended consequences, the most challenging aspect may be developing the global capacity to make these decisions. If we do, then we will have become a new kind of entity on the planet: self-aware world changers with the good sense to work with the planet and not against it.

Home Movies

We have to get a handle on ourselves, learn to act consciously as the global entity we are. Yet look at us now. How are we ever going to get from here to there? When people express despair at the possibility of change, I like to remind them how fast our relationship with the world is changing right now, in precisely some of the ways we need. The first step is awareness, and our awareness of our world is exploding.

I've described how our visits to neighboring planets have given us needed perspective to see how climate and life have evolved with Earth, but at first our forays to worlds beyond Earth were few and far between. Most of the hardware we launched off Earth in the early space age stayed very close to home, circling our planet in low Earth orbit, the first thin threads in a thickening satellite web we've been weaving ever since. On April 1, 1960, the *TIROS* (Television and Infrared Observation Satellite) was launched and began sending back the first continuous images of our planet's weather patterns. Earth began to self-monitor. From that point on, we've photographed and measured our planet with increasing coverage, detail, and instrumental acuity.

In 1961, there were half a dozen active artificial satellites orbiting Earth. By 1970 the number had grown to 164. In the early '70s, for a teenage space geek like me, a satellite launch was still a big deal. Today, they remain fun to watch, but they've become routine, with only the occasional dramatic mishap providing enough mayhem and novelty to make the news. In 1972, the first *Landsat* was launched, beginning a continuous program of orbital surface monitoring. This gave us a baseline from which to measure changes in the planet and an ability to catch unfolding natural disasters such as floods, hurricanes, and droughts anywhere on the globe. Starting then, we have had a nonstop photo album of planetary changes. We've now had some decades to become familiar with the fluctuations Earth goes through on her own, the hemispheric patterns of weather fronts and storms, and the somewhat predictable cycles of change on seasonal and longer timescales. Against this complex and shifting, but increasingly recognizable, background, we've become more adept at picking out longer-term changes and tracking our own activities (urbanization, dam building, and deforestation), which are all recorded with unblinking orbital eyes. The *Landsats* and other Earth observers gave us more than just the ability to perceive Earth whole and in unseen, revealing colors. Over the years and decades, the unbroken archive of images from these sky cams have given us a fast-forward "this is your life" view, where we can see clearly how the land, seas, and ice of our world, partly in response to our own incitements, have been changing.

The global weaving continues; the orbital web grows. In 2014, the European Space Agency launched a rocket from its spaceport in French Guyana, carrying *Sentinel 1A*, the first satellite in its Copernicus project, which is in many ways the most ambitious Earth observation program to date. Copernicus will include a fleet of orbiting craft to be launched over the ensuing decade, which will obtain continuous coverage of the entire planet in unprecedented detail over multiple wavelengths. *Sentinel 1A* can monitor any location on the globe using radar imaging, a technique we've employed with great success elsewhere in the solar system, allowing us to map places that cannot be photographed from orbit using sunlight, because they are obscured by clouds or the dark of night. On cloud-covered worlds such as Venus and Titan, radar mapping from spacecraft has allowed us to uncover the form and history of otherwise unseen landscapes. It's like flash photography, only the illumination source is not visible light, but a microwave blast sent toward a planetary surface from a spacecraft antenna. This same antenna then captures the signal reflected back from the ground, and beams the pattern earthward, where our computers reconstruct it into new images and maps. These radar flash cameras can see in the dark, but even better, microwaves are unfazed by clouds and hazes. So *Sentinel 1A* can observe patterns and changes through all kinds of weather anywhere on the ground or ocean, day or night. This makes it an incredibly powerful new tool for monitoring sea ice or ground motion, or keeping track of long-term trends, momentary natural disasters, or even humanitarian crises.

As sensors improve and coverage expands, we, as a component of a planet, do a better job of self-observation and reflection, obtaining the data we need to understand better what effect our activities are having, and become better equipped to take care of ourselves.

By 2015 about 7,000 satellites had been launched. Of these, about 3,600 remain in orbit and around 1,000 are functional.[10] We are still in an early phase of this planetary self-monitoring, still integrating this powerful new view of ourselves. It's as if Earth has been evolving a diffuse new organ, a sky-high mycelium of floating mechanical sensors growing far above its surface, to keep a continuous eye on itself and develop a new infallible digital photographic memory of its own changing visage.

The Apollo program gave us our first widely seen self-portraits—from Apollo 8 in December 1968, the *Earthrise* image we all know and love—our planet as a deep and lustrous blue jewel, moist and alive, swathed in bright cloud, lingering over the barren, lifeless, alien lunar horizon. [See page 3 of the photo insert.] From Apollo 17 in 1972, the *Blue Marble*: our first clear view of a full Earth, the entire illuminated hemisphere overwhelming us with its vivid, kinetic beauty. These will always be treasured, the baby pictures of a newly space-faring species. They changed us profoundly.

The Moon shots gave us our first snapshots, but it was satellite remote sensing that provided our first scratchy home movies, revealing and animating the seasonal changes on the continents and ice caps and the swirling daily circulation of Earth's atmosphere and oceans. The space age brought home to every living room moving pictures of Earth's pirouetting cloud patterns. More and more, we learned to connect our local quotidian experiences of the changing weather with the global motions of the roiling atmospheric sea. A hurricane was no longer merely something that happened out of the blue, an intense storm that sometimes engulfed your hometown with terrible wind and rain, but a beautiful and sometimes menacing spiral of clouds that swept across open ocean, swiping vast arms toward vulnerable islands, and threatening a run at the coast. By now, a couple of generations into the space age, aided in no small part by the ubiquitous satellite imagery in TV weather reports—so familiar we've long since stopped marveling at its origins—we've internalized this radical shift in scale and perspective so that, in our mind's eye, we seamlessly connect these two views, from orbit and from the ground.

To see what a difference a satellite network and computer models can make, compare the Great Galveston Hurricane to Hurricane Katrina. In the days leading up to September, 8, 1900, the people on Galveston Island, which juts into the Gulf of Mexico southeast of Houston, had absolutely no idea a monster storm was approaching. They stayed put as the weather worsened to the point where most structures were destroyed and much of the populace was swept away and drowned in the massive storm surge. It was the deadliest natural disaster in U.S. history. The final death toll is not known but is estimated to be around eight thousand.

If the same hurricane were approaching the Gulf Coast today, we would see it developing weeks ahead and have good predictions of its path. People would have time to head inland to higher ground. Property would still be destroyed, but loss of life would be minimal. Katrina, which struck the Gulf Coast near New Orleans on August 29, 2005, was terrible and shocking for a great many reasons. Still, at least we knew it was coming, and evacuations were ordered days in advance. The failures were organizational, bureaucratic, and cultural. There was loss of life and massive suffering because the authorities did not get their act together and organize an adequate response, because large neighborhoods were poor and disenfranchised, and because the infrastructure of the city had been neglected. These consequences fell disproportionately on the underprivileged. New Orleans is forever changed, many lives are permanently disrupted, and some communities may never bounce back. Yet Galveston can give us some idea how very much worse such a disaster would have been in 1900 before we had the orbital imaging and meteorological models to accurately predict the path and magnitude of such a storm. Can you imagine how vulnerable modern New Orleans, with its eroded barrier islands and rising sea level, its degraded coastal wetlands and increasingly taxed system of levees, would be without a constantly vigilant network of global satellites? This comparison illustrates two somewhat ironic, and cautionary, aspects of our increased reliance on scientific sensors and models to safeguard our cities and our society—both of which are important for understanding the situation and future prospects of human civilization as we enter the Anthropocene epoch.

First, we've made ourselves more vulnerable, created new problems that make us more dependent on our new solutions. By overdeveloping coastal regions and pushing the agricultural carrying capacity of our lands closer to possible limits, we have placed ourselves more in harm's way, made ourselves more susceptible to catastrophic disruptions of daily life by impulsive natural events (which are also becoming more erratic as they become less "natural," forced by a changing climate). Yet we have also built up safeguards that we increasingly depend upon, though we easily take them for granted. Giant coastal megacities without satellite remote sensing would be a recipe for disasters that would make Galveston look like a Gulf Coast clambake.

Second, with Katrina we could see what was coming but were unable to effectively act upon this knowledge. This shows that although science and technology have become indispensable survival aids, they can provide only so much protection if we don't have the institutional, educational, and cultural resilience to defend ourselves.

The ability to monitor our planet and our surroundings continuously has changed us, and changed our planet into something entirely new. Yet this aspect of our planet is in its infancy. We take it for granted, but now our Earth-observing capacity comes and goes with the launch and failure of different satellite systems. If we are going to build the capacity to avoid dangerous climate changes and asteroid impacts, then we will need long-term, stable, space-based infrastructure. We'll need monitoring capability that can be ensured to last over the lifetime of such a project. Right now we are obsessed with innovation, but at some point we will need to focus more on stability. Not that innovation shouldn't be welcomed if better ways are found to accomplish our goals, but at the very least the baseline plan must be undertaken with technology that can be built to finish the job at hand. We will need systems built to last for centuries, and the institutions to reliably maintain them.

We've shrouded our world with three thousand satellites with which we are watching our planet, spotting previously hidden patterns and constantly firing instantaneous signals around the globe, like cells in a restless planetary superbrain. With these compound orbital eyes and this self-assembling cyborg mind, we perceive with new depth and acuity that we are deeply embedded in complex global systems and cycles of matter, energy, and (increasingly) information. Augmented by satellite senses and terabyte memories, our cave-evolved intellect, supersized with supercomputers, is just now attaining vast new abilities to comprehend the global present and to model and anticipate the future.

The Bright Old Sun Problem

Let's assume we get a handle on our self-induced climate problems in the next century, and learn to manage Earth's climate cycles gracefully over the next thousand centuries. Then over a much longer timescale, we will be forced to deal with another kind of climate change, one that also goes in the category of planetary changes of the fourth kind.

When we look farther ahead, even our star, the Sun, cannot be expected to be a reliable, steady partner. As it ages, it is slowly brightening. It's just what happens to stars like our Sun as they evolve. This is why trying to understand the ancient climate of Earth presents us with the "faint young Sun problem" I write about in chapter 1. Well, in Earth's far future we'll have the opposite: a "bright old Sun problem."

Our climate automatically adjusts to the gradually warming sun, slowly lowering the average CO_2 content in the air, converting it to carbonate rocks in such a way as to balance the slight increases in solar radiation. Earth has been doing this over its entire lifetime, but eventually it won't work anymore. All the CO_2 will be drawn down, Earth's thermostat will be stuck, and the Sun will keep warming. Then, inevitably, Earth will go the way of Venus. It will become so hot that the oceans will boil and the hydrogen will be swept into space. Left to its own devices, Earth will experience a runaway greenhouse. We'll lose our water, and all carbon-based life will perish. Earth will be left completely uninhabitable.

Ultimately, somebody is going to have to step up and deal with this if our biosphere is to survive. As the Sun brightens we'll want to interfere more strongly because, as the slogan goes, there is no planet B. If we (or our descendants, or our creations, or even somebody else who's come along and made a home of this planet in the intervening eons) want to remain here we will have to take action and engineer the mitigation of the eventual, inevitable runaway greenhouse. This is not going to be necessary for a billion years, but for long-term survival, it will be absolutely necessary.

Who knows if anybody will still be around here? This is a timescale much longer than the lifetime of our civilization or even our entire species, or even all animal life. Yet if we think of the life of our biosphere, we see that this is not such a long time. Depending on how you measure it, we, Gaia, are perhaps three billion years old. So this will become a problem when we are only 30 percent older than we are now.

If we make it that far, given billions of years of engineering prowess, we will be able to solve this problem and even help Earth's biosphere outlive the Sun. Someday we may be the best thing that ever happened to life on Earth.

5

TERRA SAPIENS

It is far better to grasp the universe the way it really is than to persist in delusions, however satisfying and reassuring.

—Carl Sagan

The Great Promotion

Understandably, the term *Anthropocene* is disturbing to many people. We are both drawn to and repulsed by the idea that we are the harbingers of a new geologic era. I've seen this in response to numerous talks I've given and public conversations I've led on the topic, and in vigorous, unending debates on social media. People recognize the validity of the concept, but many also find it deeply troubling. The notion of a time in which we are somehow responsible for the world sets off alarm bells.

This is good. If you're not disturbed, you're not paying attention because—wait, isn't the whole notion just narcissistic and arrogant? Species come and species go, shuffling on and off the stage. Who are we to impart such importance to our own sudden entrance that we name a geological epoch after ourselves? Isn't it more than a little self-aggrandizing to declare our brief ascendance to be its own chapter in the multibillion-year book of Earth? It even seems to carry the obnoxious implication that we are elevating ourselves to a godlike status, that maybe we think we deserve to be in charge of this place. The conceit that this is somehow *our* world smacks uncomfortably of a biblical worldview, a world just for us, a universe in which we are the stars of the show. Hasn't science led us in exactly the opposite direction, in an exodus from self-centered delusions about our own significance, toward freedom and clarity about our insignificance? Yes, it has, but maybe we need a slight course correction.

For hundreds of years, science has driven us toward a view in which human existence is less and less central. Our discoveries have repeatedly and persistently pushed us to the margins, revealed us to be just another kind of animal, recent arrivals on an unbelievably ancient planet that is not the focus, or the point, but an insignificant dust bunny tossed in a random corner of a nearly incomprehensible cosmic vastness. It's been a process of knocking humanity off various pedestals. Four centuries ago Galileo liberated us from the tiny prison of geocentrism, by revealing that Earth, which we've always called not just a world but *the* world, is only one of many planets. Our loss of a privileged place was compensated for by a massive enlargement of our universe. Darwin then showed us that we are not the chosen species, but close kin to all life on this planet, and that we were not made in anyone's image because biological innovation does not require an intelligent creator. Modern geology and cosmology have affirmed that the story really can't be about us, because nearly all of it happened long before we showed up, and the size of our physical domain shrinks to insignificance against the immense, expanding clusters and superclusters of galaxies.

Now the recent exoplanet revolution is finishing the physical decentering of humanity, confirming that planets around all kinds of stars are nothing special. And we feel poised on the edge of a discovery that could pull us off our last pedestal. My own field, astrobiology, is pushing hard against the notion that life is unique to Earth. Soon, in subsurface aquifers on Mars, in a global ocean on Europa, or in the atmosphere of a nearby exoplanet, we will be able to test the proposition that our universe, given liquid water, carbon, and energy, easily makes life. Confirming this suspicion would destroy one of the last refuges of geocentrism.

In this emerging scheme, we are nothing. Or at least nothing special. In *Pale Blue Dot*, Carl Sagan writes:

The Apollo pictures of the whole Earth conveyed to multitudes something well known to astronomers: On the scale of the worlds—to say nothing of stars or galaxies—humans are inconsequential, a thin film of life on an obscure and solitary lump of rock and metal.

Carl referred to these successive scientific blows to the human ego as "the great demotions," and he described this progressive shedding of our infantile, self-important pretentions as a thrilling liberation from inherited delusions and a signature achievement of science.

Now—what? We're changing our minds? In declaring the Anthropocene to be an epoch of Earth history for which we are responsible, are we reversing this grand history of great demotions and giving ourselves a "great promotion" which places us back at the center of the story?

Well, sort of—and yes, this is indeed disturbing. Yet the point of science, as Carl liked to remind us, is not to comfort ourselves with feel-good stories, but to keep finding new ways to pull back the veils of delusion and see things as they really are, and then to deal honestly with whatever we learn. With the quote that begins this chapter, he admonished us to reject the old easy answers, the egotistical narratives in which we were central to the story of Earth. Strangely, we can now turn this dictate around to argue the opposite point. If we want to grasp the way things really are on Earth today, then we need to acknowledge our special part. Science has repeatedly revealed to us that we are not unique or special—except, guess what. We are. Once, in order to grasp the vastness of the cosmos and the peripheral nature of our reality, we had to awaken from naïve ego- and geo-centric dreams. Now we have to come to grips with our own significance. We've entered an age where we are radically changing the plotline, to the point where we cannot see nature clearly if we insist on ignoring our own growing role. Not only that but, as far as we know, Earth is the only place in the universe with life and intelligence. That may soon change, but for all we know we could be determining the future of all life. The delusion we may need to shed is that we can avoid the responsibility of, in some way, running this planet. We may be uncomfortable in this role, but it certainly won't help to deny it.

Our world is changing in unprecedented ways because of a new dynamic, a new set of processes, a new motive force we are obliged to examine. We are that force, and this puts us right back in the center. The term *paradigm shift* is used far too frequently to describe changing ideas in science. However, the reintegration of human activities into our understanding of the "natural" world is worthy of this term.

We've been taught that science is objective, value neutral, and depersonalized. In order to study nature clearly, we're supposed to remove ourselves from the frame, tiptoe out of the picture, brush over our tracks, and make ourselves invisible. Yet in our new Anthropocene science, the boundaries are not so clean.

As we observe the world, the lens is part of the photograph. As I've discussed, climate modeling is hard enough even when the modelers are not themselves part of what is being modeled. Ecologists, accustomed to studying various biomes, or communities of organisms existing in specialized environments, are now studying "anthromes," where human activities have become part of ecological systems. Rather than simply ignore or deplore croplands, rangelands, parks, cities, and managed forests, we can put them in our maps and models and decide how we want to integrate them into the world.

The triumphant successes of Enlightenment-driven science have drilled into our oversize, skeptical heads the notion that we cannot see ourselves as privileged in any way, that any stories in which our species is somehow special or pivotal are not to be trusted. There's some irony here, as the very success of this hyper-Copernican worldview has sown the seeds of its downfall, or at least its necessary revision. When Galileo stuck lenses on a tube and pointed his first crude telescope at the heavens, what he saw did not comport with the dominant, and biblically official, cosmology. His observations that Jupiter was accompanied by its own moons and that Venus was orbiting around the Sun could not be reconciled with an Earth that was the center of it all. In the resultant scientific revolution, we learned that we could discover the reality of our lives and our world through observation, experimentation, and deduction. This began a dance between science and technology that continues to the present. As science progresses, it deepens our understanding of nature and also allows us to build more powerful machines, including new scientific instruments that permit us to dig deeper into nature's seemingly endless mysteries.

This tango quickened with the Industrial Revolution, and again with the postwar Great Acceleration. Now we have instruments Galileo could never have dreamed of. We've split the atom and built supercolliders to pull apart the *sub*subatomic particles. We've deciphered the chemical assembly lines of living cells. We've constructed global networks of sensors that continually measure the composition of our atmosphere, monitor ocean currents, and eavesdrop on seismic vibrations that whisper to us of hidden motions and structures deep within the Earth. To analyze this flood of data, we have microprocessors that double in speed every two years and solid-state memories increasing in capacity even faster than we can fill them with information. Ever speedier computers give us a new way to interrogate the universe, by simulating everything. In addition to the classic twosome of experiment and theory, simulation has become a powerful third form of scientific inquiry. Up in orbit we have massive and elaborate descendants of Galileo's original telescope peering outward to the edges and beginnings of the universe, and dozens of telescopes ringing Earth, looking back downward. Yes, we're spying on our neighbors, but we're also revealing our planet to ourselves as never before. And what do we see? Ourselves—a world transformed by a powerful force that marks it, as far as we know, unlike any other in the cosmos.

When we first traveled far enough to look back upon our home and see it whole, as in the *Earthrise* portrait taken by Apollo 8 astronauts, we were simply struck by the beauty and liveliness of Earth in vibrant contrast to the lifeless Moon. Yet now that we've been able to observe our planet for half a century with ever-increasing detail, we've also started to notice that it is changing.

Look again at Earth from space. On the scale of worlds, are we really still so inconsequential? Individually, yes, but collectively we are moving mountains, altering rivers, draining seas and filling new ones, and lighting up the dark. Large swaths of tropical forest are disappearing. Coastal waters are choked with nitrogenous sediments, the signs of frantic, unsustainable agricultural activity. Space-based measurements show increasing carbon dioxide and decreased ozone in our atmosphere. The Arctic ice is visibly shrinking.

We used to say that national borders were not visible from orbit. This was one of the observations made by early astronauts in describing their feelings of unity and transcendence brought on by the orbital overview. These feelings and insights are real, but, sadly perhaps, it is no longer true that political borders are invisible from space. Differing patterns and styles of development have now inscribed some of our abstract political borders into clear, literal lines. One very stark example is the border between North and South on the Korean Peninsula, as photographed from the International Space Station in 2014. [See page 4 of the photo insert.]

We have to believe our machine-enabled eyes. When we fix our distant gaze carefully and steadily on Earth, we see it morphing under our own influence. The same technological revolution that has afforded us this vastly expanded view has also unleashed these accelerating changes.

The great demotions were a triumph of humility. Yet we're awfully proud of them. Our intellectual conquest has perhaps made us a little cocky. In liberating us from the ancient texts and constructing a beautiful new one, illustrated with undeniable wonders from spacecraft, telescopes, and electron microscopes, science has gained a sense of superiority over older, prescientific worldviews. Yet in shooting down traditional worldviews, science sometimes tends to pump up the human intellect, and our abilities and accomplishments, to godlike proportions. As Carl Sagan writes in *Cosmos*:

> Once we overcome our fear of being tiny, we find ourselves on the threshold of a vast and awesome Universe that utterly dwarfs—in time, in space, and in potential—the tidy

anthropocentric proscenium of our ancestors...We are right to rejoice in our accomplishments, to be proud that our species has been able to see so far, and to judge our merit in part by the very science that has so deflated our pretensions.

Paradoxically, science today is a strange mix of humility and arrogance. Nobody seeing and thinking clearly would reject its liberating, empowering insights. It's unbelievable what we've discovered. I'm biased, but it gets my vote for most impressive human accomplishment ever. Still, it's not enough. We wouldn't be having this conversation if science were solving all our problems. In tossing out other worldviews in our zeal for this one, did we throw out any babies with all that bathwater? Can we find valuable insights and intuitions on how to live well in, and with, this world from some of the prescientific cultures whose world was largely vanquished by the relentless steamrolling advance of our science-driven culture? Many indigenous peoples identified so closely with the natural systems within which they lived that they did not perceive themselves as separate from them. In some ways, we are rediscovering this ancient wisdom. In the words of Chief Joseph of the Nez Perce, "the Earth and myself are of one mind."

Many prescientific indigenous peoples intuited that our world is, in some deep sense, alive. When we gained the ability to see Earth from the outside, we could see that they were right. Those first views of Earth as a whole, seemingly living thing lost in the inanimate immensity of space also planted the seeds of a new scientific worldview in which life is not so peripheral. It's clear when we view our breathing Earth from afar that life is central to the nature of our planet. So, while it's true that, in the existence of the planet, our species has been here for only a shrug and our civilization only a wink, if we change our perspective and identify with the biosphere, with Gaia, then we're more than an afterthought, not so ephemeral after all. One way to look at the Anthropocene, at the coming of human influence, the "Phenomenon of Man," is as a new stage in the long life of the biosphere, one in which Gaia, experiencing the first flickering of self-awareness, is starting to wake up and look around.

The great demotions were a profound achievement, and provided an accurate narrative of cosmic evolution, up until recently, until the Anthropocene. Now, however, the rules have changed and Earth has entered not just a new chapter, but a new kind of chapter in which we've inadvertently made ourselves central. Science now has to recognize the Anthropocene, the geological force of humanity, because otherwise Earth is becoming unrecognizable. We're no longer just along for the ride. We're back at the center of the story. This really messes with our heads.

Is It Official?

From a cosmic perspective there is no doubt we are in a new stage. Any fool alien watching our planet over the eons could see that Earth is going through a series of novel and dramatic changes. Yet are we *officially* in the Anthropocene?

Much popular coverage of the Anthropocene concept has been given over to a protracted academic debate over whether it should be granted status as a bona fide geological time period, and when exactly it started. Both will be ruled on by the International Commission on Stratigraphy (ICS). Stratigraphy is the study and naming of layers of rock and the science of geological timekeeping. It gives us a common temporal road map as we investigate and reconstruct the story of Earth. It's how we "synchronize watches" at different locations all around the planet and order the simultaneously unfolding subplots of Earth's complexity into one overall story.

The ICS is the official keeper of geologic time as represented by the geologic timescale, the series of layers of recognizable rock types, and their associated ages, that unite our global studies of Earth history. The geologic timescale, cobbled together over centuries by geologists all around the world, is a wonderful and precious creation, arguably one of our greatest intellectual achievements. It is not only an indispensable scientific tool, but also a marvelous social and historical construct available to the whole human race (and anyone else who wants to have at it), with which we collectively assemble our story. Armed with the stratigraphic maps drawn and refined by generations of geologists, a hike down the Grand Canyon is not just a steep, rocky descent of six thousand feet, but also an illuminated, annotated journey back nearly two billion years in the life of our planet. Or if you're hiking in Canyon de Chelly, in the Arizona part of the Navajo Nation, and you come across some fossils in a layer of reddish sandstone that your geologic map tells you is from the late Permian period, then you know they were deposited around two hundred sixty million years ago, at roughly the same time as those other fossils you read about that someone you never met dug up and recorded in northwest Namibia fourteen years ago. These global correlations allow us to put all our disparate knowledge of Earth's past in a common framework.

Just the fact that we now have a geologic timescale to maintain, update, and fuss over should serve as, if it were needed, sufficient proof that we are indeed in a unique time of Earth history. At no previous epoch could such a creation, such a tool, exist. That there is a species exploring, excavating, digging up, and piecing together a unified global understanding of our planet's history— this is itself a creation of uniquely Anthropocene activities.

You've seen this diagram before, or something like it: an idealized stratigraphic column showing the entire geologic timescale. [See page 5 of the photo insert.] It's a condensed graphical history of our planet, with time running from the origin of Earth, at the bottom, to the present day, at the top. From left to right we see time broken up into successively smaller units, called eons, eras, periods, and epochs.

Currently the youngest rocks, those at the very top of the column, including those of zero age (those being made right this second, for example, by active volcanoes), are classified as part of the Holocene, an epoch that began 11,700 years ago, at the end of the Pleistocene, when the last ice age gave way to the current warm interglacial. In 2009, an "Anthropocene Working Group" was convened by Jan Zalasiewicz of the University of Leicester, to study establishing the Anthropocene as a formal part of the geologic timescale, an epoch that would follow the Holocene, marking the time of human influence. Their proposal will be put forward to the ICS to become recognized internationally. After that, it will surely be subjected to multiple rounds of revision, and it remains to be seen if the Anthropocene will ever be formally adopted as an epoch with an agreed-upon start date. The process is inherently conservative, which it ought to be. You don't want to casually mess with the geologic timescale.

In 2011, a cover story in *The Economist* declared, "Welcome to the Anthropocene." The term has gone mainstream, and the fact that the human presence on Earth has drastically altered many (previously) natural systems is no longer controversial. Yet does this qualify our time to be a new, formally named geological epoch? Some geologists regard this as more of a stunt than a serious matter of stratigraphy. They worry that it is designed to help marshal concern over environmental issues, and is thus motivated more by politics than science. Some scientists expressed alarm when they saw the term being adopted by the mass media to describe a geological concept that the geological community had not itself formally adopted. A 2012 Geological Society of America article by two stratigraphers asked, "Is the Anthropocene an Issue of Stratigraphy or Pop Culture?"

In part, they argued that our presence is as yet too brief a perturbation to merit a named place in the stratigraphic column. The amount of ocean sediments laid down since World War II is less than a millimeter. And, of course, nobody knows how long the Anthropocene will last. That depends crucially on how we, collectively, respond to the novel realization that we're now in such a time. Our presence, however, is already indelibly inscribed in Earth's stratigraphic history, and as long as (or whenever) there are geologists, it will be identifiable.

Whether or not it becomes "official," the term *Anthropocene* has struck a chord, and it helps frame our current and future challenges in deep time, in geological time, in Earth time. The stratigraphers gave us an enormous gift in starting these contemporary discussions of the concept, but they don't own it anymore. If the Anthropocene started out as a question of stratigraphy, it has long escaped that domain. It is now being debated by scientists from a wide range of fields—not just earth science but also biology, anthropology, archaeology, ecology, and even astrobiology. The discussion has also spread far beyond the halls of science to include historians, philosophers, ethicists, literary scholars, artists, and poets. If the stratigraphers conclude their debate and decide that no, sorry, there really is no Anthropocene epoch, I doubt anyone will take too much notice. The wide informal usage of the term has already outstripped the pace of the debate over its status.

When Did It Start?

The biggest kerfuffle has been over when, precisely, this new era began. Was it in the late eighteenth century, at the beginning of the Industrial Revolution, when James Watt's steam engine first began revving up the wheels of industry and began adding excess CO_2 to the atmosphere?[1] Or was it thousands of years ago when, through the spread of agriculture, we first began systematically modifying the landscapes of our planet, burning and cutting down forests and replacing them with cultivated fields?

Another popular choice that I have previously advocated dates the start of the Anthropocene to the first atomic bomb tests in the 1940s. This seems correct both scientifically and poetically. These explosions left clear isotopic and geologic signatures that will forever uniquely mark a moment in the rock record of Earth. The hundreds of above-ground nuclear tests conducted from 1945 until the atmospheric test ban in 1963 left an identifiable trace of long-lived plutonium and other radioactive isotopes in sediments around the globe. This provides what geologists call a "golden spike," a unique marker associated with an event or transition that can be identified widely around the world.[2] This nuclear horizon also conveniently coincides closely with the beginning of the postwar Great Acceleration, in which so many human impacts began to surge.

Fallout from the first nuclear explosions records a historical moment when an awareness of our awesome and fearsome new powers as world changers became unavoidable. As Robert Oppenheimer memorably remarked, reflecting on the Trinity explosion, the first nuclear bomb test on July 16, 1945,

> We knew the world would not be the same. A few people laughed, a few people cried, most people were silent. I remembered the line from the Hindu scripture, the Bhagavad-Gita..."Now

I am become Death, the destroyer of worlds." I suppose we all thought that, one way or another.

If there was ever a time when we started to see we were all in the same boat, shooting holes in the hull, it was the dawn of the nuclear age. It really is an excellent choice for the beginning of the Anthropocene because it's the moment we saw what we'd become. It also left an indelible isotopic marker. It's a point where geology and history merge.

The best argument against the nuclear test horizon for the Anthropocene beginning is that humans had already radically altered the planet thousands of years before. Geographer Erle Ellis and several colleagues have forcefully argued for what they call the "Early Anthropocene." They point out that massive planetary-scale changes had already been caused by human activities long before the industrial era. These include the extinction of most very large land animals (including woolly mammoths, giant ground sloths, woolly rhinos, and cave bears) between fifty thousand and twelve thousand years ago, the clearance of forest regions and the spread of agriculture across our planet's arable lands seven thousand years ago, which caused CO_2 levels to rise. Methane levels rose because of rice agriculture and livestock around five thousand years ago, and many scientists believe that the earliest episode of human-induced global climate changes can be traced to this period.

Another provocative proposal is to start the Anthropocene in the year 1610, when the amount of CO_2, as found in ice cores, actually decreased noticeably. This has been linked to the Columbian Exchange, the massive biological disruption that occurred after Europeans reached the Americas and trade across the Atlantic caused a sudden interchange of species between hemispheres. In terms of the swapping and mixing of biological information that occurred, it was as if the two hemispheres had sex. You cannot say it was consensual. The sudden drop in CO_2 at this time has a surprising and tragic origin. Roughly fifty million indigenous people died, largely from the introduction of pathogens to which they had no immune defenses. The collapse of Native American civilizations allowed the reversion of massive areas of abandoned cultivated fields back to forests. This caused the observed drawdown of CO_2 between 1570 and 1620, and a consequent global cooling.[3] As with the nuclear bomb horizon, this proposed golden spike, the climate signal of genocide, serves as a permanent marker of the destructive potential of human ingenuity.

A lot of ink and electrons have been spent on arguing over when the Anthropocene began, but in a certain way, this is like having a protracted argument about what day your daughter was finally grown up. Was it her bat mitzvah? Her first day of high school? Her first unchaperoned date? The day she got her driver's license? Her high school graduation? Likewise, although the argument over the origins of the Anthropocene has been framed as a debate over when it started, what's really at issue is what "it" is, which is a more interesting question. The debate is most valuable if we read it as a protracted dialogue on how humanity has journeyed from being just another hominid species in East Africa to the global force we are today. Regardless of the outcome, of who wins the argument over a start date, it has gotten us all talking about the history of our planetary-scale influence.

What was the question again? Is it when human activity first became detectable in the geological record? When humans first left a clear global marker, or golden spike? When human-induced change to the landscape became obvious? When we became an important agent of climate change? When we started to be the dominant geological force or pushed multiple environmental variables outside their normal ranges? Each of these would give us a different answer.

When we look at it in this way, we see there is no one beginning. None of the proposals is wrong. Each marks a different stage in the "hominization" of the planet. So rather than insist on a choice, I like to view them as a set. Together they describe a series of interesting waypoints in the development of the changing and increasing human influence.

Entering our time officially into the geologic record begs us to imagine these rocks being dug up and studied someday in the far future—but when and by whom? Imagine a geologist or astrobiologist coming to Earth a few hundred million years from now and digging up the rocks of our time, or finding them conveniently exposed in an area that has been uplifted and cut by a steep river valley. What would those scientists find? They might see novel "plastiglomerate" rocks, the isotope signatures of nuclear tests, or sudden changes in fossil species marking the disappearance of most large land mammals or coral reefs. They might denote the time span of our influence as a specific colored band in their maps. Truthfully, the choice between our various proposed Anthropocene markers may really not matter because, geologically, it's all the same moment. It's all right now. Ten thousand years is nothing. It's less than the width of the line drawn between most of the epochs in the geologic timescale. Many geologic transitions are not resolved within millennial timescales, and almost none of them is resolved within centuries. Those far-future geologists mapping out our time won't care which date we choose.

Who are these future geologists? Are they human beings? That would mean our species did not cause our own extinction or create machines that made us obsolete. And if in that distant time those scientists are still using our current stratigraphic system, that would mean our culture and our science somehow survived to be passed down, intact, through the ages. Can it? By implicitly posing these questions, the idea of including the Anthropocene epoch in our geologic timescale has value. This symbolism may be its main benefit. Recently some scientists proposed that rather than formally adopt the Anthropocene, the scientific community should simply continue to use it as an informal term. This seems to be what is happening anyway. We can break it up into subdivisions that make clearer which phase we're talking about: the "paleo-Anthropocene" or "early Anthropocene" or "industrial" or "modern" Anthropocene.

Now let me offer my own modest proposal for a "golden spike" to mark the Anthropocene. If we must choose a geological deposit to mark our time, one that is uniquely human-born, I would suggest the area of Mare Tranquilitatis, the Sea of Tranquility, on the Moon, where the Apollo 11 astronauts first stepped onto the soil of another world, hopped about, did experiments, took rocks and soil, and left behind machines, flags, and footprints. Those boot marks will fade in a few million years as micrometeorites grind them into the surrounding dust, but the overall disturbance of this site, including the alien artifacts we left there, will surely be detectable for as long as there is an Earth and a Moon. This could not have been produced by any other species. In symbolic potency, it matches the isotopic signature of the first nuclear bomb tests. These two markers, one terrestrial and one lunar, are tied together, stemming from the same conflict, because it was the competitive rocketry of the Cold War that carried us to the Moon. Yet our lunar excursions were also fueled by our age-old wandering spirit, the same drive that caused some of us to first leave Africa seventy thousand years ago. These disturbances are the product of our human propensity to explore in teams, to develop new tools to expand our domain to places that are not part of our "natural" habitat.

This could not have been done by a species that had not developed world-changing technology. The unique evolutionary significance of this step was captured by Arthur C. Clarke in his short story "The Sentinel" (which became the basis for the novel and film *2001: A Space Odyssey*), in which advanced extraterrestrials had left a signaling device on the Moon to serve as a kind of tripwire to alert them when Earth life reached a level of technological maturity worthy of their notice.

This altered lunar landscape also captures the moment we first looked back and saw the unity of our home and our common destiny with all Earth life in the great cosmic loneliness. The beginning of the space age is at least as good a contender as any other for the beginning of the Anthropocene.

Of course, as an actual proposal for correlating geological events on Earth, these Anthropocene lunar disturbances are

A 3.6-million-year-old footprint of *Australopithecus* in Tanzania, next to an astronaut's footprint on the Moon.

ridiculously impractical, completely useless. But so what? There is nothing practical about the decision to formalize the Anthropocene epoch. Those geologists millions of years in the future will surely have science and technology that (as Clarke also said) would seem to us like magic. They may wish to detect and map out our early influence, but somehow I don't think they will need our help. So, I know I'll lose, but I vote for Tranquility Base.

The Mature Anthropocene

When did our world gain a quality that is uniquely human? Many species have had a major influence on the globe, but they don't each get their own planetary transition in the geologic timescale.* When did we begin changing things in a way that no other species has ever changed Earth before? Certainly making massive changes in landscapes is not new. Beavers do plenty of that, for example, when they build dams, alter streams, cut down forests, and create new meadows. Even changing global climate, and initiating mass extinction, is not a human first. What really distinguishes us from the other life-forms that have come along before and profoundly changed the world? Clearly it has something to do with our great cleverness and adaptability; the power that comes from communicating, planning, and working in social groups; transmitting knowledge from one generation to the next; and applying these skills toward altering our surroundings and expanding our habitable domains. This has been going on for tens of thousands of years, and has resulted in the many environmental modifications proposed as different markers of the Anthropocene start date. Yet, up until now, those causing the disturbances had no way of recognizing or even conceiving

---

* Although the appearance or disappearance of some is used to mark geological stages.

of a global change. Yes, humans have been altering our planet for millennia, but there is something new going on now that was not happening when we started doing all that world changing.

To me, what makes the Anthropocene unprecedented and perhaps fully worthy of the name is precisely our growing knowledge of what we are doing to this world. Such self-conscious global change is a completely new phenomenon on this planet. This puts us in a category all our own, and therefore, I believe, it is the best criterion for the real start of the era. The Anthropocene begins when we start realizing it has begun. This also provides a new angle on the long-vexing question of what truly differentiates humanity from other life. Perhaps more than anything else, it is this self-aware world changing that really marks us as something new on this Earth. What are we? We are the species that can change the world and come to see what we're doing.

By this alternative criterion, the true Anthropocene, what we might call the "mature Anthropocene," is just now beginning. And what we have experienced so far, all these different stages that have been suggested as start dates (the early, or paleo, or Columbian, or industrial beginnings), all these stages of the unconscious human remaking of Earth collectively form a kind of preamble.

Defined this way, the true Anthropocene hasn't fully begun in earnest, or it is just now in the process of getting started. It starts when we acquire the ability to become a lasting presence on this world. The mature Anthropocene arrives with mass awareness of our role in changing the planet. This is what will allow us to transition from blundering through global changes of the third kind to deliberately making global changes of the fourth kind. It starts with the end of our innocence.

How do we lose our innocence? By developing "situational awareness," by becoming cognizant of how we are behaving on a planetary scale, in space and time, and integrating that knowledge into our actions. This will not require altruism or idealism or self-sacrifice, only accurate self-perception and "enlightened self-interest." Responsible global behavior is ultimately simply an act of self-preservation of, by, and for the global beast that modern technological humanity has become.

There's never before been a geological epoch brought about by a force aware of its own actions. Humanity has at least a dim, and growing, cognizance of the effects of its presence on this planet. The possibility that we might integrate that awareness into *how* we interface with the Earth system is one that should give us hope. No force of nature has ever *decided* to change course before. If we do not like some aspects of how this epoch is getting started, its outcome is not set in stone.

What makes the Anthropocene a new type of geological age is our growing knowledge of it. So, I propose that we call this time we've been living through so far, in which we've been accidentally tinkering with planetary evolution, the "proto-Anthropocene." We can regard this phase, where the world is largely being altered by our unconscious planetary changes of the third kind, as a first step in realizing our lasting role on Earth. It may be a necessary prelude to the mature Anthropocene, when we fully incorporate our uniquely human powers of imagination, abstraction, and fore-sight into our role as an integral part of the planetary system. The mature Anthropocene differentiates conscious, purposeful global change from the inadvertent, random changes that have largely brought us to this point.

Viewed this way, the Anthropocene is something to wel-come, to strive for.

One Galactic Year from Now

We pace our lives to astronomical tempos: We rise and sleep with Earth's spin; count our days in months derived from the

Moon's orbit; and we plant and harvest, measure our lives, and record history by Earth's orbit around the Sun. Other cosmic cycles have shaped us in profound ways that were completely invisible to our prescientific ancestors. Through the Milankovič cycles, Earth's climate, and (it turns out) human evolution have been pushed and pulled by the motions of the planets. We are also traveling, along with our Sun and the rest of the solar system, on a much longer orbit around the center of the Milky Way galaxy.[4] We don't feel this motion—good thing, since it is carrying us along at about 490,000 miles per hour—but we can measure it. In our current portion of this circuit, we are speeding almost right toward the star Vega, the second-brightest star in the northern sky, and away from Sirius, the brightest star in our sky. This orbit defines another timescale that was hidden to us before the twentieth century: we complete one lap around our galaxy about every 225 million years.[5]

We can assemble a scrapbook of our cosmic history measured out in these galactic, or "cosmic," years. Our universe seems to have been around for about sixty-one of them,* and Earth has almost reached the galactic age of twenty-one. As a biosphere, we're still a teenager of sixteen or seventeen galactic years.

We find ourselves embedded in an evolutionary sequence encompassing the entire history of the universe, which has gathered itself over billions of years into a succession of forms, increasing in both size and complexity. The story of cosmic evolution is one that natural philosophers and scientists have been telling and refining for centuries, and historians have recently joined in, adding a more human focus and calling it "Big History."[6] Cosmic evolution describes the entire history of the universe in terms of the major transitions it has experienced.

---

* Though the Sun was not here, in galactic orbit, for the first two-thirds of that time.

Out of the Big Bang came a dense fog of racing, swarming particles. As the universe expanded and cooled, they clumped into small atoms of hydrogen. Gravity pulled these together into stars, where they fused into a wider diversity of atomic elements. Stars burned and exploded, spewing their guts into vast, diffuse clouds of dust enriched with heavier, chemically reactive elements, where simple molecules formed. Gravity again gathered these clouds into new stars, and planets. On some of these planets, more complex molecules formed.

Then, on at least one planet, the story kept going, and a whole new saga started in some dirty, warm ponds where organic molecules experimented upon themselves. At some point, around seventeen cosmic years back, one of these assemblages of chemicals hit upon the ability to publish exact copies of itself. This was a newfound kind of stability, a way to preserve survival of type through successive generations, so the design persisted even as the individual expired. This step was the origin of life, a method of not only perpetuation but improvement. The designs get better because not all copies are exact. Typos in the genetic code ensure that mistakes are made, some of which are keepers, with increased ability to survive and propagate. By the next spin around the galaxy, complex molecules had begat life. After that, it was more than ten more galactic years before cellular life formed into complex and giant (compared to bacteria) plants and animals.

If we were more attuned to this galactic cadence, we might now be commemorating the anniversary of a great tragedy. Just over one galactic year ago our biosphere was reeling from its greatest trauma: the Permo-Triassic extinction, or the "Great Dying," which I write about in chapter 3.

And now, in the latest ticking of the last cosmic year, as the Sun completed the last couple of degrees in its most recent arc around the Milky Way, another type of evolution evolved here on Earth, one perhaps again as profoundly new as the origin of life: the origin of culture, a meta-biological way for ideas to propagate, accumulate, persevere, and evolve. Under this influence the biosphere has been morphing into something else entirely: something with electric lights and angst about the future; something that does comedy, chemistry, and cosmology and asks a lot of questions.

One of those questions is whether, in similar fashion, the story has kept going elsewhere, on planets in other warm stellar pockets. If not, what is so freakishly special about Earth? If so, what do we have in common with our sister stars?

Another is what makes us so different from the rest of the life of this planet? Since our last galactic go-round, matter has woken up in human form. This world is now seething with machinery and cattle, surrounded by thousands of satellites, searching its own history and the rest of the universe for answers. A new mechanism of change has appeared and taken over the planet, and yet another is beginning. Against the backdrop of our slow galactic turning, how should we regard this new transition where, by our own hands, this planet has been metamorphosing, first unconsciously and then, perhaps, starting to take notice. Might this, too, be a type of transformation that sometimes happens to other planets? We'll return to that question in the next chapter.

Will this new kind of planetary change catch on, take root, and persist? Will there be any sign of it the next time the Sun comes jogging around this way? One galactic year from now, when our star, orbiting in its nearly circular path, rhythmically bobbing slightly above and below the flattened plane of the galactic disk, returns again to this quadrant of space, what will the Anthropocene be then?

The Noösphere

At the inception of the Industrial Revolution, when the first steam engines began adding the tiniest wisps of CO_2 to the air, nobody had any way of seeing where we were going—or almost nobody. Some forward-looking individuals in the late nineteenth and early twentieth century seemed to see what was coming. In 1873, Italian geologist Antonio Stoppani proposed that the growing influence of humans was causing the "Anthropozoic era," but this was largely ignored by scientists of his day. In 1877, physiologist Joseph LeConte described a similar concept, calling it the Psychozoic era. In the 1920s the French Jesuit priest Tielhard de Chardin spoke of the rise of the "noösphere" (pronounced "NEW-o-sphere"), the sphere of thought, the nascent realm of human influence enveloping Earth. The word comes from the Greek *nous*, meaning "mind." Tielhard may have used it first, but some accounts suggest he heard it from mathematician Edouard Le Roy, who had been listening to lectures by pioneering geochemist Vladimir Vernadsky, who had been toying with the same concept.

When I learned about the Gaia hypothesis of Lovelock and Margulis, I was drawn to the history of ideas about life as a planetary process, and that led me to Vernadsky, the Ukrainian geochemist and polymath who, in the 1920s, developed the concept of the biosphere, which predated and anticipated the Gaia hypothesis and today's earth system science. When they first formulated and published the Gaia hypothesis, neither Lovelock nor Margulis was aware of Vernadsky's work in this area, but later they recognized him as a seminal contributor to the theory of Earth as a living system.

The Iron Curtain and the world wars that preceded it kept many in the West ignorant of Vernadsky's contributions, but in Russia he is rightfully mentioned alongside Darwin and

Einstein as one of the architects of the modern scientific world-view. In 1996, I was asked by the publisher to provide a prepubli-cation review for the first full English translation of Vernadsky's influential seminal book *The Biosphere*, which was originally written in 1926. I was astonished to learn that this important book, which I had read in abridged form and seen referenced and excerpted so many times, had, until then, never been trans-lated in full.

Vernadsky described life as a geological force that creates a continuous living layer, enveloping the Earth, transform-ing solar energy into the motion and organization of matter. He recognized that oxygen, carbon dioxide, and nitrogen in the air all result from biological processes, and described how these and numerous other properties of Earth, including the chemical composition of the oceans and continents, are com-pletely transformed by life in ways that differ radically from the qualities we should expect on a nonliving world. He saw that the role of life is so deeply embedded in the physical functioning of Earth as to be inseparable.

In other words, fifty years before Lovelock and Margulis, Vernadsky largely described Gaia. Furthermore, he hinted at a fundamentally new stage in the life of the biosphere that was being brought about by the actions of humanity. In his later life he became obsessed with the idea of the noösphere. In a 1938 essay entitled "Scientific Thought as a Planetary Phenomenon," he wrote:

> The rise of the central nervous system has increased the geo-logical role of living matter...
>
> You might say that within the last five to seven thousand years the continuous creation of the noösphere has proceeded apace, ever increasing in tempo, and that the increase of the cultural biogeochemical energy of mankind is advancing

steadily...There is growing understanding that...it is an elemental geological process.

In this same essay, he referred to this new stage of Earth history as "the anthropogenic age." Vernadsky's vision of the noösphere concerned not just the human transformation of Earth but our transcendence of Earth. He recognized that the movement of our activities into the surrounding space and out to other worlds would be a vital part of this transition.

Vernadsky and de Chardin both saw the Noösphere as a fundamental new phase of the evolution of matter, toward which geological and biological processes had been tending. The lithosphere had given rise to the biosphere, and now the biosphere had birthed the noösphere. Earth had become alive and then developed a mind.

The noösphere is the planetary realm of collective human action. Appropriately perhaps, the origin of the word cannot be clearly traced to any individual. The concept almost seems to have sprung from the Earth at a certain time when the zeitgeist was right, as a melding of de Chardin's cosmic Christian worldview with Vernadsky's cosmic geochemistry, with the whole steeped in the philosophy of Russian cosmism.

As a college student studying the history and philosophy of planetary science, I discovered the Russian cosmic philosophers, who, in the earliest decades of the twentieth century, developed a sweeping, scientifically informed, spiritually inspired view of human existence as a stage in the development of planet Earth, itself a small step in the development of the cosmos from diffuse, inchoate origins and toward a more organized, complex, and sentient state of existence. The most forward-looking of these cosmists was Konstantin Tsiolkovsky (1857–1935), a reclusive, near-deaf, self-taught rural schoolteacher who, working alone and having almost no contact with the wider scientific community, invented ingenious engineering designs for multistage rockets, orbiting space colonies, and interplanetary craft.

Though he had no formal education, for a time as a teenager Tsiolkovsky lived in Moscow, spending his free time at the public library. There he met and was tutored by the brilliant and eccentric philosopher Nikolai Fedorov, one of the founders of cosmism. Fedorov, who urged a merging of ethics and natural philosophy, wrote:

> Philosophy must become the knowledge not only of what is but of what ought to be, that is, from the passive, speculative explanation of existence it must become an active project of what must be.*

Tsiolkovsky is remembered today almost wholly for his prescient engineering designs and ideas about space technology, but he also developed a cosmic, evolutionary philosophy in which he saw both human life and intelligent alien life as part of the inevitable development of matter from simple structures to more complex forms producing biology and then mind.

Fortunately, he recorded his philosophical ideas in expansive writings, but during his life his main audience was the students in the remote country schoolhouse where he earned his living as a teacher. One of his former students recalled walking home from school with his teacher as he raved enthusiastically about interstellar travel:

> He would say goodbye to us beyond a bridge where, in impossible mud, lay the street at the end of which his house stood. Rain

---

* Which reminds me of the inscription on Karl Marx's tombstone in London's Highgate Cemetery: "The philosophers have only interpreted the world in various ways. The point however is to change it."

poured, dust thickened, but we were reluctant to start back for our homes, for this meant no more of that afternoon's inspired miraculous monologue about mankind's future.

In his own life, he experienced poverty, isolation, and cruelty. He was not impressed with the state of human civilization, yet he viewed it as a necessary stage, something to endure for the sake of what was to follow. Tsiolkovsky envisioned early spaceflight, which he helped to realize but did not live to see, as the first step in the advent of the era of "Star Culture." This, he believed, would carry our descendants far beyond Earth and eventually allow all sentient beings to become free from suffering through an ultimate understanding of the physical universe, which includes the understanding of one's own nature.

In 1928, he summarized many of his philosophical ideas in *The Will of the Universe*, writing:

> Some day, time itself will make man a master of the Earth. He will be in command of the lives of plants and animals, even of his own destiny. He will transform not only the Earth, but living beings as well, not excluding himself.

He saw humans as an early, primitive stage in the evolution of intelligence but was certain that, in the future, our descendants

> will control the climate and the Solar System just as they control the Earth. They will travel beyond the limits of our planetary system; they will reach other Suns, and use their fresh energy instead of the energy of their dying luminary.

Tsiolkovsky and his fellow cosmists believed that technological progress would merge with spiritual progress to the betterment, and ultimate perfection, of mankind. He suggested that the reason we had not been contacted by advanced aliens was that in all likelihood they would not see us, in our current form, as worthy of their attention.

> We are brothers, but we kill each other, start wars, and treat animals brutally.
>
> How would we treat absolute strangers?...Mankind, in its development, is as far from more perfect heavenly beings as lower animals are from people. We would not visit wolves, snakes or gorillas...In the same manner, higher beings are not able to communicate with us for the present.*

Tsiolkovsky was the first one (as far as I know) to express a thought that has become commonplace among futurists and space enthusiasts of the twentieth and twenty-first centuries: that we ultimately face a choice between spaceflight and extinction. It is a thought echoed by H. G. Wells, Arthur Koestler, Arthur C. Clarke, and Octavia Butler.

That is what I believe—and not because I think we are going to leave Earth after we mess it up and find another unspoiled planet. That's a cop-out, an escapist fantasy. Rather, it is because the kind of society that will thrive sustainably on Earth is one that embraces space technology for wise stewardship, for Earth observations, for asteroid deflection, for continued planetary exploration and the Earth wisdom it brings, and eventually for resources that will allow us to stop depleting our home planet. Perhaps, in the very long run, if we manage to stick around, we'll use space resources for climate management and—yes, eventually, in several billion years, when even the Sun proves to be nonrenewable—for escape. Tsiolkovsky's most well-known

---

* Here Tsiolkovsky was addressing a question that many decades later would become known as the Fermi paradox, which I'll discuss in chapter 6.

quote expresses this sentiment: "The Earth is the cradle of mankind, but one does not stay in the cradle forever."

Awakenings

Scientists, philosophers, naturalists, and science-fiction authors have been pointing out for a century at least that we were approaching a time when human activity would be remaking Earth faster than any "natural" process. The visionary rocket designer Tsiolkovsky, the evolutionary priest de Chardin, and the biospheric geochemist Vernadsky—they all saw what was coming. In the mid-1960s, Polish science-fiction author and polymath Stanislaw Lem wrote about the "psychozoic era," the era of human influence, and also discussed the "psychozoic era of the galaxy," in which he imagined Anthropocene-like transitions occurring on many planets around different stars, perhaps communicating and helping one other through the transition.[7] So the Anthropocene is not really a new concept, but a new name, useful for sparking widespread conversation on an idea that has been around, within some communities, for quite some time. Yet now there is an important difference. Throughout most of the twentieth century it may have seemed like just a conceptual reframing, an approaching state, a cool way to think about the future. Now, sixty-five years into the Great Acceleration, it is something that is happening, something that we need to deal with. Smart people have been warning us for a while, but it is our generation that is approaching so many global limits. What is new is a sense of urgency.

Today the noösphere is not just an abstract concept. The transgenic chickens have come home to roost. The seas are rising and acidifying. The Arctic ice is melting. Mass awareness of the vulnerability of natural systems to human perturbations has often arrived in reactive waves of shock: to *Silent Spring* and pesticide poisoning, to the smogging up of our biggest cities; to nuclear fallout, acid rain, ozone destruction; and now to climate change. Note that for every item on this list there is a common progression: the issue sneaks up on us, an initial alarm is sounded, followed by public distress and activism, and then mitigation through education, innovation, treaty, and regulation. None of these problems is completely solved, all are works in progress, but all show this pattern of movement following propagation of widespread concern.

These are stirrings of a mature Anthropocene, when our actions (meaning here humanity's collective actions) become modulated by a realization of global consequences. We may have a long way to go before this mode dominates our behavior, but there are many examples that show we have this capacity.

One watershed moment was the limited atmospheric nuclear test ban of 1963, when most of the major nuclear powers, even in the midst of their life- and civilization-threatening competition, took their hands off each other's throats long enough to sign a treaty eliminating an obvious threat to everyone's health and safety.[8] Another came two decades later, with the publication and dissemination of the theory of nuclear winter. That was the moment when it became undeniable that warfare, aided by technology, had transitioned from being local or regional to global, when we realized that, because the "theater" of conflict is finite, in vanquishing our enemy we would be vanquishing ourselves. When destruction becomes globalized, all fire is "friendly fire," and the whole world is the front line. Disarmament then became less a matter of pacifism, idealism, or altruism and simply one of enlightened self-interest.

The TTAPS nuclear winter paper (discussed in chapter 1) was published in *Science* in December 1983. At the time, my father, Lester Grinspoon, a psychiatrist interested in the neuropsychological aspects of the big issues of our times, was the scientific program chair of the American Psychiatric Association. For their annual meeting in May 1984, he organized a large symposium, Nuclear Winter: Some Psychosocial Implications, held at the Los Angeles Convention Center. This gathering was moderated by Carl Sagan, and participants included biologist Stephen Jay Gould, psychologist Erik Erikson, psychiatrist Robert Jay Lifton, and Catholic priest Bryan Hehir (who won a MacArthur "Genius Grant" that same year for his work on the ethics of nuclear strategy). It was a heady group and not exactly a "fun" topic, but due to Sagan's star power, the giant convention center was packed. Everyone gets accustomed to the trappings of their upbringing, however atypical, so I was used to seeing my dad on TV and to being around such people. Still, that was an occasion of real pride for me, seeing my father gather this impressive bunch to discuss such a vital topic in front of that massive crowd. Sagan began with a typically masterful presentation of the science of nuclear winter. It was a teachable moment, and he seized it, presenting a little climate science and a little comparative planetology and laying out the case, clearly and rigorously, for how, even with the inevitable uncertainties, we knew enough to know that nuclear war would be mutually suicidal. The other speakers discussed the moral, psychological, and spiritual dimensions of this realization.

This was during my second year of grad school, when I was just starting to learn how to model the climates of dusty greenhouse worlds, and it really sank in that this science was also connected to our understanding of human survival.[9] It was still a decade or so before climate change started to become the highly visible topic of public debate that it is now, but the speakers that afternoon made it very clear that the human power to alter the habitability of our planet marked a fundamental shift in our role on Earth—one that we had not yet integrated into the institutions wielding that power.

Sagan ended his presentation with this appeal:

Our talent, while imperfect, to foresee the future consequences of our present actions and to change our course
appropriately is a hallmark of the human species and one of
the chief reasons for our success over the past million years.
Our future depends entirely on how quickly and how broadly we
can refine this talent. We should plan for and cherish our fragile world as we do our children and our grandchildren; there
will be no other place for them to live. It is nowhere ordained
that we must remain in bondage to nuclear weapons.[10]

Another of these noöspheric stirrings came in the late
1980s, with the signing of the Montreal Protocol on Substances
That Deplete the Ozone Layer, which I describe in chapter 3. Yet
another extremely important example is the work of the IPCC,
the Intergovernmental Panel on Climate Change. Starting in
1990, this group, which was set up by the United Nations and
relies on volunteer labor from scientists around the world, has
released five assessments of published climate-related research,
along with various other special reports on specific topics.

It seemed as though the Fourth Assessment, released in
2007, hit especially hard. It came at a time when climate change
was very much on the radar, at least among those people not
fully occupied with simply trying to stay alive another day. The
question had already been raised to the highest levels of public consciousness, and the Fourth Assessment gave an unambiguous answer. It stated that "Warming of the climate system
is unequivocal" and is "very likely" (greater than 90 percent
probability) due to human activities. This release caused a huge
splash. Around the world, it generated a moment of focused
attention on this problem.

Given what I know about how hard it is to get even a small

group of scientists to agree on exact wording, and how hard it is to forge an international agreement that has any teeth at all, I have been inspired by the substantive depth of these papers. The fact that a document can be so thoroughly vetted and still say anything meaningful at all is nothing short of amazing. It is the product of the collective intellectual and diplomatic exertion of thousands of scientists working to forge a consensus statement on a complex, highly loaded subject on which our understanding of many of the most salient aspects is still shifting. There has never been anything like it. History will look very kindly on this effort, perhaps even as one of the greatest achievements of our time—not because it is scientifically flawless or beyond reproach in any other way, but because it has allowed the global scientific community to find a voice and speak coherently, and thus has focused the world's attention on the climate change that humanity is causing.

Another event that, after some time, may stand out as one of these moments of noöspheric awakening is the international climate agreement forged in Paris in late 2015. It has been praised as a historic and promising breakthrough in global cooperation and also criticized as an insufficient Band-Aid slapped on the festering wound of climate change. Both these views are potentially correct. When nations come together to acknowledge and address an issue of global concern, it creates an opportunity to move our global actions from inadvertent changes of the third kind to intentional changes of the fourth kind. The Paris Agreement is clearly a necessary step in the right direction, but will it be sufficient? Right now it's too early to tell. To know for sure, we would have to accurately predict the interaction of two horribly complex nonlinear systems: the physical climate system and human cultural/political dynamics. Simple linear extrapolations would suggest it is not enough. Even if all the commitments made in Paris are honored, it looks as though it will not keep the world from experiencing more than 2 degrees Celsius of warming—a dangerous prospect. Even with our most sophisticated models, this amounts to, at best, an educated guess, even if we assume the climate system will not pass any hidden tipping points that send it hurtling faster toward an unsafe state. Yet history shows us that human culture can also experience tipping points, sudden and surprising phase changes of awareness and action. These are even harder to predict. If the Paris Agreement turns out to be an initial step in an accelerating series of actions, facilitated both by evolving values and technological breakthroughs, then it may come to be seen as a significant point in our turn toward the mature Anthropocene.

Regardless of the trajectory of the climate and other changes we are inducing, it is encouraging that there is now more broad discussion of our role on the planet. The news on any given day can be scary, but the trend over the last few decades is toward global awareness and action. We are slowly waking up.

A Good Anthropocene?

There are those who are strongly opposed to any suggestion that the Anthropocene can be regarded as in any way a positive development for humanity, or for Earth. A raging debate has sprung up among scholars and environmentalists over whether there is, or can be, such a thing as a "good Anthropocene." It's an argument that has been simmering for a while, but it heated up in late 2014 after Andrew Revkin wrote a piece in his *New York Times* blog, *Dot.Earth*, entitled "Exploring Academia's Role in Charting Paths to a 'Good' Anthropocene." He even put "good" in scare quotes, indicating that he knew it was a loaded word. The response was intense. In a blog post entitled "The Delusion of the 'Good Anthropocene': Reply to Andrew Revkin,"

Australian author and ethicist Clive Hamilton wrote that "the 'good Anthropocene' is a failure of courage, courage to face the facts." Hamilton writes very eloquently on the danger humanity has gotten itself into and the need to change our economic and political systems to meet the crisis. He does not take kindly to any wording that hints at optimism, or anything less dire than acknowledgment of an impending, inescapable apocalypse.

New Yorker writer Elizabeth Kolbert responded by tweeting: 2 words that probably should not be used in sequence: "good" & "anthropocene." When I saw this tweet, it caused me to wonder, "Why not?" What else would you wish for? If there can be no "good Anthropocene," what can there be? What is so bad about trying to turn the Anthropocene in a direction we won't regret, or will regret less? Are you saying there is no Anthropocene? Or that it simply must be bad?

The "good Anthropocene" debate has been productive in refocusing attention on the wider moral and cultural dimensions of a potential human role in managing Earth. The downside is that it easily becomes polarized into a simplistic thumbs-up or thumbs-down on humanity. That binary thinking that plagues us: we're good or we're bad. As the Johnny Mercer song goes, "Don't mess with mister in-between."

Some are suspicious that somehow the whole notion of a good Anthropocene is a plot to get us to accept global warming. Whereas I see it as the opposite: as an attempt to get us to own up to the role we are now playing. Nobody can credibly deny that we are in a time of rampant human influence on Earth. Defined in this crude way, the Anthropocene obviously exists, so why insist it must be bad. What do you propose? That we convince everyone to feel bad about their rotten species? Beyond this, I can see little point. The point is to work for a good Anthropocene.

We may romanticize the planet without human influence.

Even if that were our wish, however, we'd be ordering something that is not on the menu. Our choice is over what kind of human-influenced Earth we will have. We may lament this truth, but we no longer have the option to choose not to be geological change agents. So let's get over that, because we do have options as to *how* to be geological change agents. How to do it right—*that* should be our concern. So rather than yearn for some innocent days of yore, we need to take stock of where we are now and proceed, with eyes open, into the future.

How, then, do we reconcile the valid horror at what we've done, at what we've awoken to find ourselves doing, with the hopeful, optimistic, resourceful spirit that will serve us best in changing course? Here I find it helpful again to differentiate between the proto-Anthropocene and the mature Anthropocene.

If the Anthropocene is seen solely as the pattern represented by what we have done so far (the proto-Anthropocene) and we simply project that pattern into the future, then yes, it is "bad." If the Anthropocene is what we can, should, and *must* do, applying our awakening awareness of planetary processes and human influence to the problems at hand, curbing the worst trends and finding a way to live well with the Earth (the mature Anthropocene), then it can be "good."

Does just the fact that we've altered Earth make us inherently evil? You would have to detest so many life forms if that were the case. What actually distinguishes us from other creatures who've changed the world and caused mass extinctions is that we alone have the potential to prevent ourselves from following through on foreseen disaster. Might we even become a big plus for the biosphere? I would say that the cyanobacteria, despite their many victims, ended up improving the world. They enabled oxygenic life and habitable continents. We, too, can improve the world, even learn to protect and ensure life's future. Let's work toward that. So, while some environmental philosophers have written about the Anthropocene as a topic of fear or shame, I see it instead as a hopeful step, albeit one we're still trying to achieve.

Traditionalists versus Ecomodernists

Where does our need to preserve wilderness come from? Is it for "the environment" or ourselves?* Several writers have proposed that we have a psychological need for wilderness, or at least for wildness. I know I do, but this comes in part from a privileged youth running wild in New England forests and on Cape Cod dunes. We didn't go to church or temple, but our yard was adjacent to a protected bird sanctuary, so right out back was "the woods," an expanse of wetland swamps and dense, ferny New England forest, where my brothers and I explored what seemed an infinite vastness. It was not virgin forest, and this added to the enchantment, for there were ancient artifacts to stumble across: the occasional old farmer's stone wall running incongruously straight through some tangled, disorderly thicket; or the decaying hulk of a 1940s vintage car, a treasure to be explored, played in, and left to rust in peace.

Those woods also provided streams for our hydrologic projects. Building dams out of rocks, sticks, leaves, and mud was a favorite activity. At one point, new ditches started appearing that had apparently been dug illegally in order to try to drain the woods (protected wetlands) so they could be developed. Our dam building became not just fun but principled, which made it more fun. Of course, this protected sanctuary surrounded by suburbia was a far cry from actual wilderness, but I did get a big early dose of wildness. I wish it for every kid.

---

* Or do we err again in making the distinction?

As an adult set loose in the mountains, I have been able to wander realms free of any obvious human stamp. In thirty years of living in Arizona, California, and Colorado, I hiked and rafted through plenty of places that felt much as they must have hundreds or thousands of years ago (although my nylon backpack, hi-tech boots, and Neoprene raft, not to mention the occasional contrail stretching high overhead, were a dead give-away). As much as I have one, the wild western coasts and high mountain valleys are my temple.

So I find myself naturally sympathetic with the frequent appeals to the spiritual value of wilderness that come from one corner of the new debate roiling the environmental community. The skirmishes over a "good Anthropocene" have added fuel to a long-raging battle within conservation over which values and goals should guide our interactions with the rest of the world.

The current flare-up has divided the conservation world into two camps. In one corner are the ecopragmatists, or ecomodernists, who ask: since the human-altered aspect of our landscapes is obviously not going to disappear, how do we find ways to integrate it more successfully with ecosystems? They believe we should acknowledge that we are now managing planet Earth, and that our best way forward is to intensify our use of technology in order to reduce our land usage. They see farming, energy extraction, forestry, and settlement as activities that can be made more efficient, and thus compressed into smaller areas, to allow the rest of the world to heal. By increasing agricultural yields, moving toward more centralized and powerful energy sources, and continuing the concentration of population into cities, we can relieve pressure on natural landscapes. Ecomodernists also emphasize economic growth, including expanding energy use, particularly in developing countries, as of prime importance for both moral and environmental reasons. They are unashamed to promote human needs first, believing that smart development will both ease poverty and preserve nature. For example, if people have electricity and well-heated homes, they don't need to pollute the air and deplete the countryside with fires from wood and dung. Alleviating poverty also leads to falling birth rates, which cascades to help innumerable environmental problems, including climate change. The ecomodernists are blatantly optimistic, believing that within the next century, human impact on the environment will peak and poverty can be ended. They also believe that nature will be in better shape, that the world will be greener and wilder. This point of view was asserted in "An Ecomodernist Manifesto," published online in April 2015 by eighteen prominent academics, authors, and scientists.[11]

On the other side are traditional conservationists, who are generally horrified and offended by the notion that humanity is in any way managing the planet. They feel that wilderness is intrinsically, spiritually valuable for its own sake, and that our increasing intrusions into its realm are a desecration. They are more interested in restraint than growth. They believe that we need to scale back our technological impacts and that our main goal should be to protect what is left of wilderness from human intrusion. The traditionalists are decidedly more pessimistic about the prospects for positive human interventions or technological solutions.[12] One of the main tensions between the two sides is over economic development. The traditionalists accuse the ecomodernists of a willingness to trash wilderness in the service of economic gain. The ecomodernists accuse the traditionalists of valuing trees more than people, of a willingness to sacrifice the lives and well-being of poor people in order to protect a vanishing illusion of unspoiled nature.[13]

Many traditional conservationists are deeply suspicious of the idea of the Anthropocene. They see the word itself as an illegitimate claim on power. To them it is not just a neutral name for a geologic epoch, but code for a threatening and dangerous agenda. They describe their enemies in this war as the proponents of the "Anthropocene worldview." Some writers have caricatured a belief in the Anthropocene as synonymous with cheerleading for development, celebrating human hegemony over Earth, and believing that human needs justify destroying other species and that technology and capitalism will just take care of everything, so there is nothing to worry about. Of course this is a straw man argument, a cartoon view of the Anthropocene and ecomodernism constructed for the purpose of knocking it down.

At their worst, some ecomodernists do live up to this stereotype. They can be so enamored of technical innovation as a way to solve our global problems that they sometimes advocate that we can ignore natural limits or suggest that there are no real physical limits that cannot be overcome by our inventiveness. They often reject the notion that overpopulation is a problem because, in their view, the idea that Earth has a finite carrying capacity is a myth.[14] They seem to relish, even delight in, tweaking the noses of the environmental and scientific establishments. They undercut their own effectiveness with hectoring diatribes against those whom they judge to have the wrong shade of green, tweeting antagonistic rants such as "The sky may be the best waste dump we can imagine!! Nobody lives in the sky! If our shit has to go somewhere, why not there?"[15] This is not "modernism" but a call to the prescientific era, when we thought the world was infinite and imperturbable, when we could just throw things "away" and they would stay there forever, and when the sky seemed like some azure dome hanging above and separate from the world. This is just wrong. "Nobody lives in the sky"? Actually, we *all* live in the sky. The sky is simply the atmosphere, which is finite and increasingly subject to modification by human industrial activity. As I write in my book *Lonely Planets*, "The Earth has many lands but only one atmosphere, and we are all in it together."

One need not go to such alienating extremes to make a compelling case for a pragmatic, technology- and innovation-friendly approach to our environmental challenges. Yet true pragmatism must include dealing with reality, and you cannot simply wish the capacity of the atmosphere to absorb effluent, or any other aspects of our planet, to be infinite. So the radical technophilic approach ("To hell with limits—We'll just invent new mathematics and new laws of physics!") is not any more reality-based than an extreme technophobic approach in which all new technologies are rejected.

The ecomodernists offer up their own straw man caricature of the traditionalist's viewpoint, accusing them of a naïve attachment to completely nonexistent pristine wilderness, and of ignoring the real world, and the needs of struggling people, in favor of romantic nostalgia.

When you pit these two opposing straw men against each other it seems like a vast difference in worldview. People love a fight, love an enemy, to feel righteous, to rally troops and raise funds. Beyond the bluster, however, this is largely a trumped-up debate, an epic battle of straw on straw. The truth is that the traditionalists have long been evolving in a direction that largely aligns with the ecomodernists. When you listen to the most thoughtful voices on either side, you discover that there is much common ground. When they are not busy ridiculing and caricaturing each other, both sides are actually saying many of the same things. Many of the traditionalists do in fact recognize that there is a continuum of wildness, from the most distant, roadless almost-wilderness to—I don't know—downtown Newark? Yet they still argue strongly for the preservation or expansion of the most untrammeled areas.[16]

If we drew a Venn diagram for these supposedly opposing philosophies, and blur out the extreme voices on either side, in the rather large area of overlap you would find the outline of a wise approach for application of human technical ingenuity in promoting the health and longevity of our indivisible wild and human Earth. You would find many shared concerns and ideas, and a few interesting substantive differences.

One honest disagreement is over how resilient nature really is. The ecomodernists emphasize that nature adjusts and recovers. They can point to numerous success stories where this resiliency is on display, where disaster zones have grown from the brink back into lush, diverse biomes. There are so many important and inspiring stories of restoration.

Part of the backdrop of my youth was the Charles River, a scary toxic mess running through our town in suburban Massachusetts toward Boston Harbor. Five years before I was born, an article in *Harper's* described it as "foul and noisome, polluted by offal and industrious wastes, scummy with oil, unlikely to be mistaken for water." When I was old enough to walk, I was being warned to stay away from it. In the early 1960s, cleanup efforts began. Now, though it is still not perfect, it is swimmable and improving, on its way to being again fully healthy for children and other living things. It is undoubtedly true that ecosystems, if not pushed too far, have shown an amazing ability to bounce back.

Yes, Gaia is tough, and will ultimately be fine with or without us. Yet it is also true that everything we treasure, both the creative works of humanity and a large number of the species we share Earth with today, is threatened by the worst-case climate change scenarios and other self-inflicted existential risks. The traditionalists emphasize that so many species are threatened and that ecosystems can collapse. They make the almost unquestionable assertion that extinct is forever. Neither side denies the tragic potential of the current moment, but the ecomodernists emphasize, rightly, that nature is brilliant at carrying on by forming new alliances and adaptations. They suggest, therefore, that with enlightened management we can create and encourage new "anthromes" that have the qualities we most treasure about Earth's ancient biomes, but that accommodate human needs and preserve the bulk of the species we care about.[17]

The traditionalists are eloquent on what we have lost and are losing. They are very clear in describing what they are against, what they fear, what they mourn. They are long on what they don't like (exploitation, destruction of nature, capitalism, and, sometimes, human beings), but short on where we are going and what we should seek. It's not always clear what they want, beyond vague notions of relinquishing our privileged place on the planet. When they do give prescriptions, they generally involve reducing the human presence on Earth. I have to admit I do like this idea. I think, in the long run, to some degree this will happen, in combination with the smarter use of technology advocated by the ecomodernists. Yet it is not clear to what point they would scale back. Some suggest ending industrial food production altogether, and that we should simply become "plain members and citizens in the community of life."[18] What does this mean? Abandon agriculture and go back out on the hunt?[19]

There is a persistent message that we should not get comfortable with the Anthropocene but should fight it. This sounds to me like saying we should not get comfortable with adulthood but should resist it. Growing up was a bad deal. Look at all we gave up: innocence, infinite wonder, a sense of immortality, and, for some, a sense of safety and endless worry-free days. Why would we ever accept anything else?

Though I am often annoyed by their tactics, my heart is with the ecomodernists, because of their lack of sentimentality, their acknowledgment that we do have a role to play on this planet, and their clear-headed appraisal of our choices. Their vision is not perfect, but at least they have a vision for the future. However, there are a few areas where I think they, and we, would benefit from listening more to the traditionalists.

Traditionalists argue persuasively that wildness has very important value that cannot be measured in economic or utilitarian terms. Even as I question their quasi-animist notion of "self-willed" jungles and rivers, I also think they have a very important point here. There is a spiritual, emotional, and cultural value to wildness, something irreplaceable.* Wilderness as a pure entity, a platonic ideal, may not be helpful. Maybe it never existed. Yet what of wildness as a quality to mourn and treasure and protect and foster? An appreciation of wildness as something that we need, and do not fully comprehend, can also encourage humility, restraint, gratitude, respect, and wonder at the world. These are values we will need to cultivate, along with innovation, courage, confidence, and curiosity.

Both sides in this debate are motivated by the same consternation and concern with how we are going to manage ourselves and our planet in the critical century ahead. Each accuses the other of being attached to the status quo, but at least they share a realization that the status quo is unstable. Both recognize the climate emergency we are in, and that business as usual is not okay. Both support efforts to move our energy supply rapidly away from fossil fuels, though ecomodernists would do it with more nukes and fewer windmills. Do we try to go back to nature or move forward to our own reinvented world? It will have to be some combination of the two. To deal with the world as it actually is today, we'll need a synthesis of these approaches.

We need invention *and* self-restraint. The most useful inventions will result from paying more attention to limits, not less. Even as we find better, less intrusive systems to power and feed

---

* An ecomodernist might ask, "Do you care more about your self-willed reserves than helping the poor in Africa achieve a decent standard of living?"—to which I would answer that we must have both. There is a global population level at which we can. I don't know exactly what it is, nor does anyone else, but I would wager it's a hell of a lot less than eleven billion.

ourselves, we need to recognize that even if we wanted to, we cannot replace or do away with Earth's massive ancient cycles. The innovations we need most will help us find ways to work with our planet's properties, not ignore them. That will include deepening our understanding of ecological and biological systems so that we can learn to be better collaborators.

Respect and reverence for the integrity and mystery of wild nature remind us that our innovation must be tempered with caution, because our knowledge of planetary functioning is still dwarfed by our ignorance. I see awareness of limits and the need for restraint as embedded deeply within the notion of the Anthropocene—the moment in Earth history when the world stopped being functionally infinite and we had to start dealing with that fact.

Human Nature

The traditionalists, the purists, remind us of the essential value of large robust parks and reserves, connected wildlife corridors, and healthy forests. Yet what about the environments where most of the world's people actually live and where more will be living in the future? How should a conservationist regard a city: simply as something to try to have less of or try to mitigate? Or should she concern herself with the health of what is in that environment?

I'm a huge fan of urban nature, not as a substitute for more wild places, but as a marvelous and evolving thing in and of itself; not just a consolation prize, a retreat or last stand, but as something to be celebrated and cultivated. I love the dark skies of the Arizonan and Australian deserts, but I am also fond of urban astronomy, where the stars are muted and the skies are ruled by the dance of the moon and planets, where Venus looms out of a twilight sky sometimes tinged with a lovely red glow. Knowing that this glow comes from particulate smog, I'd like to see a little less of it, but it can still be quite pretty. I love both nature and cities, and I love them together. I love the fact that here in Washington, I often don't get in a car for days on end (something that was much less likely when I lived in the "wild" West, in Tucson and Denver, where the post-highway cities are spread out on huge square grids, with all that vast pseudo-wilderness just out of reach, just a half hour's drive out of town...).

Right now, looking out my kitchen window on a summer day on Capitol Hill, I see a complex, shifting scene composed of about 50 percent brick and 50 percent trees. It's lovely, a riot of organic forms bouncing in the wind. The brick is festooned with lichen, ivy, and moss, its rigid geometry softened and blemished by hundreds of years of wind, rain, and life, and illuminated by splintered sunlight refracted through blowing branches and leaves. A squirrel skitters along a power line, balanced, at ease, "natural," as if he's been evolving to do this for a hundred thousand years. The trees are diverse, some deciduous and some evergreen. They look happy, at home, healthy, and strong. They are permanent residents, compared to any people. The birds and rodents that nest, chase, chatter, and squeal among them seem at home as well. I know that here I am reducing this to my superficial impressions. What about the trees themselves? I do believe they have impressions, experiences, and feelings of a kind, but they will never write books about it. Some of them may become books. Well, not these trees in my landlord's backyard. It's nature of a different kind: call it "human nature." It's not so wild, but it's beautiful and valuable, too.

When I have time and inclination I find a little more wildness. On a hot summer day, I can drive half an hour, or bike two hours upstream, and scramble through woods and across the mossy shore to wade into the cool Potomac, with no buildings or people in sight. Even closer at hand, within the city, is the Kenilworth Park and Aquatic Gardens, where I can sit in a field of blowing cattails on the muddy edge of an expansive tidal marsh and watch herons land and turtles crawl. Jet planes on approach to National Airport snap me out of my reverie, reminding me of the surrounding city.

As we measure the extent and direction of our influence, it's important to remember that human-made does not have to mean ugly. It can also be beautiful and inspiring, especially when built in concert with the nature around and within. Who can deny the beauty of Machu Picchu or Angkor Wat? Who hasn't been moved by the magnificent sight of the Golden Gate Bridge elegantly joining San Francisco to Marin County, its parabolic cables swooping between bold vermillion Art Deco towers? This is a handsome addition to the geography, not a compromise. For me, such a structure contains a message of hope: that humans can collaborate with the land and build with skill and beauty, that even our giant works can be integrated gracefully into Earth without diminishing it. Today, meaningful conservation must concern itself with the ecological health of places that include human activity within them—and the whole world is now one of those places.

Scientific and technical ingenuity allows us to discern and work with the laws and parameters of the universe, but not to change them. Some of those parameters include the geometry, chemistry, and physics of our planet. Our brilliance will succeed by discovering, embedding, and entraining ourselves within our world's ways. The ecomodernists are all about embracing innovation and the power of human ingenuity. I'm with them as long as that also includes embracing innovation in reimagining our global economy and moving it, in the long run, toward one that encompasses, rather than ignores, the reality of finite global geometry.

We do need to transition to a sustainable economy that incorporates ecological limits. What economists now call macroeconomics is really very micro, ignoring the finite physical capacity of our planet. In this area the traditionalists have it right. Embracing the Anthropocene cannot mean embracing the economic status quo that considers future environmental costs as "externalities" to be ignored. A mature Anthropocene will require the opposite. Recognizing our role requires awakening to the reality that business as usual has no future and accepting the responsibility of finding a new way. This need not require us to abandon or overthrow capitalism, merely to adapt it in smart ways to encompass our increasing awareness that there are constraints on an economic system that is growing within a finite planetary system.[20]

Three Futures

It's often said that we humans learn best through disaster. An endless supply of tragic historical anecdotes supports this claim. Yet that is not the whole story, because we are also uniquely gifted with foresight. This is a great strength because we can do much of what would otherwise be trial-and-error learning by running scenarios inside our heads. As individuals, we're always considering outcomes and making choices. We picture ourselves skidding off the road, so we step on the brake. We've developed mechanisms to do this in larger and larger groups, with mixed results. Our big evolutionary challenge now is to take this to the global level more effectively. If we can learn from disasters foretold, we can steer clear of danger.

In 1979, for *CoEvolution Quarterly*, the underground comic artist R. Crumb drew "A Short History of America," a series of twelve sequential panels portraying the path of progress (such as it is) over the twentieth century in a random spot, a typical American town. [See page 6 of the photo insert.]

The first panel shows a bucolic scene of a rolling meadow at the edge of a deep forest, with a large flock of birds flying overhead. The next view is nearly the same, but now with a train running through on newly laid tracks. In the third panel a telegraph line and a dirt road have appeared alongside the tracks, along with a lone, modest farmhouse. The scene can still be described as bucolic. The decades pass, and the pace of development, which started innocuously and gradually, quickens. Paved roads, sidewalks, streetlights, electrical and telephone wires appear. The trees decline in number. Billboards and signs grab at our attention. In the final three panels there is no more grass, and nothing but the sky that is not human-made. In the last scene even the sky is visibly altered, dull and smoggy, the backdrop to a clogged, slightly decrepit urban corner where life goes on. A couple rides past on a motorcycle, going about their business. Although the difference from the first panel to the last is jarring, each step seems minor, and each view looks unexceptional, normal. The final panel contains the only text: the caption "What next?"

In 1988, Crumb answered his own question with a follow-up drawing, entitled "Epilogue." [See page 7 of the photo insert.] It depicts the same scene again in three more panels, showing three possible futures. The top one, labeled "Worst Case Scenario: Ecological Disaster," depicts a postapocalyptic wasteland. The shells of buildings and cars are decaying in abandoned, weed-grown streets under a frighteningly bright sun and a diseased-looking sky.

The second panel, "The Fun Future: Techno-Fix on the March," is a much happier place. A clean sky is traveled by sleek hovercraft and flying cars. A sign reads, "NO GROUND VEHICLES IN THIS SECTOR." The buildings look prefab, molded, efficient. Some green has returned, in the form of well-manicured grass and carefully planted, well-spaced evergreens.

In the third panel, "The Ecotopian Solution," the trees have grown back tall and wild. The roads are dirt again and are traveled by healthy-looking people who smile and wave as they pass by, riding bicycles and pulling wagons. These are the only machines visible. A roadside stand sells healthy-looking produce. Wooden tree houses and geodesic domes are sheltered in the forest. People sit beneath a tree, playing acoustic instruments. It looks like rural Oregon in the early 1970s, but it's supposed to be the future after humanity has chosen to loosen our technological grip on the world.

It strikes me that many of our arguments and discussions about how to handle ourselves in the Anthropocene are summed up in these last three panels. We all want to avoid "Worst Case Scenario." To do so, however, we need to make some big changes and lower our output of greenhouse gases. The "Fun Future" is the ecomodernist fantasy in which our problems have been solved by nifty new inventions. The ecotopian solution caricatures the traditional conservationist fantasy where we have lowered our numbers drastically and returned to a simpler, more low-tech lifestyle.

Either of these fantasy worlds looks pretty pleasant, but neither is a realistic depiction of our likely future. It's going to have to be a combination of these two. In order to avoid "Worst Case Scenario," we need an artful and wise combination of "ecotopia" and "technofix." There is a reasonable path forward for those who love wild nature, fear the consequences of its further diminishment, and yet recognize that both technical innovation and close attention to limits are going to be important. Ultimately I believe we'll choose both to reduce our numbers and to embrace our nature as engineers and innovators.

The Nature of Nature

To a thoughtful person, the distinction between "natural" and "man-made" has always been somewhat problematical.* Are we ourselves not products of, part of, nature? If so, then our own productions are as well. Yet when we were much fewer in number and much lighter on our feet, we were surrounded by a world that was mostly wild, untouched by humans. We could certainly draw a distinction between those things that were products of our own intellect and handiwork, and everything else, between "artificial" and "natural." Now this distinction is fading. There is no place in nature that does not bear at least the faint chemical and isotopic touch of human hands. The weather itself, the wind and rain, have become, in part, artifact.

Traditional conservationists describe the Anthropocene pejoratively as the age of human "dominion" over Earth, and the Anthropocene agenda as the vain belief that humans now control the planet. "Control" is a loaded term, but you can't deny that our decisions or lack thereof do now have major effects on Earth. So, do we take our hands off the wheel, duck, and brace for impact? Or do we try to learn to drive? It has to be the latter, while I would also suggest tapping the brakes, not the accelerator. Still, it is reckless to deny that you are driving, and to refuse to look if in fact you are.

We are, in some sense, running the world now. Yet I see this much more as a reluctant realization of a fearsome responsibility than as a power grab. It's not that we think we are so great, but that we are stuck managing this world that we have accidentally pulled off the rails. Some have interpreted this as tantamount to declaring ourselves gods. I don't like the eco-pragmatist phrase coined by Stewart Brand: "We are as gods

---

* And not just because of the inherent sexism of "man-made."

and might as well get good at it." I see the truth in it, but I don't like it. We're not gods, or if we are, we're pretty lousy ones: reluctant, incompetent gods who have been "volunteered" by fate to assume a leadership position we know we're not really qualified for. We're more like teenage hackers realizing that we're flying a 747 without instruments or instructions. We have no choice but to figure out how to land the thing.

Let's face it, it's disquieting to think we are even sort of running Earth. Nobody in their right mind loves this or is cavalier about the dangers and challenges we face. Who appointed us? We're like actors who play a starship crew on TV finding themselves suddenly, accidentally, in control of a real starship.* How do you work this thing? We don't know how to run a planet. Good luck with that. Worse still, we're only just realizing that we're driving this contraption, which would likely be a good first step in gaining control.

Nevertheless, however reluctantly, unconsciously, and incompetently we are doing it, we are now managing this planet. Many people don't want to admit this, don't want it to be true. Well, of course we don't want it to be true. Everyone prefers the carefree innocence of childhood to the weighty responsibilities of adulthood. Still, it is far better to grasp things the way they really are. When we deny that humans are now to some degree in charge of Earth, aren't we persisting in a comforting illusion? If we want to get to work solving the problems we've created, the first step is seeing clearly who we are.

Though we are not gods, we did create nature—meaning we invented the concept, and made it necessary. Before us, there was no nature to be despoiled or protected, there was just the tangled bank of the world with its profusion of crawling species growing upon and adapting to one another, living, dying,

* This is the plot of the movie *Galaxy Quest.*

and evolving. Once we started pondering things, we saw that we inhabited a world, and asked where we and it came from, and whether we had always been part of it. With that question, we invented nature as something apart from us. Now we're realizing that we may have fallen for our own illusion.

How about realizing that we can be neither saviors nor external, outside managers but must recognize ourselves as integral? If we are to feel truly not apart but a part of the biosphere, then we must recognize exactly what sort of part we are: a thinking, communicating, planning, engineering part. We need to use those unique capacities to see more clearly the situation we're in. We're not separate entities on an infinite, nonliving world. The nonhuman world was doing just fine before we got here, but it needs us now, to clean up after ourselves and to chart a path forward. We need to use our human qualities to be better partners, better spokeshumans, better colleagues with the nonsentient and nonliving parts of this planet—because we need them, because we are them, because it is right.

Promise

One popular response to our current predicaments is to insist that the human race is inherently ignorant, destructive, and divisive. It's an unimaginative, defeatist reaction. We need to look clearly at our record, but we also need to see our promise, because our sense of ourselves can help guide our choices.

Certainly it's not difficult to find examples that point to a record of carelessness and casual disruption and destruction of natural systems at the hands of humans. There are books full of them being published as fast as trees can be felled to print them. However, it is equally true that humanity is uniquely creative and constructive. If we weren't, we wouldn't be in this mess. Right now we are certainly facing a situation that is new for us. For so long we behaved as if the world were infinite, and compared to the scale of our activities, it basically was. Then we spread around it so thoroughly that we ran back into ourselves. Our great cleverness has reached clear around the world to bite us in the ass.

I want us to differentiate between the proto-Anthropocene, the time of unconscious technological transformation of our planet, and the mature Anthropocene, not so that we can escape responsibility for what we have done, but so that we can fully and clearly assume responsibility for what we need to do.

In order to lose innocence, you need awareness. Awareness of a certain kind seems to be something we humans are blessed (or cursed?) with as individuals. Yet is it something we possess as a global entity? Humans are aware, but is humanity? Complex global problems require intelligent global responses. Intelligence entails the ability to sense changes and consequences and modify behavior. Yet humanity has rarely acted collectively, or consciously changed anything to meet collective needs on a planetary scale. On this level, we have not acted with intelligence, with sentience. Now the challenge of climate change is forcing us to face this fragmented aspect of our nature. That's why, in chapter 4, I referred to the "twisted gift of global warming." It may turn out to be the wake-up call we need, and the resulting gradual mass realization of our planetary role could represent a turning point, a critical juncture in the evolution toward intelligent life on this world, as currently manifested by the bumbling progress of the human race.

In becoming dependent upon energy sources that endanger the future of our civilization, we've stumbled into a trap. Yet don't assume we're stuck there. Our fleets of Earth-observing satellites, our climate-modeling supercomputers, and our tightening mesh of global communications are all very new phenomena.

Many people have pointed out that the Internet looks a lot like Vernadsky's noösphere. Is the world developing a kind of mind that might be capable of looking out for itself? Could it be that this big, slow creature is awakening and has just started to notice, and is slowly turning to face, the threat?

This transition to self-conscious, self-aware global change seems to us, trapped in the flitting pace of human lifetimes and generations, painfully slow. Yet on the cadence over which Earth goes through its ages and stages, bobbing along in its orbit around the galaxy, a time frame in which a century is nothing at all, it is lightning fast.

If, for a moment, we step back, way back, and seek a view that is not only less human centered, but also less focused only on our own time, instead regarding the history of the planet and the biosphere as a whole, then we see that there is something truly new and remarkable happening here. Yes, there is a tragic component to it, just as there was a tragic component to the end-Cretaceous extinction that took out the dinosaurs. As I've discussed, extinction, even mass extinction, is a familiar refrain in the story of Earth. Yet this time is different because we have some awareness and therefore some sense of responsibility and agency. We have the potential not only to forestall the worst of the wave of extinction that has accelerated on Earth in our time, but also even to prevent future mass extinctions that would occur inevitably without our influence.

Self-aware global change is really so new on this planet that, arguably, it has not had time to catch on and catch up with the careless changes that have dominated our impact. Looking at it this way, I believe there is something profoundly different, unprecedented, and ultimately incredibly *promising*, about the Anthropocene. The mature Anthropocene begins with our mass awareness of our role as world changers. So it starts right now, in our discussions about the global environment, in our growing realization of ourselves as a planetary-scale entity with a need to start behaving like one. All attempts to spread the word are, in effect, efforts to get the Anthropocene started in earnest. Every book, lecture, discussion, online argument, flame war, and bar fight about climate change, the global economy, and the Anthropocene itself is a part of this beginning. Even the voices seemingly pushing us backward, arguing against the obvious reality of global warming and attempting to thwart efforts at global responses to our global problems—like those recalcitrant neuronal circuits in your head firing in protest of your clear need to get out of bed in the morning—are a manifestation of our collective global brain starting to pay attention more and more, to think these thoughts and wake up to the new reality. In this sense, the Anthropocene is just beginning, and it's something not to fear but to welcome. Let's get it started. Let's do it right.

Humanity: An Event, a Phase, or a Transition?

Scientists debating the Anthropocene have been so focused on locating its beginning. It might be more important to ask where it is going, and when will it end?

What will be the significance of the Anthropocene rock layer and the ultimate legacy of the human race when, in another 225 million years, our star, having completed one more dance around the black hole at the center of our galaxy, returns to this quadrant? Will we simply leave a thin layer rich in refined metal and Twinkie wrappers, underlying a layer bereft of coral reefs? Or will we leave more lasting changes on this world, or even never leave it at all?

In the scientific literature, you see the Anthropocene referred to sometimes as an "event" and sometimes as an "epoch." "Event" implies it will all be over pretty quickly, whereas "epoch" implies some more prolonged phase of human influence. From the standpoint of Earth evolution, which will we be, a moment or a phase? There are plenty examples of each in Earth's history. The end of the Cretaceous period was an *event* caused by a disruptive impact, marked by a centimeter-thick layer of clay that represents a few years of accumulation. What immediately followed was an epoch, the Paleocene, which lasted more than ten million years and finally ended with the Paleocene-Eocene Thermal Maximum, one of the most rapid and extreme periods of global warming the world has ever seen.

Is it audacious to think that, geologically, the changes we bring might be anything more than an event on this world? After all, species come and go; why should we be any different? Notwithstanding the great longevity of certain species—sharks have been here, pretty much unchanged, for more than one galactic year—why should we expect to be spared? This frequently stated fatalistic opinion ignores the central observation of the Anthropocene: that, all value judgments aside, human civilization has brought new forces to bear in the dynamics of Earth and of evolution itself. The past is no longer the key to the present because the rules of the game have changed, but not necessarily in our favor. The adolescent stages of a young technical species appear to be fraught. We need to find a way to power our growing civilization without wrecking our environment, but this is merely the first in a string of challenges brought by an expanding human population and the increasing global reach of our technology on a finite planet.

So, *event* or *epoch*? There is actually a third possibility: a *transition*. The origin of life, the Great Oxygenation Event of 2.3 billion years ago, and the Cambrian explosion are examples of unreversed transitions that left Earth dramatically and permanently transformed. The Anthropocene could mark the beginning of a transition of similar importance in the history of Earth. If we make it past the next few centuries, it will be because we've honed our survival skills to make them work on a planetary scale. Once we achieve that, we have done much more than ensure our persistence against near-term self-induced challenges. We will have unleashed the power of reason and foresight in defense of Earth's biosphere—but first we have to get through a bottleneck.

The Bottleneck

We are blessed and cursed to live in interesting times. Not that there have been any uninteresting times, but both cosmic evolution and its local subplot of biological evolution seem to have gone through long periods of cruising along in relative stasis, like a drive across Kansas and Nebraska. Then they hit the Rockies. Naturally we should be suspicious of the impression that there is anything special about our current era. It's easy, I'm sure, for the people of any time to think that theirs is the most crucial in history. This could just be another pre-Copernican type of fallacy: "*Ours* is the crucial time!," as suspect an opinion as "Ours is the most special place." We've escaped from the parochial trap of assuming Earth to be the center of everything just because it's where we are, and we don't want to fall into the trap of thinking that now is important just because it's when we're here. Yet even looking carefully for this type of bias, I still come to the conclusion that major changes are afoot in the very mechanisms of evolutionary change and in the relationship between life and Earth. Such transformations do not happen every day.

All those upward-arcing curves defining the Great Acceleration described in chapter 3 cannot continue. That would be physically impossible. Global warming is only the most visible of several accelerating and interconnected challenges confronting humanity now, others being issues of energy, population, environmental pollution, and global supplies of food and potable water. Add to these several developing technologies that have been credibly proposed as existential threats for humanity. Among these are the nexus of artificial intelligence, genetic engineering, and nanotechnology, which together suggest the possibility of dangerous self-reproducing and evolving agents being loosed on the world by accident or malice. Combined, all these risks define the "twenty-first-century bottleneck" we must pass through, either learning to achieve a sustainable balance with our own expanding population and technology or suffering dire consequences. This is the test that will determine whether our time will ultimately be just a strange, thin layer in the strata, or the early stages of something lasting and wondrous.

In the last century, we've increased our speed of transportation and communication by huge factors. Life expectancy has risen sharply around the world, and levels of poverty and malnutrition have declined. Many terrible infectious diseases have been put in their place. By several meaningful measures, we're doing pretty well. We may have already passed "peak child," that is, the number of children being born is no longer increasing, and we are approaching "peak human," after which the global population will start to decline. By that time, we will surely have largely transitioned off fossil fuels, though we'll have an atmosphere with substantially more carbon dioxide. I strongly suspect that by century's end we will be in the process of removing CO_2 from the air, drawing it down toward preindustrial levels. So, if we can just get through this century...

As I described in the previous chapter, we are on the cusp of having the ability to prevent certain terrible natural disasters that until very recently we would have been helpless to avoid.

A composite image of Earth assembled from data taken with NASA's
Suomi satellite in 2012.

Planetary changes of the first kind—an asteroid impact causes a mass extinction. Art by Don Davis.

Earthrise, as seen by Apollo 8 in orbit around the Moon.

The Korean Peninsula, taken by the Expedition 38 crew of the International Space Station, reveals that some national borders are now visible from space.

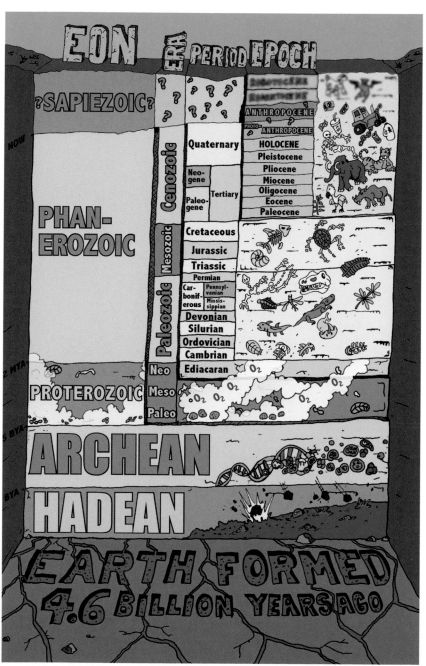

The geological time scale. Art by Aaron Gronstal.

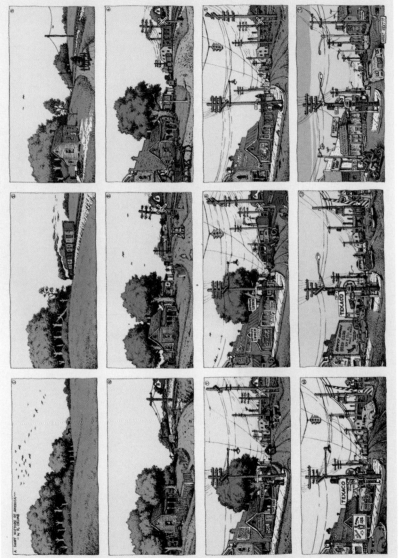

"A Short History of America." Art by Robert Crumb.

"Epilogue." Art by Robert Crumb.

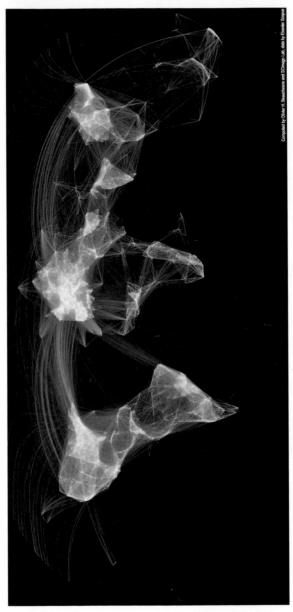

A global map showing four years of the geography of scientific collaboration.

Compiled by Olivier H. Beauchesne and SCImago Lab; data by Elsevier Scopus.

At the same time, we are creating the potential for new, equally devastating disasters. All this locates us at a curious and frustrating step in our evolution. We can see how technology could be used to ensure, rather than threaten, survival, and we can conceive of the idea of a truly intelligent, sustainable society. Yet we don't know if we can get there from here.

Not just one but several transitional moments are near or upon us, any or all of which could be pivotal in life's future. Our accelerating development of science and technology carries both great promise and great peril. There are many ways in which we could quickly doom ourselves. Yet these same abilities have improved life for vast numbers of people, and if we can learn a little self-control, will also offer us possibilities for survival never before open to any other species on Earth.

The twentieth century may have been a build-up toward a cusp in human history that will come to a head in the twenty-first or twenty-second century, where our accelerating technological development may become either a threat to our survival or, alternatively, a tool for great longevity.

Biologist Edward O. Wilson, in his 2002 book, *The Future of Life*, described it as follows:

> [T]he immediate future is usefully conceived as a bottleneck. Science and technology, combined with a lack of self-understanding and a Paleolithic obstinacy, brought us to where we are today. Now science and technology, combined with foresight and moral courage, must see us through the bottleneck and out.

British astronomer Sir Martin Rees, president of the Royal Society, in his very sobering book *Our Final Hour*, surveyed a range of existential threats and gave humanity a 50 percent chance of surviving to the end of the twenty-first century. He is particularly concerned about the fact that biotechnology and interconnectedness are potentially making it easier for a malicious actor, or an error, to cause huge worldwide damage. As he put it,

> In our interconnected world, novel technology could empower just one fanatic, or some weirdo with a mindset of those who now design computer viruses, to trigger some kind of disaster. Indeed, catastrophe could arise simply from technical misadventure—error rather than terror.[21]

Others rebut this pessimistic view by pointing out that defenses against biological and computer viruses are advancing as fast as the potential to do harm. So while the point remains valid that mischief makers or careless experimenters will have increasingly powerful tools at their disposal in the coming century, the doomsday predictions involve a lot of guesswork. They should concern us but not paralyze us with fear.

As I discuss in chapter 3, our Anthropocene dilemma is that right now we have global influence without global self-control. What if technological intelligence is actually an evolutionary dead end? In small doses, it may be a liability and a threat. Like bird's wings, you wonder how it ever evolves to the full thing, as a partially formed version seems like an awkward disadvantage.

H. G. Wells, the Arthur C. Clarke of the paleoindustrial age, saw this all coming in 1920, when he wrote, "Human history becomes more and more a race between education and catastrophe." Wells also said, "There is no way back into the past. The choice is the Universe—or nothing." Einstein, in 1946, noted that powerful new technology "has changed everything save our modes of thinking and we thus drift toward unparalleled catastrophe." Einstein was referring specifically to the threat of nuclear weapons, about which he no doubt felt some special personal responsibility. Both these brilliant men recognized humanity's problematic and dangerous combination of great cleverness with limited vision, and foresaw the accelerating collision of these tendencies, which confronts us today. A few years later, in his short story "The Sentinel" (1950), Clarke, describing why the aliens had left a monolith buried on the moon, instead of placing it on Earth, where we would find it much sooner, explains that its builders

> were not concerned with races still struggling up from savagery. They would be interested in our civilization only if we proved our fitness to survive—by crossing space and so escaping from the Earth, our cradle. That is the challenge that all intelligent races must meet, sooner or later. It is a double challenge, for it depends in turn upon the conquest of atomic energy and the last choice between life and death.

A decade after that, as the Cold War simmered, Arthur Koestler predicted that within the foreseeable future we would either destroy ourselves "or take off for the stars." Subsequent events and discoveries have only made this choice more clear than H. G., Albert, or either Arthur could have foreseen, and they foresaw a lot.

An evolutionary perspective reminds us that the human race has survived bottlenecks before. Just prior to the emergence of modern humans (which seems to have occurred around 190,000 years ago), Africa was in a warm climate phase, and many different human populations all around the continent were migrating and interbreeding with one another. Then there was a very cold, glacial phase, and humanity suffered an apocalypse. With the changed climate, populations of many game species disappeared, and we nearly did, too, because most of Africa could no longer support our hunter-gatherer ancestors. We almost went extinct.

DNA analyses of modern people suggest that the human population declined dramatically, to perhaps only a few hundred. When we talk about all men and women being brothers and sisters, and people of all types being closely related, we are not just being poetic and romantic. Compared to other species, humans exhibit very little genetic diversity, a trait that stems back to that time, not too long ago, when a small band of survivors had to repopulate humanity.

We don't know exactly when this "genetic bottleneck" happened, or where in Africa the survivors lived and hung on. They may all have lived in one small region of the southern African coast. Curtis Marean, a paleoanthropologist at the Institute of Human Origins at Arizona State University, has for many years been leading a team excavating an area called Pinnacle Point, near the southern tip of present-day South Africa, and has come to believe that this is the site where modern humans emerged. When most of Africa had become uninhabitable due to climate change, the people at Pinnacle Point survived largely on shellfish and coastal vegetation. They had to adopt a new lifestyle, and they developed sophisticated new technologies, including heat treatment of rocks in specially made and controlled fires to make spear points for hunting. This is an example of a "long-chain complex recipe technology," something that requires a sequence of steps to be performed in just the right way in order to produce the desired product. It's the kind of process that makes you say, "How the hell did they figure that out?" because there is no obvious connection between the starting materials, the details of the process, and the finished product. Such methods would have required language to pass detailed knowledge between generations. The people of Pinnacle Point also apparently made art and pigments for symbolic self-adornment. So they had language and arts, and were creators and teachers of a sophisticated material culture. They may not have been the first modern humans, but they seem to have been among the earliest.

We were once an endangered species almost completely wiped out by climate change. Some would say that we are again today (although this confuses the plausible threat of societal decline and collapse with the much more unlikely threat of human extinction). Back then, we avoided extinction by utilizing and honing our unique skills at cooperation, group innovation, and technological cleverness.

Now, strangely, we've been so successful at those skills that we've accidentally triggered a new era of climate change that again challenges us to evolve. We may be near the beginning or the end of the story of present-day human civilization—or both. We may be nearing the end of a ten-thousand-year adventure and beginning a much longer one.

Welcome to the Sapiezoic Eon

Some see the concept of an Anthropocene epoch as grandiose. They wonder if the coming of humanity is really so important, seen against the immense backdrop of Earth history. They worry that we are thinking too big. Yet I think the opposite is true. Maybe in thinking of it as *only a new epoch* we are thinking way too small. A shift to a new epoch is not that rare. An epoch typically lasts for a few million years, which, for Earth, means it's no big deal. Yet this is not merely another geological shift among many in Earth's long, ever-changing history. The advent of conscious agency as a force of change on Earth is a major inflection point in the way the planet functions. What we are witnessing, and manifesting, is something much more significant than a shift to just another epoch. It is more properly regarded as the potential dawning of a new *eon*.

Just as we divide our days into hours, minutes, and seconds, we divide geological time into different-size units. Look again at the diagram of the geological timescale shown on page 5 of the photo insert. The column on the right shows the epochs, typically lasting a few million years. These form convenient markers for scientists studying the detailed and tangled history of different locations around Earth. They generally don't represent major directional Earth changes so much as fluctuations in the complex interplay between changing geography, oscillating climate, and evolving species.

As you go left on the chart, you come to larger and larger segments of time: periods, which last for several tens of millions of years; and then eras which are much longer, each encompassing several periods. Finally, at the very left of the chart, we have the largest time divisions, the eons. There have been only four eons. Each represents a completely different phase of planetary history. During each, Earth was a fundamentally different planet. The transitions between them were revolutions in the relationship between life and the planet.

Let me summarize briefly the four eons of Earth history. Here I will gloss over a lot of detail in favor of some essentials. In very crude terms, to which any trained geologist will howl with a million caveats and complexifications, the four eons of Earth, and the transitions between them can be described as follows:[22]

<u>The Hadean Eon</u>, roughly the first half-billion years, was so named because it was pure hell, or would have been if anyone had been around to experience it. My friend and former officemate Kevin Zahnle, one of the galaxy's top experts on planetary origins and evolution, has referred to the Hadean as "a world of exuberant volcanism, exploding meteors, huge craters, infernal heat, and billowing sulfurous steams; that is, a world of fire and brimstone punctuated with blows to the head."[23]

The Archean Eon is often said to have begun 3.8 billion years ago with the origin of life. Some have attached it to the appearance of the first rock. Either definition would mean that the date will keep moving earlier in time as we explore Earth's past more fully, or, as Zahnle states it, this "puts the Hadean into the same category as the fastest mile or the tallest building." Sometimes its start date is simply pegged at 4 billion years ago.

The Proterozoic Eon began 2.5 billion years ago, at the time when oxygen from photosynthesis began building up in the atmosphere, generating the Great Oxygenation Event that I write about in chapter 3. This is arguably when life took over Earth's atmosphere, and so may also have been the origin of Gaia, or the time when life became a global entity.

The Phanerozoic Eon began 542 million years ago, with the Cambrian explosion. This was when life suddenly became complex and macroscopic. Up to that time, life had consisted almost entirely of single-celled organisms. Suddenly Earth became a world crawling with animals and plants.

So, to simplify further, and summarize each with a one-liner:

Hadean: origin of Earth
Archean: origin of life
**Proterozoic: Great Oxygenation Event, origin of global
    biosphere**
Phanerozoic: origin of complex life

When trying to put our planet and ourselves in the context of cosmic evolution, and asking what parts of the Earth story might possibly be universal, it's best to ignore the epochs and periods on the right side of the geologic timescale. Our eyes are drawn toward the major eon transitions shown on the left side of the diagram. [See page 5 of the photo insert.]

It seems obvious to all but the writers of grade B science-fiction films and some of the more gullible UFO conspiracists that the evolution of life, however it may have unfolded elsewhere in the universe, will not, in the details, resemble what has occurred on Earth. Consider a roughly Earth-size exoplanet around a Sun-like star thirty light-years away. You can bet there will not be any Jurassic there—no dinosaurs. There will not be a Pleistocene. No woolly mammoths. The epochs, on the right side of the timescale, even those lasting tens of millions of years, represent the unique, random details in the meandering path of life on our planet.

Yet what about the left side: the eons, the major life events of our planet? These are of particular interest to an astrobiologist because each represents a fundamental transition in the relationship between life and Earth. And each, I believe, is something we could legitimately search for on other planets.

One of our main strategies for seeking life elsewhere operates on the suspicion that other atmospheres have been oxygenated by photosynthetic life. Perhaps this is geocentric thinking, but it's hard for us to imagine that opportunistic evolving life would not plug into the powerful and ubiquitous solar energy source. So we half-expect something like a Great Oxygenation Event elsewhere.

Likewise, it is entirely reasonable to ask if other planets had a Cambrian explosion, or something like it. Indeed the question comes up all the time in astrobiology: should we expect only simple microbial life on other worlds, or larger and more complex organisms?

One of the other big questions is whether we can find intelligent, technological aliens. For reasons I'll delve into in the next chapter, we won't be able to find intelligent life if it is only a brief phase, a flash in the pan. If we find other civilizations, it will be the ones who have made it through the bottleneck of technological adolescence. If technological intelligence is able to take hold, and become a lasting part of a planet's functioning, this would arguably be at least as great a transition as any in Earth's past. The moment where cognitive processes become a dominant mechanism of change is easily as significant as the oxygenation of the atmosphere or the advent of animal life. If global intelligence becomes a lasting planetary force, then I believe it is more appropriate to regard this as the beginning of Earth's fifth eon.

What should we call it? I propose the Sapiezoic Eon.

Terra Sapiens

When Carl Linnaeus, the great Swedish botanist who started our modern system of naming species, was coming up with a name for humans, he sought a quality that differentiated us from other, similar animals in the genus *Homo*. He chose *sapiens* from the Latin *sapientia*, meaning "wisdom." Here he was echoing Charles Darwin, who wrote, "Of all the differences between man and the lower animal, the moral sense of conscience is by far the most important… It is the most noble of all the attributes of man." Perhaps giving us this lofty title was a bit of overreach, of self-flattery, but it can also be seen as an aspirational goal, something we wish to be. And we alone are capable, it seems, of wishing or imagining ourselves to be something we are not, or not yet.

If technological intelligence is to become, as life did long ago, a permanent part of the workings of our planet, it will be because we learned to integrate our activities gracefully with the ancient and deep dynamics of Earth. In chapter 2, I describe the debate over teleology in the Gaia hypothesis: the fraught question of whether the biosphere could possibly be said to be *seeking* some kind of homeostatic balance for its own good. With the Sapiezoic, we enter an era where Earth processes are unambiguously conscious and teleological, where life is obviously seeking, and finding, a balance among biological, geological, and cognitive processes, and where the boundaries between these, long identified as indistinct, fade to insignificance.

In chapter 3, I make the distinction between cleverness and wisdom. Sapience (wisdom) differs from intelligence (or cleverness) in that it is not merely a kind of cognitive skill, but includes the ability to act with judgment born of experience. You can be intelligent but lack wisdom. You may be the best in the world at answering questions on IQ tests, or the cleverest solver of puzzles, but still lack the judgment to make good decisions. Wisdom is clearly something that some individual humans and teams of people (and possibly orcas and elephants) carry in abundance. Yet can we, humanity, collectively, globally, find wisdom? We're in a phase characterized by our strong global influence, largely untempered and unmodified by our observations of the effects it is having. Can we make the transition to a phase of considered global control, where our actions result from an incorporation of all the knowledge we are gaining about the planet and its history, our history, and our growing awareness of the interactions between the two?

This would be a changed world, and I have a name for it: Terra Sapiens, or Wise Earth. This is an Earth where we and the planet have both changed to come to grips with each other; where we've learned to live comfortably over the long haul with world-changing technology, applied with a deep understanding of planetary function; where intelligent and wise application of our engineering skills has become smoothly integrated into global processes. It's a vision for the planet, but it's also an aspirational name for ourselves, for who we must become to manifest this world. On Terra Sapiens we won't differentiate between the two because we'll identify deeply with the planet. We'll understand that wise self-management and wise planetary management are one and the same.

This may be wishful thinking. No, it *is* wishful thinking. I don't see it as a forgone conclusion that we can achieve this new world where we've learned to apply our cleverness with wisdom. Yet it is by no means certain that we can't. It's valuable to have a vision of where we think we could conceivably be going. With the halting beginning of our mature Anthropocene, we tentatively dip our toes into this new eon.

In calling such an era Sapiezoic, I don't assume it will be humans (Anthropos) who will attain the wisdom to run a planet or, more accurately, to run with it. I suggest it as a more generic name for an eon when cognitive processes become a stable part of a planet's functioning. It is when life, any life, realizes what it is doing and incorporates that knowledge into the operation of its planet.

If we humans do not make the leap to sapient planetary management, other species may develop this ability in the future. Or, if we blow it this time, another human or hominid civilization may rise again, perhaps learning from our mistakes.* Earth may have a Sapiezoic Eon that does not begin with us. The term also need not refer just to Earth. Minds and machines are powerful and world changing. I believe they will sometimes appear on other worlds. Sometimes they will last, and these worlds will be transformed by them. Such planets will have a Sapiezoic Eon. When I think of SETI and the question of what it takes to achieve a long-lived planetary civilization, I imagine not just clever species with lots of nifty technology, but species who have

---

* No, *not* like in *Planet of the Apes*!

also navigated the hurdles we are now facing, which have to do largely with discovering and redefining our relationship with our planet. I imagine worlds where technology has become an integral attribute of a biosphere, one that ensures longevity.

I've argued that one way to approach the question "What is life?" might be to think of it as a property that a planet sometimes takes on. Similarly, mind may be a phenomenon that sometimes can become a property of a planet. If planets can have Sapiezoic Eons, then there is a form of stable intelligence, manifested on some worlds, that has not quite yet appeared on Earth. This might even help explain why we haven't heard from anyone.

INTELLIGENT WORLDS IN THE UNIVERSE

All is but a woven web of guesses.

—Xenophanes

We are scatterlings of Africa, on a journey to the stars.

—Johnny Clegg

A Common Story?

Right now, we seem to be sleepwalking participant/observers in some new kind of planetary transformation. Could this be a local version of a stage in cosmic evolution where some planets start to wake up, look at themselves self-consciously, and at the same time look outward and wonder if they are unique? Maybe this perspective can help us in our effort to figure out what is going on here, and what our role is in this transition, and to awaken fully before we do too much damage.

If we view the Anthropocene as the beginning of an

unprecedented and significant transition in Earth history, then we naturally wonder if this has happened to worlds beyond. Yet when we try to imagine this same thing possibly happening to other planets, what exactly are we picturing? Obviously there are not going to be people. We don't expect bipedal creatures driving metallic cars on asphalt highways between concrete cities, internal combustion engines suffocating their planet's air with regurgitated primordial carbon. This isn't a *Star Trek* universe with scattered almost-Earths peopled by almost-humans with tattoos and funny ears, possessing slightly different craniums, customs, fashions, spaceship designs, and architectural tastes. Evolution does not stamp out cheap replicas. As biologist Stephen Jay Gould famously surmised, if we could rewind and replay the tape of evolution on Earth, nothing like the modern biosphere, at least in detail, would emerge. To me this metaphor always seems strained, since a tape recording plays back more or less the same every time. Evolution is more of a crapshoot, or a long series of them, run over and over again. What Gould meant is that it's a random, contingent process, and the outcome is not neatly determined by the starting conditions. Given the chance, even on identical worlds, evolution would never repeat itself exactly. Now add in the fact that there surely are no identical worlds. As I describe in chapter 1, planets, even in the absence of life, are complex balls of feedback and chance, and in detail no two will be the same. So, the game of life will never be played with the same preconditions, and run in the same environment, even twice.

There will be no people, but is there some essence of what people are that nonetheless can be expected, sometimes, to slither out of the mud of other home worlds, compelled by the same forces that brought us here? Are there common trends, inventions that evolution is bound to stumble upon? Life is so incredibly resourceful at finding ways to exploit environments to thrive and self-perpetuate. Natural selection is a ceaseless and thorough trial-and-error search for survival solutions. Certain innovations just work so well that they will always be found sooner or later. Many times in Earth history, life has independently hit upon the same survival tricks more than once. The similar aerodynamic swimming shapes in dolphins and fishes evolved completely separately. The octopus eye and the human eye are constructed in much the same way. Winged flight evolved separately in insects, bats, and birds. It seems the relentless, industrious incubator of natural selection has certain go-to solutions. Such convergent evolution suggests there may be a vaguely predictable quality to evolution that even spans worlds, not in the specific steps, but possibly in some overall trends.

We've made the mistake many times before, of thinking that Earth was unique in various ways. And we've made the opposite mistake: assuming it to be completely typical. When we first realized Earth was one of a handful of worlds, many seventeenth-century astronomers and natural philosophers imagined the surface environments of other planets as essentially identical to Earth, populated by similar creatures and peopled with minor variations on European societies. When we don't have any details, our minds tend to go with what we know, filling in gaps with the familiar. Today still our thoughts and calculations about alien life and civilizations are unavoidably rife with assumptions based on terrestrial evolution and history.

When we talk about extraterrestrial species and civilizations being at "our level" or "more advanced," we are adapting a narrative where life, intelligence, and civilizations proceed through some common, predictable, or directional progression of stages, one logically following another, like an organism going through phases of growth, or a soldier being promoted up the ranks. We don't know how good an assumption this is. SETI might help us find out.

Could some version of this Anthropocene transition have happened elsewhere?

Now we know that planets are common. We don't yet know about biospheres, but we suspect they too may be common, and we wonder to what extent they will also fall into classifiable types and follow well-worn evolutionary and developmental sequences.

We imagine that, on the right kind of worlds, there will be an origin of life that will then organize itself into some kind of recognizable cellular and organismic structures. Sometimes we get even more specific and self-referential, imagining that, elsewhere, life will use familiar carbon configurations,[1] evolve photosynthesis, poison its atmosphere with oxygen or methane that we might find in our spectrometers, make ozone layers, and/ or fill its continents with something like forests that may betray themselves by their coloration.

On Earth, in order to make forests and trees, and animals to swing and crawl among them, life had to advance beyond the stage of individual cells, and organize itself into larger, more differentiated organisms. Here that jump in complexity did not follow closely on the heels of cellular life. Rather, it took life billions of years to make this next leap. Why the wait? Does this long delay mean that such a step was difficult or unlikely, and so should be rare on other worlds?

Based on this, some astrobiologists have concluded that most planets with life might have only microbes. Yet the usefulness of these "timescale arguments" is much debated. How much can we really say about the likely paths and probabilities of life elsewhere based on our one example? We don't know why life here paused for so long (over a billion years between oxygenating the atmosphere and building the larger animal bodies fueled by that liberated oxygen), so we really shouldn't overinterpret this fact. Yet we can't help it. We can't stand not knowing, so people tend to make up their minds prematurely. For every scientist who is sure the probability of complex animal life elsewhere is close to zero, you can find another who is sure that it is 100 percent.

Then, what about the transition from complex animal life to creatures who tell stories, do science, make art and machines, and struggle with awareness of their world-changing potential? Several biologists have argued, on the basis of life's history on Earth, that evolution of human-style intelligence is a freak occurrence that would be highly unlikely to be found on other planets. If it is a likely or easy development, why did it take another five hundred million years to go from animals to zoologists?

Not only that, but the brightening sun will make Earth uninhabitable (assuming nobody interferes) after only about another one or two billion years, so it actually took almost all Earth's useful habitable lifetime for evolution to produce technological intelligence. Given this, it is easy to imagine that it might never have happened here at all. So, some say, big-brained technologists must be rare. However, this conclusion ignores many facts: There are innumerable stars out there with lifetimes much longer than our Sun, many with potentially habitable planets. If Earth were farther from the Sun, out toward Mars, its habitable lifetime would be significantly longer. Finally, we really have no idea what determined the evolutionary timescale for the appearance of intelligent technological life on Earth. Given all this, there is little reason to conclude, on the basis of our single data point, that what has happened here is rare.

It took 4.5 billion years for Earth to go from dead rock to space walk, from molten ball to shopping mall, from sea to me, from goo to you. This is probably the most overinterpreted number in all of science. Timescale arguments are really pretty weak because we must assume our planet is typical. We can't account for the role of luck. Are we reading huge significance into the results of a careless roll of the dice? If we want to know the likelihood of something like us evolving on other worlds, we have to go out there and find the answers.

And what of the "postbiological" parts of the story starting to unfold on Earth? Here, intelligence of a certain kind made the leap to culture, global civilization, and world-altering technological activity. If we accept that these qualities of human agency (group problem solving, cumulative knowledge, technology, and some small degree of foresight and intentionality*) are now altering our planet's evolution, we can ask if these could be influencing other worlds as well. It's a trick question because either way you answer, yes or no, you can be accused of arrogance. Either you're putting Earth on a pedestal, saying our planet is so special and gifted as to possess potentials and developments found nowhere else in the universe. Or you're putting humanity on a pedestal, daring to presume that our human qualities are cosmically significant and would be shared by "advanced beings" elsewhere, by the shapers of other worlds, which seems to elevate us to some kind of role model for galactic intelligence.

Crown of Creation?

The idea that our evolution is so much of a freak occurrence as to be potentially unique in the universe, not only in detail but in its overall functional attributes, has most recently been argued by my friend and colleague, the brilliant and reflexively contrarian Australian astrophysicist Charley Lineweaver, with his "Planet of the Apes hypothesis."[2] This revives an argument made many times by eminent biologists, including, among

---

* Just enough to design and drive factories and cars, not enough to seriously consider how much of that is a good idea.

others, naturalist Alfred Russell Wallace in 1904, paleontologist George Gaylord Simpson in 1964, and biologist Ernst Mayr in 1995. All have argued that the fossil record reveals human-type intelligence to be such an unlikely chance occurrence that it has probably never evolved on any other planet.

In 1964, Simpson published, in *Science*, "The Nonprevalence of Humanoids," a scathing diatribe against both SETI and planetary exploration. He railed against belief in extraterrestrial intelligence, for reasons that closely recapitulated the Earth-centric views voiced six decades earlier by Wallace, codiscoverer with Darwin of evolution by natural selection. They argued that the evolution of human intelligence required too many accidental and lucky factors to represent a process that could occur anywhere else in the universe.

Wallace wrote, in 1904:

The chances against such an enormously long series of definite modifications having occurred twice over, even on the same planet in different portions of it...are almost infinite...

Simpson's 1964 paper concluded:

I shall close this chapter with a plea. We are now spending billions of dollars a year...on space programs. The prospective discovery of extraterrestrial life is advanced as one of the major reasons, or excuses, for this. Let us face the fact that this is a gamble at the most adverse odds in history. Then if we want to go on gambling, we will at least recognize that what we are doing resembles a wild spree more than a scientific program.

To some it seems that the reward could be so great that facing any odds whatever is justified. The biological reward, if any, would be a little more knowledge of life. But we already have life, known, real, and present right here in ourselves and all around us. We are only beginning to understand it. We can learn more from it than from any number of hypothetical Martian microbes.*

Lineweaver has renewed this argument. By comparing them to the authors of the schlocky but fun film *Planet of the Apes,* he is offering a mocking critique of those who feel that evolution of smart, inventive creatures may be possible on other planets. Yet one does not have to literally believe that creatures on other worlds would be hairy humanoids wearing medieval costumes, riding around on horses with guns and whips, speaking English, and acting like evil fascists to think that other biospheres might eventually find their way to producing some sort of life-forms that are clever, social, sophisticated thinkers who develop technology.

Brains are good survival tools. If you evolve to sense your environment and behave in response to it, then you will need some sort of nervous system to process this information and coordinate behavior. Better-functioning nervous systems will be selected, and so some species will get smarter. How far this can go is anyone's guess. It seems irrational to think that it can't in some locales go a lot farther than it has on Earth. So rather than argue over whether "human-level intelligence" can ever be reached elsewhere, it might be more reasonable to wonder to what degree it would be surpassed.

What does the fossil record say? It depends whom you ask. It is easier to make the case that human intelligence is a convergent feature, one that might represent some common or inevitable trend in evolution, if you accept the published evidence

---

* Based on this paper, astronomer Frank Drake described Simpson as "...perhaps the most eminent person ever to have misunderstood totally the foundations of SETI."

that brain size has increased over time within groups of species. There is a history of disputes over the quality and significance of this evidence. In 1995, biologist Ernst Mayr publicly feuded with Carl Sagan on this question, with Mayr arguing that SETI programs were an irresponsible waste of money,

> since it can be shown that the success of an observational program is so totally improbable that it can, for all practical purposes, be considered zero.

Sagan countered that

> the notion that we can, by *a priori* arguments, exclude the possibility of intelligent life on the possible* planets of the 400 billion stars in the Milky Way has to my ears an odd ring. It reminds me of the long series of human conceits that held us to be at the center of the universe, or different not just in degree but in kind from the rest of life on Earth, or even contended that the universe was made for our benefit. Beginning with Copernicus, every one of these conceits has been shown to be without merit.

Charley's arguments are not *a priori*—meaning not based on any relevant data or experience. He uses data from Earth evolution and argues that isolated populations of prehumans on several continents had plenty of time to make the transition to human intelligence, but only in East Africa did this actually happen. He feels that this shows that our "cognitive revolution" was a rare and unlikely type of development.

Charley's an awfully smart guy who has made many

* If Sagan had written this ten years later, he could have deleted the word "possible," and shown yet another conceit to have fallen.

important contributions in cosmology and astrobiology, but I think here he overstates the case, because the reality is that our evolutionary history is still quite poorly known to us. The details of what happened to make us what we are today are largely still buried in the dirt, lost in the murk of time. Just in the couple of years while I've been writing this book, several major discoveries have been made that require us to rewrite substantially the provisional drafts of our human memoir. A 2.8-million-year-old jawbone discovered in an Ethiopian hillside is the oldest human fossil known to date, and pushes back the origin of the genus *Homo* by nearly 500 million years. A cutting tool found under a 15,800-year-old ash layer in Oregon may be the oldest evidence of human occupation of North America. A new discovery of stone flakes near Lake Turkana, in Kenya, may push back the known date of toolmaking and use by 700,000 years, back far before the evolution of the genus *Homo*, meaning that our non-human ancestors manufactured tools. Also, we have discovered that, very recently in Earth history, there were in fact many different human species living simultaneously. These previously unknown contemporaries of modern humans belie the notion that *Homo sapiens* has been the only tool-making human species on Earth since the Neanderthals died out.

In 2003, remains were found of extinct species of humans on the Indonesian island of Flores. *Homo floresiensis* stood three and a half feet tall, used sophisticated stone tools, and lived until at least twelve thousand years ago. In 2010, a forty-one-thousand-year-old finger bone was found in Denisova Cave in Siberia. It belonged to a young female who was human (genus *Homo*) but not *Homo sapiens*. DNA analysis indicates that these Denisovans ranged widely over Europe and interbred with both Neanderthals and modern humans, and that traces of their DNA can be found in several modern populations, including Melonesians and Aboriginal Australians.

Yet another recent and distinct species of humans, the Red Deer Cave People, lived in China until at least around eleven thousand years ago. Hints of other recent human species have also been found, and it's clear that we don't really know how widely peopled Earth was, until very recently, with alien humanoids. Twenty thousand years ago, Earth may have been home to a wide range of distinct human species. We don't know what happened to all these close cousins, but it seems likely that in various ways, they fell victim to the great success of our species as we spread around the globe. What a strange and different world it would be if multiple species of humans had survived to the present day. Perhaps in such a world we would be less prone toward thinking of ourselves as the crown of creation, and thinking of our own intelligence as something so unique and rare in the universe.

It's easy to fall into the illusion that modern science has figured most things out and just the details need to be filled in. Every generation of scientists thinks so, always falsely. The pace of discovery suggests we have much to learn about our own origin story and its meaning.

Charley feels that, as we chart the paths of evolution, and speculate on what we might find on other worlds, we put too much emphasis on searching for "human type intelligence," and specifically, too much emphasis on the evolution of big brains. He regards this as a form of self-aggrandizement, suggesting that we fixate on brains out of pride, simply because they are our own most unique anatomical attribute. As a thought experiment, he suggests that a race of intelligent elephants might instead measure and chart evolutionary progress based on the size of noses of various species, and speculate on the existence of advanced species with massive trunks elsewhere in the universe. This seems to me to be a silly argument, and not just because of the delightfully goofy image of a bunch of intelligent elephants sitting around discussing evolution. The phrase "a race of intelligent elephants" contains the flaw in the argument. How are these putative intelligent elephants able to have this imaginary conversation? What gives them the ability to do science, to know something of biology, to study their planet's natural history, to make graphs of evolutionary progress and discuss them with one another? Wherever these abilities come from, these elephantine evolutionary biologists seem to have a great deal of "human-type intelligence" which they use to study life on their own planet and speculate on its paths elsewhere. Whatever it is that gave Charley's imaginary elephants the ability to do science, they are doing something that, on Earth today, is being done only by human beings. Anybody who can ask such questions about evolution, whatever their anatomy, is demonstrating enough "human-type intelligence" to put the *I* in SETI.

No, brains are not just some random feature that could be swapped out for any other in an evolutionary analysis, and the Planet of the Apes hypothesis seems to willfully ignore what is actually happening to our planet right now. It is our nervous systems, our cognitive abilities, amplified by our social skills and our propensity for toolmaking and storytelling, that makes us such a powerful and unprecedented force of planetary change.

I always enjoy talking to Charley and reading his provocative papers and talks. It's valuable to be reminded to question our assumptions, and that we don't know what we mean when we use words such as *intelligence*. He reminds us to "check our privilege" as cognitive creatures searching for like minds elsewhere, and to be aware of geocentric, egocentric, or homocentric assumptions. Once every few decades or so, some well-spoken individual comes along and refurbishes this argument. It is worth listening to. Then, having done so, we can nod our outsize heads and turn our attention back to searching the cosmos.[3]

Just as astrobiology has challenged our ideas about the

universal qualities of life, and SETI has provoked much thought about the nature of intelligence and technology, we can ask, assuming these cognitive qualities do exist elsewhere, how they may be interacting with planetary evolution. By seeing the Anthropocene as a particular example of something that can happen to a planet, we are forced to imagine a view beyond the random particulars of terrestrial history.

Accuse me of being bio-, ethno-, geo-, or egocentric all you want. There is something so strange and beautiful about the presence of human minds on Earth, about the fact that this one bit of biosphere is awake and aware and curious about everything—stuck in the gutter, perhaps, but looking up at the stars. Yes, we are perhaps unduly impressed with ourselves. But how can we help but wonder if any other part of the universe has this same sense of aliveness, this same urge to look out at the stars, and also to wonder if somewhere under a different sun some other growing things began sensing and responding to their environment, evolved memory, cognitive faculties, and— who the hell knows—maybe social behavior and language and culture? And technology? Have they, at some point, been surprised and troubled by their ability to change their world, or doomed by their own cleverness?

Are we reading too much of ourselves into this generalized cautionary fable? Perhaps, but I don't believe there is strong evidence from hominid evolution on Earth to convince us one way or another, and as long as they might be there, then we have to try to find them.

The Interstellar Age

Sixty years ago a strange glowing object, unlike anything ever seen, suddenly appeared in the sky above Earth, scooting purposefully among the stars. A polished metal sphere twenty-two inches in diameter, with four spindly antennae, it circled our planet for three months. *Sputnik 1* transmitted for three weeks, its simple, high-pitched beep, beep, beep causing delight and wonderment among those who had the right gear to receive it, and provoking awe, hope, and fear among everyone else. This portentous message from the sky, lacking content but rich with meaning, announced the arrival of the space age and heralded the rapidly approaching interstellar age of SETI.

This age started on September 19, 1959, when a persuasive paper appeared in *Nature* advocating for the feasibility of radio communication between the stars, and suggesting that humanity now possessed the tools to search for signals from alien civilizations. Radio telescopes, at the time, were a brand-new kind of instrument—born of wartime advances in radar technology. Astronomers were still figuring out what you could do with them when two Cornell University physicists, Giuseppe Cocconi and Philip Morrison, realized they could be used to communicate with planets orbiting distant stars.

Morrison came to SETI circuitously. Before he started speculating on the existence and longevity of other civilizations, he became intimate with the possible destruction of our own. A gifted nuclear physicist, he trained with J. Robert Oppenheimer and, in 1943 at the age of twenty-eight, was recruited, despite a huge FBI file on his youthful Communist activities, to join the Manhattan Project. Motivated by valid fear of a Nazi bomb, he made key contributions to the design and testing of the first nuclear weapons. He calculated how much plutonium was needed for a successful device, and designed the shape of the nuclear triggers. For the first atomic test, Trinity, Morrison rode from Los Alamos out to the test site in the backseat of a Dodge sedan alongside his plutonium bomb core, which he later described as "slightly warm, like a small cat." He watched the blast at a distance of ten miles and was seared and stunned by its frightening heat. He helped assemble the bombs that would destroy both Hiroshima and Nagasaki, and helped load both on the planes that would drop them.

Shortly after the war, he visited Hiroshima. Shaken by what he saw there, he realized that nuclear weapons could trap humanity in a disastrous arms race. He wrote that "the public must realize that the bomb opened a door to fear, expense, and danger rather than just end the war," and became a passionate lifelong activist for peace and nuclear disarmament. The Manhattan Project physicists knew they had unleashed a threat that could bring about the end of human civilization. Morrison was instrumental in organizing them into a powerful and authoritative brain trust for the disarmament movement.

I first met him thirty years later, when I was a young teenager and my parents socialized in the Boston academic scene with Philip and Phylis Morrison. They were an academic power couple in the nicest possible way: gentle, mega-erudite, and seemingly knowledgeable and curious about everything. As a kid, I encountered some adults who made scientists, teachers, and writers seem cooler than athletes and rock stars. This put ideas in my head. The Morrisons were definitely in this category. Another decade later, I reached out to Philip when I was working with a student disarmament group at Brown University. I was thrilled when he accepted my invitation to come speak in Providence. He was our big fish for the Brown University Conference on Nuclear Disarmament in April 1981.* We attracted a huge audience, and Morrison didn't disappoint. He had a lifelong disability from contracting polio as a child, needing a leg

---

* I spoke to my parents on the phone while writing this section, reminiscing with them about the Morrisons. When I mentioned the Disarmament Conference, my mother said, "Oh yes, that's when you got a haircut and took your earring out!" These were not the details I remembered, but I guess it speaks to how seriously I took my role in this event.

brace, a cane, and eventually a wheelchair. He approached the dais slowly, hunched over his cane awkwardly, his gait lopsided and deliberate. He could barely see over the lectern, and as he bent the microphone downward, it creaked discordantly. Yet, as his soaring contralto voice and radiant intellect filled the hall, the auditorium quickly fell to a hush. He spoke directly to the students in the room: "I believe," he began,

> if we can solve the problem of maintaining the peace against the claims, the all-too-urgent claims of nuclear warfare, then I think we have a chance, an excellent chance, for solving the rest of those problems which you all know so well, the last generation or two has bequeathed you.

After World War II, Morrison had joined the faculty at Cornell, where his research interests veered into astrophysics, and he founded the field of gamma ray astronomy. He started to wonder if hyperenergetic gamma rays might be used by extraterrestrial civilizations to communicate across the galaxy. He and his Italian colleague Cocconi decided to look into it. In working out the physics, they became convinced that gamma rays would not work well for sending information, and that instead radio waves would be the preferred medium for interstellar greetings. Out of this came their landmark 1959 paper, which provided the theoretical basis for modern SETI.

In 1964, Morrison left Cornell for MIT, where he remained until his death in 2005. There he became known as a riveting lecturer on physics, biophysics, and astronomy, and a committed, energetic, and increasingly visible public communicator of science. He and Phylis Morrison, a gifted writer and educator, collaborated on books and radio and television shows.[4] Throughout his storied career he maintained his involvement in both SETI and the disarmament movement. He was the nexus between these topics, having played key roles in the origins of both the atomic bomb and SETI, and he often connected the two, arguing that nuclear war, or a more general tendency toward technological self-destruction, might limit the lifetimes, and therefore the numbers, of civilizations in the galaxy.

The word *seminal* is used to describe many important papers, but when Cocconi and Morrison published their *Nature* paper "Searching for Interstellar Communications," they planted a perfect seed into fertile ground. Three months later, on winter solstice 1959 (the day I was born), the cover of *Life* magazine carried a banner headline: "Target Venus: There May Be Life There!" As the 1960s began, voyages to the planets were imminent, and scientists were predicting that plant life would be found on Venus and Mars. The heady optimism of these times was reinforced when chlorophyll (the green stuff in green plants) was detected on the Red Planet, or so it was reported in *Science* magazine.* A 1961 National Academy of Sciences panel concluded that "the evidence taken as a whole is suggestive of life on Mars." The news was also peppered with flying saucer reports. The beeping of *Sputnik* and its successors had made the concept of high-tech messages from above seem less fantastic. Space was wide open, and the world was ready. The SETI idea germinated and became rooted in the scientific consciousness, and soon sprouted into public visibility. Cocconi and Morrison presented the case so clearly, elegantly, and convincingly that, quite suddenly, the notion of exchanging messages with technically advanced alien civilizations was no longer an outlandish speculation. They somehow made it seem like a reasonable proposition by treating it as a physics problem, working out the equations and determining the size of telescopes, energy

---

* Later found, like so many "life on Mars" discoveries, to be a mistaken interpretation born of wishful thinking.

of transmitters, and sensitivity of receivers needed to succeed. They showed that even a young and relatively primitive scientific civilization, just like late twentieth-century humanity, would, in the course of investigating the physical universe, likely develop equipment that could easily be used to communicate across the interstellar void. Their paper concluded:

> The reader may seek to consign these speculations wholly to the domain of science-fiction. We submit, rather, that the foregoing line of argument demonstrates that the presence of interstellar signals is entirely consistent with all we now know, and that if signals are present the means of detecting them is now at hand. Few will deny the profound importance, practical and philosophical, which the detection of interstellar communications would have. We therefore feel that a discriminating search for signals deserves a considerable effort. The probability of success is difficult to estimate; but if we never search the chance of success is zero.

Who could argue with that? The Cocconi and Morrison paper became the theoretical basis for an unprecedented kind of experiment that began a year later, in 1960, and continues to the present day. For all we know, it may just be getting under way. It is an experiment that a scientist can undertake only if she is comfortable with the idea that it may not succeed within her lifetime, if ever.

The Green Bank Dolphins

The time was ripe. In a stunning example of convergent intellectual evolution, a few hundred miles (or 0.001 light-seconds) to the south of Ithaca, New York, Frank Drake, a brave and gifted young radio astronomer, had independently produced the same calculations as Cocconi and Morrison, and reached the same conclusion. Not only did Drake determine that a radio search for alien messages had become possible, but he decided to do something about it.

In 1960, he began Project Ozma, aiming the eighty-five-foot dish of the National Radio Astronomy Observatory in Green Bank, West Virginia, toward two nearby Sun-like stars, Tau Ceti and Epsilon Eridani, tuning the frequency dial by hand to search one possible alien radio "station" at a time. Lo and behold, Drake immediately picked up a strong signal! Could it be that easy? No. The signal was coming from a radar installation—on Earth. The intelligence was military, not extraterrestrial. From April to July of that year, for six hours a day, Drake searched for a signal. Nothing turned up that was both artificial and extra-terrestrial, but the era of experimental radio SETI had begun.

In 1961, Frank Drake hosted the first SETI conference at Green Bank. There were only eleven in attendance, includ-ing Philip Morrison; Carl Sagan; Melvin Calvin (who, during this conference, received a call awarding him a Nobel Prize!); astronomer Su-Shu Huang, who invented the notion of habit-able zones around stars; and neuroscientist John Lilly. Swept up in optimism and camaraderie, the participants formed a whim-sical organization called the Order of the Dolphin, after Lilly's work toward communicating with these sleek, bright creatures who seemed to encourage our hope for conversing with other intelligent species. The presence of all these notables helped to establish SETI as an endeavor that reputable people could take seriously—or, as Sagan described it, they "crossed the ridicule barrier." Maybe so, but one wonders why there are no published proceedings or photographs from this meeting.

In putting together the agenda, Drake wrote down all the different factors that must be considered in any attempt to estimate the number of broadcasting civilizations in the galaxy. How often are new stars born? What fraction of them have planets? What portion of these have suitable conditions for life? What fraction of suitable planets actually develop life, and on what portion of these does intelligence develop? What fraction of these develop the capacity and desire for interstellar communication? How long do they last?

Drake realized that these agenda topics could also be combined into one equation, with each question, each unknown probability, represented by a variable, and the output, N, representing the estimated number of civilizations in the Milky Way galaxy. Some of the variables were well known, such as R^*, the formation rate of stars. Some were wild guesses, such as f_i, the fraction of biospheres that would develop technological intelligence. One important variable, f_p, the fraction of stars with planets, was a wild guess at the time and has since become a known, observed quantity. Each fractional probability was given a value between zero (no chance of occurring) and one (inevitability, 100 percent chance). This mathematical tool became known as the Drake equation. The formula was not meant to provide a single, definitive, correct answer for N, the number of civilizations, but to organize our collective thoughts and discussions. It is a "heuristic" equation, useful for thought experiments. It allows us to ask questions such as What if we assume that life always forms complex communicating societies on a planet after two billion years, but these last for only an average of five thousand years? How would this affect the total quantity of civilizations? We can plug in numbers and get answers for our hypothetical situations. As we slowly learn more about the universe, and replace wild estimates with observed constraints, the range of reasonable hypotheticals narrows down. The Drake equation serves as an evolving quantitative framework for our continuing assessment of the paths and outcomes of cosmic evolution. Or, as Frank Drake describes it, the point of his equation is "to organize our ignorance." It has served this purpose brilliantly, and has become a ubiquitous, time-tested tool, the iconic encapsulation of our efforts to suss out the population density of thinking species in the galaxy.

Historically, others had made a similar calculation but had not connected it to the practical consideration of an actual search program. In chapter 1, I mention a pioneering meeting on climate change organized in 1952 by Harvard astronomer Harlow Shapley. There, Shapley discussed "The Abundance of Life Bearing Planets" and wrote out an equation that is nearly identical to the Drake equation.[5] His calculation led him to conclude that

> the life phenomenon is widespread and of cosmic significance. We are not alone. And we should admit, of course, that the animal, vegetable, or other organisms on other happier planets may have far "surpassed" the terrestrial forms. There is no reason whatever to presume that *Homo sapiens*, *Apis mellifera* and *Corvus americanus** are the best that biochemistry and star shine can do.

Yet there was an essential difference between Shapley's treatment in 1952 and Drake's nine years later. Shapley stopped short of discussing complex or intelligent alien life as something we might ever actually observe. Perhaps in the early 1950s such a thought would still have seemed too outlandish even in such brave and forward-thinking company. In the intervening years, with the advancement of radio astronomy, SETI had become possible as an observational science. In Cocconi and Morrison's

---

* Here, by mentioning in passing honeybees and crows, Shapley ingeniously pointed out that evolution on Earth has produced more than one kind of complex life with some version of intelligence.

theoretical framework and Drake's calculations, these estimates made the leap from idle speculation to testable prediction.

At the Green Bank meeting, Drake encouraged the participants to come up with a consensus estimate for every factor in the equation. They estimated that half of all stars have planets (a little low, we now know) and that the average number of habitable planets orbiting each of these stars is between one and five (probably a little high). They figured the probability of life forming on one of these planets is 100 percent. This seems optimistic, but sixty years later, we still don't have any empirical purchase on this number. As I've described, a high probability is (weakly) supported by the quickness with which life established itself on Earth. Then they made their most controversial assumptions: the percentage of biospheres where life, once established, will evolve to intelligence is also 100 percent, and of these, the fraction that will develop the technology and desire to communicate with radio telescopes is 10 percent. These estimates, reflecting the zeitgeist of those times, now seem wildly optimistic. Yet, of course, nobody knows. The point of SETI, after all, is to try to find out.

The Dolphins concluded it was definitely worth a look.

Russians

SETI was born as the Cold War was heating up. While the nuclear powers were aiming missiles at each other across the Bering Strait, devising strategies for mutual assured destruction, scientists on both sides were looking out into the galaxy and wondering about establishing friendly relations with other civilizations.

At the time of Green Bank, a growing community of Soviet scientists was becoming interested in SETI, thinking along nearly parallel lines as the Americans, but with a Russian twist. The idea of more advanced civilizations that should evolve from early industrial ones fit very well into a tradition of Russian thought going back at least to the early twentieth-century cosmists such as Fedorov, Tsiolkovsky, and Vernadsky, who all saw contemporary human civilization as a passing phase in cosmic evolution. They imagined that the Anthropocene (or the noösphere, as they called it) was something just getting under way on Earth and that it might one day extend beyond this planet and make contact with its counterparts.

Soviet scientists also tended to see the subject within the political/philosophical frame of Marxist dialectical materialism, in which human history is unfolding according to inevitable historical laws. It was only reasonable to suppose that if such laws existed they would, after the current epoch, produce more advanced stages of civilization, and that these same patterns would be playing out elsewhere in the universe. Many of the boldest and most visionary ideas about galactic civilizations and the far-future evolution of societies (or past evolution on planets that got the jump on us by millions or billions of years) have come from Russian scientists steeped in these philosophical traditions.

The Soviet counterpart to the Green Bank meeting, the First All-Union Conference on Extraterrestrial Civilizations and Interstellar Communication, was held in May 1964, at the Byurakan Astrophysical Observatory in Soviet Armenia. Perhaps less concerned about the "ridicule barrier" and any public threat to research funding in their more centralized scientific enterprise, the Soviet scientists were eager to publish their papers. So this is the first SETI meeting for which published proceedings exist.

The father of modern SETI in the USSR, the Soviet Frank Drake, was Ukrainian astrophysicist Iosif Shklovsky, from the Sternberg Astronomical Institute in Moscow. Shklovsky, already an eminent pioneer of Soviet radio astronomy, had been inspired by the Cocconi and Morrison *Nature* paper in 1960. When he turned his career toward the theoretical study of SETI, it started a vital and sustained tradition of important Eastern Bloc contributions to this field.

He opened the Byurakan conference with his talk on "Multiplicity of Inhabited Worlds and the Problem of Interstellar Communication," proclaiming,

> We are witnessing the inception of a new science, which occupies a boundary position between astrophysics, biology, engineering and even sociology.

His logic was similar to that outlined at Green Bank, but his conclusions were less optimistic:

> Is intelligence an inevitable consequence of the long and tortuous evolutionary process of life forms? In my opinion, this is by no means so.

He acknowledged the important work of Cocconi, Morrison, and Drake, but offered this critique of the American efforts:

> In our opinion, the main deficiency of Cocconi and Morrison's idea and of its realization by Drake is in the assumption that the extraterrestrial civilizations are approximately on the same technological level as the Earth civilization. An inevitable conclusion which follows from this assumption is that the power of the receivers and the transmitters available to the alien civilizations is roughly the same as that of our equipment. But this proposition is inherently fallacious. It is well known that the time scale for the technological development of civilizations is exceedingly short. Consequently if there are civilizations in the

Universe, they should greatly differ in the degree of their development. The great majority of civilizations should be technologically much more advanced than we are. We are only infants as far as science and technology are concerned.

This notion that other civilizations would be much more highly advanced than ours was further emphasized by Nikolai Kardashev, a brash young astrophysicist and star student of Shklovsky's. Only two years out of grad school at this meeting, Kardashev quickly established himself as a leading voice in global SETI, with bold ideas about the properties of advanced civilizations and how to find them. At Byurakan he proposed an enduring scheme (which became known as the Kardashev scale) for classifying technological civilizations.

Kardashev presented an engineer's view of the salient features of human history. The most quantifiable and, he thought, predictable quality of human societies over time was a steady increase in the use of energy. He proposed that this would be the nature of all technical civilizations in the universe. He showed that if our current annual increase in energy use continued at the same pace, that it would be less than a century before we would be utilizing the equivalent of all the solar energy available on Earth, only a few thousand years until we would expand into the solar system and consume the entire energy output of the Sun, and less than ten thousand years until our energy needs would rise to equal the output of all the stars in the galaxy. Based on these considerations, Kardashev proposed that all civilizations be classified into one of three types:

- **Type I civilizations** are masters of their planets, and have learned to make use of the entire energy resources of their home worlds. Kardashev suggested that humanity was currently approaching type I.

- **Type II civilizations** are masters of their solar systems. They would have expanded beyond their home planets and learned to use the entire energy resources of their home star.
- **Type III civilizations** are masters of their galaxies. They have built an interstellar society and learned to harness the total energy resources of their entire galaxy of hundreds of billions of stars.

Kardashev concluded that type I civilizations like our own should represent a brief transitional phase and thus would be relatively rare in the galaxy. Therefore (echoing Shklovsky's complaint that the Americans were too eager to assume that ET was at our level), rather than listen for signals from type I civilizations, we should look for the massive engineering works of type II or III civilizations.

British American physicist Freeman Dyson had been thinking along similar lines in 1960 when he published a paper in *Science* entitled "Search for Artificial Stellar Sources of Infrared Radiation." Dyson proposed that an advanced civilization will ultimately surround its star completely with solar collectors to catch all its energy. For building materials, they might take apart fallow planets or use a multitude of asteroids, rearranging their solar system to be more biofriendly. Such a "Dyson sphere" surrounding a star need not be a single solid structure; it could be a cloud of smaller orbiting objects that, together, create a thick spherical shell, capturing all the star's light in the service of life and civilization, and also providing plenty of surface area for living space and for whatever it is that superadvanced post-planetary civilizations do for fun.

Dyson pointed out that if such structures existed, they would be observable across space at great distances. The high-energy solar radiation would be absorbed by the sphere, which would radiate at cooler infrared wavelengths. So, to find Dyson spheres created by advanced civilizations, we could search the galaxy for stars emitting an unusual excess of infrared radiation. Such a search for advanced societies by looking for their observable engineering projects carried an important advantage: they did not have to be sending out signals in order for us to discover their presence. The Soviet astronomers at Byurakan were aware of Dyson's work. Kardashev suggested that Dyson spheres could be the observable architectural creations of type II civilizations. Science-fiction writers, picking up on this idea, have since generated many riffs on Dyson spheres and the varieties of Kardashev civilizations, helping us flesh out the possible paths of long-lived star cultures.[6]

What the L?

The most solid quantitative conclusion reached by the Soviets at Byurakan was the same as that reached by the Americans at Green Bank: that the distance between technical civilizations, and therefore the possible success of SETI, rests most crucially on the factor Drake called L, the average longevity. If L is small, because most technical civilizations last less than, say, one thousand years, then the equations show that communicating civilizations are few and far between, and the chances of SETI succeeding are nil. If, however, L is large, say, millions of years, then the galaxy should be full of chattering sentience, some quite near and easy to contact. Why? Imagine that technological intelligence existed on another planet for only, say, five hundred years. What chance would we have of ever observing it at random? That's less than 0.0001 percent of a planet's lifetime. Blink, and you've missed it. You'd have to stare at that world continuously and attentively for billions of years, or you'd never catch it in the act. It would be as if every cherry tree in

Washington, DC, instead of blooming for a week or two each spring, flowered only once, randomly, for a few hours, during its fifty-year lifetime. If you stood watch over enough trees for long enough, you might see this occur, but if you observed just a few thousand trees for only a few minutes each (the equivalent of our current SETI searches), you'd almost surely know nothing about these episodes. Civilizations must make their presence known for many millennia, or they will be effectively invisible.*[7]

Although they did not use the word, the SETI scientists at Byurakan spent much time discussing the duration of the Anthropocene age, or its equivalent, on other planets. As Shklovsky put it in his talk,

> Another problem of cardinal significance is the duration of the Psychozoic era (the age of intelligent life) on any given planet. Here we are on fairly uncertain ground.

He summarized some of the possible threats to survival of technical civilizations, including "self-destruction as the result of a thermonuclear holocaust or, in general, a discovery leading to unexpected and uncontrollable consequences," or "a crisis connected with the creation of artificial intelligence."

It is hard to discuss the question of longevity without projecting our fears or expectations for our own future. Wondering whether other geek civilizations could survive for long periods is an excellent way for us to think, from a slightly different perspective, about our own prospects. If technological civilization is something that cannot last long, or usually does not, then are we doomed? Conversely, if we cannot imagine our own civilization navigating through our existential threats and surviving for

---

* Another way to think of it: Imagine if the birth rate stayed the same but people lived for only one month on average. Earth would be much more sparsely populated. Think how little e-mail you would receive.

many more thousands of years, why should we think that others typically can? If L is tiny, less than ten thousand years, then our attempt to hear the brief, ephemeral pings of scattered technical life before they fade out is akin to a search for the black box of a missing airplane in a vast ocean when the battery life of the transmitter is limited. Will anyone hear *our* ping before we fade to silence?

In the early 1960s, while SETI was getting off the ground, the threat of nuclear annihilation was ever present. I was aware of it as a young kid growing up in those times. Bomb shelters and nuclear dread, along with the Space Race and the Beatles, were part of the landscape. My first political action, I am told, was being pushed in a stroller in a Ban the Bomb march in Boston when I was three years old. Because of my parents' concern over radioactive iodine from atmospheric tests, my brother, Danny, and I didn't drink fresh milk. Our breakfast cereal was served with powdered milk. With cruel irony, childhood leukemia took Danny's life a decade later.

During the tense peak of the Cuban Missile Crisis, my parents happened to be hosting a delegation of Soviet psychiatrists who were visiting Boston. Family lore has it that while they were making polite conversation in the living room, Danny and I were noisily roughhousing, crashing around in the bedroom above. When a particularly loud boom shook the ceiling, one of the visiting Russians remarked, "Sounds like an atome-ic bome!"

My recurring childhood nightmare was of hiding in shelters, and trying to find my family, while the bombs fell. In the daytime, hopes and dreams of space and an enlightened future beyond Earth provided an escape and a direction.

Even as the Cold War simmered, scientists across the Iron Curtain exchanged ideas about SETI. One international partnership, in particular, between Iosif Shklovsky and Carl Sagan, served as a catalyst and conduit for exchange between the separate but nearly parallel efforts. In 1962, for the fifth anniversary of *Sputnik*, Shklovsky published a magnificent SETI manifesto that drew penetrating conclusions about the evolution of life and civilizations on other worlds, and the prospects for using current and imagined science to find it. Its Russian title translates as *Universe, Life, Mind.* Upon learning of this work by the middle-aged Russian scientist, which contained ideas so similar to his own, the twenty-seven-year-old Carl Sagan, who had just begun his appointment at Harvard, wrote to Shklovsky and suggested a partnership. An English translation of Shklovsky's book, significantly expanded by Sagan, was published in 1966 as *Intelligent Life in the Universe*, with Sagan and Shklovsky listed as coauthors. Many years later, in his posthumously published autobiography, Shklovsky wrote that he had misunderstood Sagan's proposal, believing the offer was to have his book translated, with Sagan adding some introductory material. He was shocked when he saw the published work with both authors' names printed on the cover. He also felt that the deal provided him with a best seller without any of the large royalties usually associated with one. As he summed it up,

> With his American business sense, Sagan effectively used the "Soviet-American book" as the springboard to a dynamic pop-science career, the apotheosis of which was his thirteen-part TV series *Cosmos*. Now he's a very progressive millionaire, an active fighter against the threat of nuclear conflagration, and a scientist out on the rosily optimistic flank of the spectrum on the question of extraterrestrial civilizations. I have no grievance against this businesslike, cheerful, and congenial American: at my request he did all he could to help my brother when he fell sick in Paris.

The two authors had never met in person when the book came out, and (though it might be hard for wired young people to imagine today) their communication was restricted to mailed packages and letters. Somehow this sub-light-speed exchange across the space between their worlds allowed a misunderstanding to persist. Yet, however clouded its origins, the book became, and remains, the Bible of SETI, bursting with prescient, penetrating insights and fearless speculations from these two luminous minds working together across the gulf of the Iron Curtain.

Many of the most riveting ideas from this work, presented in a more casual, readable style, mixed with personal anecdotes and packaged with appealing and captivating artwork, formed the basis for *The Cosmic Connection*, Sagan's first genuine best seller, which launched his career into orbit.

Parts of the book read like a dialogue between the two astronomers, and although they saw eye to eye about the evolution of life, the universe, and almost everything, and shared a cosmic outlook on the human condition, each was a product of his own society, and there are places where the Cold War tension leaps off the page. Shklovsky declares that planets where capitalism exists will always be at great risk of self-destruction, and therefore advanced extraterrestrial civilizations will all be based on communism. Sagan responds:

No one today lives in a society which closely resembles Adam Smith capitalism or Karl Marx Communism. The political dichotomies of the twentieth century may seem to our remote descendants no more exhaustive of the range of possibilities for the entire future of mankind than do, for us, the alternatives of the European religious wars of the sixteenth and seventeenth centuries. As Shklovsky says, the forces of peace in the world are great. Mankind is not likely to destroy itself. There is too much left to do.

Their lively collaboration, embodied in their coauthored book, was the corpus callosum joining together the two hemispheres of SETI thought in a dangerously divided world.

Each was critical of the excesses and failings of the other's government, yet, steeped in awareness of possible self-destructing civilizations, each advocated for the abolition of nuclear arms. Shklovsky was especially scornful of the scientists who continued to assist the nuclear complex, referring to Edward Teller and his Soviet counterparts as "the cannibals." In his memoir he spoke of being moved by a visit with Philip Morrison:

> I know personally one American scientist who displayed real heroism and civic courage in his relations with the cannibals. He is Phil Morrison, who is now one of America's leading theoretical astrophysicists. Seriously ill, to all intents and purposes a cripple, even back in the far-off 1940s he realized that a scientist's probity and honor are incompatible with the service of Beelzebub...Sitting with him at a table in a small Mexican restaurant in the old section of Albuquerque, some hundred miles from Los Alamos, I looked into his deep blue, childlike, clear eyes—the eyes of a man with a crystal-pure conscience—and my heart was uplifted.

Sagan's later work on nuclear winter, first published in 1983, fifteen months before Shklovsky's death at age sixty-eight, is thought by many to have contributed to the end of the Cold War by rendering undeniable the futility of global nuclear conflict.

Although they acknowledged other possibilities, Shklovsky

and Sagan agreed that advanced civilizations were most likely to be peaceful, reasoning that any communicating civilizations out there must have been around much longer than we. Therefore, they would surely have learned to live with technology that, if accompanied by aggressive or warlike behavior, would have been incompatible with survival. They would have solved the existential problem humanity was facing, for the first time, with the Cold War. The continuing tension of that conflict, set against the optimism of SETI, formed a poignant backdrop for the first international SETI conference.

After the success of their joint book, Sagan and Shklovsky decided it was time to bring their two communities together for an international meeting. The result was the Soviet-American Conference on Communication with Extraterrestrial Intelligence, jointly sponsored by the National Academies of Science of the USSR and the United States, and held, at Byurakan, in the same location as the first Soviet meeting (and thus sometimes referred to as Byurakan II).

Sagan enlisted Drake and Morrison to help round up the delegation of American (and British and Canadian) scholars, which included a who's who of late twentieth-century science, including visionary physicist Freeman Dyson, Nobel laureate Francis Crick (codiscoverer of the DNA double helix), and artificial intelligence pioneer Marvin Minsky. They also invited noted historian William McNeill[8] and others representing sociology, psychology, linguistics, and anthropology.

The Soviet participants were organized by Shklovsky and Kardashev. Kardashev wanted to include only physicists and astronomers, as he felt that scholars from the humanities were merely "windbags," but Shklovsky insisted that the Soviet delegation include linguists and philosophers. The proceedings, edited by Carl Sagan, and illustrated with candid black-and-white portraits shot by Phylis Morrison during the conference, include transcripts of the sometimes contentious discussions after the talks.

Sagan began the meeting by writing the Drake equation out on a blackboard, and again it served as an agenda to organize the week of discussions, the topics beginning with the astrophysical and ending with the sociological. He presented estimates suggesting that humans would need to examine about one million stars to have a decent chance of finding one alien civilization, and "thus the probability of success of all efforts to date is $< 10^{-4}$"—which was Carl's nerdy way of opining that they would need to multiply previous efforts by more than 10,000 before success could be considered likely.

There was much discussion of longevity and whether we could really say anything about the properties of long-lived civilizations. Historian William McNeill acted as an outsider to the SETI community and questioned many of the assumptions made by the astronomers. He pointed out the tragic misunderstandings that have occurred when human civilizations (which presumably have a lot more in common than we would with aliens) have encountered each other. Morrison countered that, given the distances and times involved, our encounters with alien civilizations would be inherently indirect, less like Cortés and the Incas and more like the way contemporary societies have learned from the culture of Ancient Rome. We could exchange information without fear of exploitation.

Another topic was the possibility of establishing a common language. Here again McNeill played the skeptic. He did not share the scientists' confidence that they would be able to communicate with ET. The SETI scientists shared a strong conviction that whatever immense dissimilarities we may have with our alien cousins, we will understand the same math and physics, and could use both to develop the basis of a common language.

Marvin Minsky, reflecting his faith in the convergence of mathematical minds, and giving a nod to the early '70s "generation gap," offered that "It is probably easier to communicate with a Jovian scientist than with an American teenager." McNeill was dubious, feeling that although our math and physics may seem universal to scientists, they may well be social constructs, peculiar to our own society, and therefore completely incomprehensible to aliens. The scientists answered that he simply did not understand mathematics and physics well enough to grasp why they must be universal. Near the end of the meeting, McNeill stated,

> I must say that in listening to the discussion these last days, I feel I detect what might be called a pseudo or scientific religion. I do not mean this as a condemnatory phrase. Faith and hope and trust have been very important factors in human life and it is not wrong to cling to these and pursue such faith. But I remain, I fear, an agnostic, not only in traditional religion but also in this new one.

Some of the scientists could take only so much of these challenges. They weren't really there to talk with "windbags" about whether SETI made sense. They wanted to talk about how to go about doing it. At one point, impatient with the digressions into such obscure questions, Freeman Dyson got up and declared, "To hell with philosophy. I came here to learn about observations and instruments and I hope we shall soon begin to discuss these concrete questions."

The Inevitable Expansion Fallacy

There is a consensus narrative, based on an extrapolated interpretation of human history, that ascribes properties of aggressive growth and relentless territorial expansion to "superior" alien cultures. At Byurakan II, the discussions on the nature of advanced civilizations were centered on Kardashev's three types. This classification system had been given a big boost when Shklovsky and Sagan adopted it in their best-selling classic book. Sagan had decided, based on energy use, that human civilization (in the 1970s) was currently at 0.7—not quite a type I, where we would have mastery and control over our entire planet. What a difference that 0.3 makes: such a small quantity, separating our current conflicted, confused state from our perceived destiny as wise, confident planetary masters. Represented like this, it seems such a minor deficit. Only 0.3? We're almost there. Sigh.

The basic premise (that civilizations would proceed inevitably along the path to greater population and energy use until they had godlike command over entire galaxies) appears to have been widely accepted by the scientists gathered at Byurakan II. It seems strange, from the vantage point of the twenty-first century, that nobody in this gathering of insightful scholars questioned the assumption that such endless expansion would be the obvious path of advanced societies. Rather, several speakers mentioned the existential danger that would befall any civilization that stopped expanding. Perhaps it reflects the zeitgeist of the time, smack dab in the middle of the Great Acceleration, when the notion of human progress was still connected, in such an untroubled way, with the idea of endless expansion. I suppose it fit well with both the Soviet belief in inevitable historical progress and the capitalist ideal of endless growth.

All these brilliant minds saw the problem of survival through a lens shaped by the anxieties and hopes of their time. Early discussions of L always mentioned the likelihood that civilizations would invent nuclear weapons around the time they discovered radio technology. If this were the case, then inevitable nuclear holocaust might ensure that L was only decades long. SETI pioneers such as Drake, Morrison, Kardashev, Shklovsky, and Sagan imagined that if L was short, it was because most civilizations might "blow themselves up" in a nuclear holocaust. Between Green Bank and Byurakan, the Cuban Missile Crisis flared up, and annihilation seemed like a real possibility.[9]

Humanity's biggest challenge was seen as avoiding nuclear war and achieving a peaceful society, which could then put its resources into progress and development for the betterment of all mankind. Progress was defined to a large extent in terms of reworking the landscape with huge engineering projects. Both Freeman Dyson and William McNeill, in their personal accounts of Byurakan II, noted what wonderful progress the Soviets were making in diverting water from the huge lake near the Byurakan Observatory and irrigating the previously fallow landscape, allowing the small town of Byurakan to grow into a thriving city. Of the dozen largest dams in North America today, more than half of them were built within six years of this gathering. The Great Acceleration was in full force. Apollo 8 had just given us our first look in the mirror, and the first Earth Day was held a year and a half before Byurakan II, but the global environmental movement was still only nascent.

Today, this idea about progress, that there must be a universal pattern of unending population and energy growth, is still ingrained in contemporary discussions of SETI. The Kardashev narrative, classifying civilizations into types I, II, and III, based on an assumption that the longer a civilization survives, the larger and more power-hungry it will become, is everywhere in the SETI literature.

Yet, these days, this ubiquitous assumption—let's call it the "inevitable expansion fallacy"—seems more dubious than it must have in the 1960s, now that it is becoming clear that by defining our own progress simply through continued exponential growth in population and energy use, we may be limiting the longevity of our civilization.

Today nuclear weapons still threaten our survival, much more than we generally acknowledge. Yet the existential threats most on our mind are related to the runaway exploitation of resources, the destruction of key natural life-support systems, and the unintended consequences of mindless, unplanned development. Given current anxieties, some present-day discussions about L and the galactic prevalence of intelligent civilizations are beginning to focus more on our Anthropocene existential threats of climate change or resource exhaustion and the challenges of sustainability. What connects these concerns is the overarching question "How can an advanced technological species develop a long-term, stable relationship with world-changing technology?"

Clearly our own proto-intelligent civilization is confronting the limits and dangers associated with an ethic of unquestioned growth and reflexive implementation of powerful technology for its own sake. So we could look a little differently at the assumed qualities of "advanced" civilizations and question whether achieving true planetary intelligence, that capable of creating a civilization built to last, might require a different set of guiding values. It may be that the Kardashev scale and its variants amount to an assumption that intelligent civilizations must act in a way that is, in fact, not very intelligent.

The idea that relentless expansion will be a universal drive is often supported with the argument that any technological species must be the product of a Darwinian evolutionary process, and therefore all will have the commandment to be fruitful and multiply bred into their bones or exoskeletons. Endless expansion is a genetic obligation deeply embedded in organic life. It served us well throughout four billion years of biological evolution, when increasing numbers could help ensure species survival. Since we are still struggling to free ourselves from this animal imperative, we tend to project it onto our future selves and other sentient races. We picture aliens as technologically powerful but still mindlessly driven to expand their population and their control over resources at all costs.

Yet this imperative is clearly becoming a threat to our survival, and it seems quite likely that advanced civilizations will have discarded it. This period of multiplying like rabbits cannot last long. It is not just a good idea to stop this behavior. It's physically impossible to continue it for very long. We'd soon run out of planet, and even the prospect of interplanetary or even interstellar expansion does not help. At best, it slightly delays the consequences. If we maintained a population growth rate of 2 percent per year, in less than a thousand years we would experience devastating overcrowding, no matter what. It is easy to show that even if we learned to expand off the planet and increase our domain *at the speed of light* (a pretty safe theoretical upper limit), and if we managed to colonize all available stars and planets within a sphere expanding at this impossible speed, then we would still run out of planets and perish in our own waste within a few thousand years. There is no alternative to limiting population growth and resource use in the not-very-long run.

Because of this, it is reasonable to suppose that truly successful, long-lived species have all discarded the expansion imperative, and replaced it with an ethic of sustainability, of valuing longevity over expansion. If technological intelligence has a true and lasting form, one of its basic properties must be that it moves beyond the exponential expansion phase (characteristic of simple life in a petri dish or on a finite planet) before it hits the top of the S-curve and crashes. For us, achieving this kind of planetary intelligence will require critically examining our inherited biological habits and shedding those that have become liabilities. Planetary intelligence would mean thoughtful control over one's self, escape from the mindless drives to multiply, to expand, to lay waste, kill, and drown in your own waste. Perhaps this is why we will not find what Shklovsky called "miracles," the highly visible works of vastly expanded super-advanced civilizations. Because advanced intelligences are not stupid.

This realization could also influence our SETI strategies. It is an often unquestioned assumption in SETI theory that civilizations become more obviously visible the longer they exist. However, I wonder if, after a short period of proto-intelligent flamboyance, the opposite may be true. Yes, our presence on this planet has certainly become more and more obvious the more "advanced" we have become. If we define advancement by the power of our technology to rework our environment, which we often do, then this is a tautology. Yet what if it's in the nature of advanced intelligence to undergo a transition toward a less obvious imprint? If every planetary civilization encounters a crisis comparable to our Anthropocene dilemma, then the logical outcome may be that the truly intelligent species become harder and harder to observe. Awakening to the reality of planetary existence may involve becoming more aware of, and thoughtful with, one's long-term patterns of development. Older and wiser civilizations, those who've learned how to use technology in the service of long-term survival, may be less wasteful and therefore less clearly visible.

In 2015, a group of Penn State University astronomers published the results of an "Infrared Search for Extraterrestrial Civilizations with Large Energy Supplies." In a very thorough search, no civilizations were found. As they summarize their results: "We show, for the first time, that Kardashev Type III civilizations (as Kardashev originally defined them) are very rare in the local universe." The authors' assumptions about advanced

civilization include the following: "Detectably large energy supplies can plausibly be expected to exist because life has potential for exponential growth until checked by resource or other
limitations, and intelligence implies the ability to overcome
such limitations."

Does it? I wonder. Perhaps true, lasting technological intelligence may imply nearly the opposite: the ability to overcome the
biological need for exponential growth.

Natural selection weeds out those who are unfit to survive.
This will be true on a planetary, civilizational scale as well. What
if an essential part of becoming a very wise species, equipped
for survival with powerful technology, is to realize and internal-
ize the advantage of living more in accordance with the natural
systems within which your existence is embedded? What if one
characteristic of really advanced intelligence is to become less
and less distinguishable from natural phenomena? That would
certainly explain why we have not seen the predicted "miracles"
created by type II and III civilizations.

The Continuity Criterion

It is said that Mahatma Gandhi, when asked to comment on
Western civilization, remarked, "I think it would be a good
idea."[10] That's how I feel about intelligent life on Earth, especially when I wonder about truly intelligent life.

Intelligence, like life, is hard to define. To search for it
elsewhere we need a working definition. Among the radio
astronomers of SETI, it's often tacitly assumed that the hallmark of intelligent life on any planet is the ability to do radio
astronomy. This "radio definition" is often offered with tongue
firmly planted in cheek, acknowledging the self-serving irony.
Yet we also use it pragmatically, circumventing tricky questions

of comparative evolution and human uniqueness by choosing to search for something concrete and recognizable.

Certainly we can imagine intelligent technological life that does not use radio. Yet what of the opposite: could there be nonintelligent life that *has* developed radio? Well, look at the one planet we know of where radio telescopes have been built. Does Earth host intelligent life? Are we intelligent in a way that would be recognized as such by other sentient technological creatures? We could quibble for centuries over what this means, but at a bare minimum, consider this potential criterion: It's not intelligence if it cannot solve the puzzle of how to survive on a planet. Legitimate technological intelligence as a significant planetary phenomenon must be able to sustain itself for some nonnegligible length of time. I say this for two reasons:

First, what is the use of all this cleverness and ability to control one's environment if self-preservation is not possible, or self-destruction is inevitable? If "intelligent life" is stupid enough to ensure its own rapid destruction, then perhaps it should be called something else. We humans are obviously brilliant at certain kinds of problem solving, but arguably if we cannot surmount the existential hurdles we place in our own path, then on the galactic stage, whether or not anyone else is watching, we could not be considered truly intelligent.

Second, as with the "radio definition" of intelligence, there is a pragmatic angle. Intelligent life[11] that doesn't last long would be extremely hard to detect, even if it existed on a large number of planets. To put this in the mathematical language of the Drake equation, the average distance between civilizations will be too great to make contact feasible, unless L is much greater than the present age of our own civilization. There is only a decent chance of finding a message over a reasonable span of time (say, within another century) if the galaxy is fairly well sprinkled with broadcasting civilizations having average life spans *much* longer than centuries. Real galactic visibility requires serious longevity.

This means that even if nearly every planet out there produced a technological civilization very similar to ours, there could still be nothing much on the galactic radio.* In order for our current searches to have a reasonable chance of success, someone must be maintaining a signal or beacon for thousands of years, something we certainly have not done. These broadcasting civilizations would be quite different from us. We don't have that kind of commitment. Will we ever? Will our civilization even be around in another five thousand years? By this minimum standard—let's call it the continuity criterion—intelligent life has arguably not yet evolved on Earth, and radio telescopes by themselves are no definite sign of intelligence.

Maybe, however, they are a step in the right direction. If aliens can build radios, surely they'll be capable of simple algebra and geometry. And probability arguments. They, too, will realize that unless someone else is broadcasting continuously for thousands of years, at a minimum, there's little point in listening. They'll understand that the problem of SETI is the same as the problem of sustainability.

To think about finding someone else out there is to look at our current technological civilization and ask: Is there a stable state achievable with this tool kit, with these qualities and potentials? Is there a developmental path leading from what we are now to one of these long-lived broadcasting civilizations? There may well be, but our Anthropocene predicament illustrates why it is not a given that such a path is easily navigable.

Could a cave painter in France eighteen thousand years ago

---

* We are not currently broadcasting signals for other listening species to pick up. Sure, there's an expanding sphere of our radio plays, game shows, sitcoms, and automobile commercials. Yet our century of leakage is quite weak, and dissipating rapidly as it expands, fading into the background hiss and crackle of the galaxy.

imagine a jet plane or a laptop computer? Can you and I imagine the technology of the year AD 16,000, or even the year 2500? My grandparents were born before automobiles, and three of them lived into the space age. You and I have never known a time when the shifting of the built environment was not keeping us on our toes. What could we possibly say about the qualities, abilities, or motives of alien civilizations that are hundreds of thousands of years more aged than we? SETI demands that we consider planetary civilizations that long ago passed through a stage of technological transition analogous to what we're experiencing now.

"Where are they?" is an ancient question, and one SETI scientists have been studying theoretically and experimentally for more than half a century. Faced with the challenge of surviving, and thriving, in the Anthropocene, we can turn the question around and ask, "Where will we be in twenty thousand years?" The answer may depend on three other questions of great relevance to both SETI and our current situation.

The first is: does the development of a technological civilization capable of SETI always precipitate global environmental crisis? On Earth at least, both were set in motion by the same explosion of technical prowess and scientific understanding. In chapter 3, I argue that the first wave of technological success to sweep a planet will always cause global "changes of the third kind," which will bring about major ecological disruption.

The second question is: can a civilization in such a crisis transition to one that employs technology well, in ways that facilitate long-term survival? At least sometimes it may arrive at a more stable state dominated by the fourth kind of planetary change, where knowledge of planetary function is integrated into technological deployment. Only a planet that has reached such a stage would have the continuity to host a long-lived communication effort that could produce decent visibility on the galactic scale.

And the third question is: what does that take? Just as astrobiologists try to discern universal principles that may shape biological evolution, SETI theorists have, for the last sixty years, been poking at this question: could there be common laws that shape cultural evolution on different worlds? We infer that we are young and imagine that any civilization we are likely to encounter might be some kind of much older version of us. If we believe in the possible existence of these long-lived societies, does that mean we could become one? What is required of us? What are the impediments? How does that vision of long-lived civilizations map into our Anthropocene dilemma and how we respond to the threat of us?

No Transition, No Transmission

When we imagine alien technological intelligences, we start with some variation of what we have here. Try as we might, it's hard to avoid taking human civilization as some kind of standard or template. Most discussions of extraterrestrial intelligence seem to accept the implicit assumption that, with the age of space travel and radio astronomy, we ourselves have now crossed some threshold and become an "intelligent civilization." Having achieved this state, we can begin to search for other intelligent civilizations. A *Scientific American* blog in February 2015 asked: "Is human-like, technological intelligence likely to be common across the universe?" This is the way the question is usually phrased. We discuss our own longevity, putting it so far at less than one hundred years, and we speculate that other technological civilizations must have achieved longevities of many thousands of years. So we tend to picture those more aged transmitting societies as a continuation, an elaboration, of what we are now. We think we are already one of them, just very young. We're charter members of the Galactic Club, intelligent, radio capable, and ready to listen, and it's just a matter of time before we graduate to becoming broadcasters. We take our own existence as proof that evolution can produce intelligence of the kind that can participate in interstellar conversation.

Yet there is inherent asymmetry in galactic radio discourse. It is much easier to listen than to transmit. A huge gulf yawns between the ability to build a radio telescope and the ability to mount a sustained multimillennial broadcasting and listening program. We cannot reasonably search for our equals. A serious SETI broadcast, the kind that could realistically be detected by young ones like us, would take a continuity and global unity of effort that is currently beyond us. We clearly have some, but not all, of the tools required for cosmic visibility. We can build fancy machines but are not quite capable of using them in any coherent, planetary-scale, long-term way. To succeed, SETI requires that more mature intelligent species will have developed not only machines that are more advanced than ours, but civilizations entirely different from what we have so far constructed. When you think it through, and you do the math, you realize that with SETI, we are not really searching for others of our own kind, but for evidence of something we hope to become. There is an aspirational aspect to the search.

Maybe the critical threshold, to the kind of intelligence that matters on a cosmic scale, is one that we haven't crossed yet. In that case, what we seek is not a slightly more advanced version of ourselves, but something qualitatively different. To get there may require more of a transformation than simply a continuation. Many rivers to cross.

As we ponder the distinction, between a proto-intelligent

society like ours that may not be ready or able to broadcast consistently or continuously, and those "wise ones" we are listening for, we might again consider a definition of intelligence that doesn't assume we are at the pinnacle of evolution. What will those long-lived civilizations be like? They would have planetary intelligence: the ability to act coherently and intentionally, on a global scale, on projects that persist for many millennia.

This maps into the four kinds of planetary change I discuss in chapter 3. If you have built radio telescopes, you are employing clever technology and are clearly capable of planetary changes of the third kind. Yet to enact a sustained program of messaging for thousands of years on behalf of your planet's inhabitants? That requires something else. This purposeful, long-form coherent activity would mean you are capable of planetary changes of the fourth kind. It would be a sign of true planetary intelligence. So any messages we receive will be from a civilization that has experienced the transition. No transition, no transmission.

As I discuss in chapter 5, we seem to be entering a bottleneck through which our increasingly powerful technology will either fairly soon (within the next couple of centuries) end the tenure of our civilization or ensure that, in some form, it survives for a long time—perhaps for much, much longer than the roughly ten thousand years since we started settling in villages. Our habits, honed for survival in the Holocene, are mismatched with the new threats we face in the Anthropocene. Perhaps civilizations elsewhere in the universe must all reach a point, similar to where we find ourselves now, when they are forced to confront the global implications of their actions. Is this an inevitable juncture for a young technical civilization? The specifics of the path will vary from planet to planet, but some waypoints could be the same. The universal nature of life (at least of nonintelligent life) is to expand, to exploit available resources, and to alter its environment. On planets where life becomes complex, evolution of intelligence may sometimes allow for the discovery and development of technology, which will accelerate these biological patterns, up to a point of excess and ecological crisis. If so, then early technological development will always cause unintended consequences and "global changes of the third kind," and these will always force a reckoning and a choice: to face catastrophe or to develop a new kind of intelligence capable of exerting the planetary self-control needed to enact global changes of the fourth kind: what I am calling planetary intelligence.

Those who make it through the technical bottleneck will have emerged as a different kind of entity. They will have learned to apply global technology in concert with the functioning of their world, augmenting it in the service of life. They will have unimaginably powerful tools and understanding at their disposal, so natural disasters will have become avoidable. Once intelligent life gets to that point, it will have become a stable, integral part of the functioning of its home planet.

In chapter 5, I discuss the possibility that one outcome of the Anthropocene could be a transition to a new eon, the Sapiezoic, where a long-lived, sustainable technological civilization becomes a component of a planet for the rest of its existence. If we get through this bottleneck, then we will be fine for at least several billion years, until the Sun swallows Earth (and that gives us plenty of time to figure out what to do then, possibly going interstellar). If such a bottleneck is a universal challenge to young technical civilizations, this has interesting consequences for the meaning of L, the average longevity of civilizations. It implies there is a *bifurcation* of lifetimes. This means that the probability distribution of lifetimes is not a smooth, continuous function—like the probable lifetime of a human being, which peaks at average life expectancy and tails off at higher and lower ages—but, rather, a choice with two possible end states. Many, perhaps most, civilizations may be short-lived ones that don't make it through, but those few (perhaps only an infinitesimal fraction) that do may be very long-lived indeed.

How long lived? Possibly immortal.

I call them quasi-immortal because we don't know if the universe is going to last forever.* They would be "immortal for a while," or until this universe ends. Even if most civilizations at our present stage are indeed doomed to self-destruct, this would not be at all incompatible with some small fraction making the transition to quasi-immortality.

This possibility was raised by Iosif Shklovsky back at Byurakan I, in 1964:

There is however, a possibility that some civilizations, having reached a highly advanced level, will find themselves past the inevitable crises and internal contradictions which plague the younger civilizations. The evolutionary time scale of these quiescent civilizations may be considerably longer, approaching the cosmogonic scale.

Shklovsky remarked on how incredibly strange it would be if the immortals did not exist:

I would like to stress that for me there could be no greater wonder than a conclusive proof that no "cosmic wonders" exist. Only an astronomer can fully grasp the true meaning of the fact that among the 1,000,000,000,000,000,000,000 stars comprising the visible part of the universe not a single one houses a sufficiently advanced civilization.

---

* Some cosmologists will tell you that we know it's not going to last forever. I think it would be more accurate to say "according to current cosmological ideas..."

Frank Drake has long maintained that it is the immortals whom we are most likely to hear from with SETI. These immortals would be another kind of entity altogether, much more different from us than you are from a mite on your skin. It seems likely that, following Arthur C. Clarke's famous dictum that "Any sufficiently advanced technology is indistinguishable from magic," they would appear to us more like gods than highly capable engineers. Although it sounds like a crazy, techno-futurist fantasy, somewhat religious in its idealism, the more you think about the timescales available to evolution in the universe, and the accelerating pace of technology, the more you realize that the result of such a process must be beyond our ability to imagine. If you allow yourself to sit with the numbers, the idea of the immortals starts to make sense.

Technosignatures

We are in the very beginning stages of an era when we will be able to observe the atmospheres, and other qualities, of a large number of exoplanets. If some planets can reach a stage with stable world-changing intelligence, we should think about what that would look like. What if the current troubles of our civilization are really the adolescent growth pains beginning the transformation to true planetary intelligence? Earth's Sapiezoic Eon may conceivably last for a significant portion of the planet's existence. Some portion of the exoplanets we observe in the future will be considerably older than our solar system, and some may have biospheres where complex life arose more quickly than on Earth. So, we might discover planets where such a Sapiezoic transition occurred long ago. If there are any "*Exo sapiens*" out there, how would we spot one?

This gives us another way to think about the possible

qualities of Earth in the Sapiezoic Eon. It also suggests a new approach to SETI, to complement our existing radio and optical searches. The "radio intelligence" search mode has been justified because it is pragmatic—it gives us something specific and well-defined to look for. Yet it may be that, with the discovery and investigation of exoplanets, we have opened up a new way to search for our cosmic company, one that does not require that aliens have any interest in contacting us. As we start to scrutinize exoplanets we should keep one eye, or one spectrometer channel, open for signs of worlds that do not seem "natural," whether or not they are blasting galactic public radio in our direction. Obviously when we find planets that seem promising for life, we should check for radio waves or laser pulses. Yet we could also end up discovering intelligent life simply by finding the atmosphere of a managed planet. In astrobiology we talk a lot about how to identify "biosignatures," the signs of life on another world that we could detect from afar. I suggest we also stay alert for technosignatures, or "noösignatures," that is, signs of a noösphere, a sphere of intelligent influence. When we define biosignatures, we are forced to question which qualities of terrestrial life might be universal. Likewise, it is worth considering which observable aspects of technological civilization might be widespread, and how these might be discernable from "mere" biosignatures.

In the 1960s and '70s, the SETI pioneers talked about looking for astrophysical "miracles," the great works of astro-engineering built by million-year-old civilizations who had progressed far beyond being simple planet shapers. Yet when SETI got started, even the idea of finding planets around other stars, as we have now done, was just a scientific fantasy. The notion of actually discovering what such planets were like must have seemed even farther off, almost akin to warp drive or teleportation—something that could almost conceivably be achieved by the science of a different time or civilization, but not by us. They didn't know that fifty years later we would be starting to observe the atmospheres of exoplanets. And they also didn't understand, in the visceral way we do now, the extent to which industrial civilizations alter atmospheres and climates. Now, more than half a century after Green Bank and Byurakan, this notion of world-changing civilizations feels a lot closer to home. We have become acutely aware that we are inadvertently altering a world, and we have no choice but to learn to engineer it to survive. The first task of this engineering project will be to stop our restless emissions and ecological wreckage. Yet it doesn't stop there.

Arguably if our civilization is to survive, or give rise to one that will survive with technology, Earth will have to stay managed. It's strange to say, but I think true, that from now on Earth will to some degree be a terraformed world. We are at a branching point, and it's interesting to consider what the distant signature of a sustainably regulated planetary climate might be.

An evolving biosphere is complex, and defies prediction. Yet, as I have argued here, there are stages that may well be convergent, and therefore universal. Inhabitants of other worlds will probably have eyes and use light to see, so perhaps at some point they will light up their nights. Their atmospheres and surfaces will undergo rapid changes as they develop science and technology and "master" their worlds. Technological life, it has been proposed, will use energy in increasing amounts. Up to some point, that should be true, but what will happen when unrestrained technical cleverness encounters planetary limits? Whatever path it takes, successful life, at its core, is homeostatic, self-stabilizing. So intelligent aliens will find ways to transcend the environmental limitations and vicissitudes of an unaltered planet. They will engineer their worlds.

The obvious first thought is "Let's look for industrial air

pollution, for global warming, for marked increases in CO_2." Yet it is extremely unlikely that we would see this. As with radio SETI, the math argues that it is nearly impossible to find anyone at our stage. Yes, fossil fuel deposits on planets with photosynthesis do seem quite plausible. But, even if we surmise that other civilizations will arise somewhat in parallel with ours and encounter the same problems, an age of rampant pollution is inherently self-limiting and has got to be brief. Any planetary civilization we have a chance to find would have long moved on, and would be altering their world in ways that are compatible with long-term coexistence.

Planetary changes of the third kind would be the hardest to find elsewhere because they can't last. A planet experiencing such a phase must be very rare, like an animal going through some quick molting or metamorphosis, and you'd have to be very lucky to catch one. This inadvertent stage will either be terminal, or a gateway to a long-lived phase where sentience becomes part of the way a planet functions.

How might we identify such *Exo sapiens*? We can imagine how we ourselves would terraform worlds, or mitigate natural hazards on a future Earth. For example, as I've discussed, in the future we will likely choose to dampen down the Milankovič cycle climate oscillations. If our descendants are still here and want to maintain their civilization, let alone a thriving biosphere, they will intervene to stop future ice ages and, eventually, prevent the Venus-style runaway greenhouse that will despoil an Earth left to its own devices. It's hard to imagine such a significant global engineering effort that wouldn't in some way be detectable at interstellar distances. If we find an exoplanet with a strange climate that is being controlled by unexpected atmospheric compounds such as chlorofluorocarbons, that should get our attention. Or if we find a world with a suspiciously unusual pattern of albedo (reflectivity) or day/night pattern of brightness, we might suspect planetary engineering with mirrors or surface alteration. We should take notice if such a world seems to be in a climate state that preserves or extends an earlier evolutionary stage, stabilizing against the aging of its star. These are just fun guessing games, though. We can't really anticipate the choices of superadvanced intelligent aliens. Yet we should stay alert for "unnatural" planetary states, and be prepared for surprises as we start to observe the properties of exoplanets.

If I am right that the kinds of planetary change I've outlined have any universal applicability, then there should be three kinds of exoplanets: dead, living, and sapient. Most of them will probably be dead. I think some will be living. These will, I predict, be fairly obvious when we find them. Certainly our own biosphere would be. From afar, the most noticeable feature of our planet is still that ridiculous surplus of oxygen in the air. So arguably, up to this point, the ancient cyanobacteria have been more effective than humans at betraying our presence to aliens. They, at least, have sent an existence proof, a calling card of life, making our planet noticeable and interesting to distant astrobiologists. If we are lucky, some living worlds may be nearby enough to find with the telescopes and spectrometers we'll build in the coming decades. Sooner or later we might also find one that is sapient.

Cosmic Optimism

I don't know if we can become one of these quasi-immortal civilizations, but I like the idea that they exist. Part of the point of SETI has always been a search for answers about our own cosmic potential and destiny. If *they* are out there, it means that there may be hope for us. It means there is a solution to this puzzle of forging a healthy, long-term relationship between a planet and a technological civilization.

True intelligence is not an easy gang to join, but once you're in, you're in for life—the life of the universe. If some civilizations, even a very tiny fraction, make it to this quasi-immortal state, then the number of sentient civilizations should be growing over time, and the universe must be increasingly permeated with intelligence. This is not an opinion, it's a calculation, and it leads me to a kind of "cosmic optimism." It is possible to be highly optimistic about the cosmic prospects for technological intelligence, while hopeful but uncertain about the human future. Even if our own civilization is somehow doomed, there may be a bright future for life and consciousness. Perhaps there is, as Franz Kafka put it, "plenty of hope, an infinite amount of hope—but not for us." I really don't know if our long-term prospects are great, but I think the prospects for intelligence in the universe are promising. There is plenty of hope for the future, but whether we get to be a part of that is up to us.[12]

There is another kind of comfort that comes with belief in a universe widely inhabited by mind. It absolves us of the weighty responsibility of possibly being the only cosmically aware life-forms in the universe. Novelist Doris Lessing began her masterful space-fiction series, *Canopus in Argos: Archives*, with this dedication:

> *For my father, who used to sit, hour after hour, night after night, outside our house in Africa, watching the stars. "Well," he would say, "if we blow ourselves up, there's plenty more where we came from!"*

If we really are alone and we somehow blow it here on Earth, wiping ourselves out through war or overpopulation, then we are not only behaving self-destructively but robbing the universe of an ability to ponder and explore itself. A widely inhabited universe lets us off the hook.

I also find it comforting that there are likely beings out there who have evolved powers of reason far beyond ours. It is exciting, and strangely encouraging, to realize that this universe is surely capable of producing "minds that are to our minds as ours are to those of the beasts that perish."[13] They may have found the answers to some of our most difficult and persistent questions, such as the basis of morality, the nature of consciousness, the mystery of free will, and how these all relate to the laws of physics. Our deepest philosophical and spiritual questions, such as "Why?" and "How ought we to live?" seem to be intractable. We are smart enough to conceive of these questions, but we cannot answer them. This may reflect that there are no definitive answers. Maybe that is the nature of reality, and we just need to live with it. Or this may have to do with the dim level of awareness and marginal intelligence we have achieved. Maybe we are just not too bright compared with true intelligence.

In the decades since Project Ozma, our searches have become orders of magnitude more powerful, and where Frank Drake once turned his radio dial by hand, our receivers now automatically sift through millions of radio channels simultaneously. Large swaths of sky have been scanned, at least for short periods, and more than a thousand individual stars, deemed promising prospects, have been targeted. The quest has been extended beyond the radio frequencies, now including optical searches for alien messages that might come in the form of laser pulses. So far, nothing has broken the frightening silence of those infinite spaces. Oh, there have been a few briefly exciting false alarms, but in all this time not a blip, squeak, or peep has been detected that can be repeatedly found by independent observers, clearly came from beyond Earth, and has properties that reveal an intelligent design. Yet, if you consider all the stars in our area of the galaxy, all the possible frequencies that could be scanned, all the possible times you could have had a telescope trained on them over the last half century, you realize that we have still barely scratched the surface.

Immersion in SETI takes the edge off the present, makes me worry a little less about the here and now. Is this escapism or an enlarged realism? The mind is diverted away from quotidian worries and toward searches and solutions that will unfold over the coming decades and evolutionary scenarios that play out over untold eons. When pondering life's efforts to connect with distant life, you can't help but gain some increased identity with Earth's four-billion-year-old biosphere, a resilient beast that is going through some awkward transition phase now but is not credibly threatened with extinction. And you can't help but wonder what, if we survive the coming century, we might become.

SETI requires of us patience and sustained attention. The chances of payoff in any given year may be low, but the potential return on investment is incalculably large. Such a project is a challenge for individuals and societies. We respond to urgency and novelty. We focus for a while and then our minds wander off, our attention responding to the next distraction or call to action. In attempting to build a sustained observation program, the founders of SETI are trying to establish an experiment that will run for as long as it takes, over multiple generations if necessary. In so doing, they are calling upon faculties we need to develop in order to begin the transition to one of those long-lived civilizations. In order to do SETI right, we need to start to become that which we seek, engaging ourselves in activities and goals that transcend individual lives. Still, it's easy to be a little impatient, to want the discovery to come on your own watch.

It's hard to be dispassionate about the prospect of making

contact with, or even becoming aware of, life elsewhere because, as the poet Diane Ackerman writes, "I am life, and life loves life." Who wouldn't want to see their faces and shake their… whatever, or even just enjoy their greetings from afar and bask in their distant radio company? I don't expect actual contact (i.e., face-to-face, or face-to-that-thing-they-look-out-of) to be easy or necessarily good. But a message seems harmless at worst, and it would be so nice to know we're not alone, to have our faith shored up by revealed knowledge. Even if contact ends up being dangerous, scary, or deeply disturbing, we all just want to know, don't we?

SETI is a hopeful enterprise. We are on some level aware of our massive double standard: that what we seek is something that has never existed on Earth, and is almost unimaginably different and more highly evolved than our own young, fragile, transitional civilization. In a sense, this search is an act of faith that such a historical outcome is possible. Scientists aren't supposed to believe in angels or miracles, but we can believe in the possibility of evolved ETs with godlike capabilities and unknown power to transform our species beyond our current troubled state.

As a grad student in 1984, I attended my first SETI conference, the Search for Extraterrestrial Life: Recent Developments, at Boston University. Philip Morrison gave a riveting retrospective on the first twenty-five years of the field that started in response to his 1959 paper. Drake and Sagan both gave inspiring talks. I was impressed by the undimmed optimism of those pioneers who were certain that the search was in its earliest stages and that eventual success was almost assured.

Twenty-four years later, in September 2008, I attended a SETI workshop entitled the Search for Life Signatures, at UNESCO in Paris. Sagan and Morrison were no longer with us, but Frank Drake was there, showing more than a touch of gray, but as sharp and optimistic as ever. This workshop was focused on new developments that gave reason for optimism. We heard about how the *Kepler* spacecraft that was being readied for launch a few months later would revolutionize our knowledge of habitable planets. (Since then, it has!)

The keynote talk was given by astronomer Jill Tarter, who combined hopeful vision with hard-nosed engineering detail. It is this combination of personal qualities that makes the SETI community so enjoyable to work with. These people are realistic dreamers with their feet on the ground and their eyes wide open, fixed on the stars. Their wild speculations are backed up with solid equations, and they are not just hoping, but searching, searching...

Tarter was a grad student studying astronomy at Berkeley in the 1970s when she became drawn up in the quest. Following Drake's retirement, she assumed his leadership role in the SETI Institute in Mountain View, California, which, after the U.S. Congress cut off research funding for SETI in 1993, has doggedly and courageously maintained American SETI as a privately supported research effort. In the years since that Paris meeting, I've had the opportunity to collaborate with Jill several times on scientific and educational projects. She is one of the most brilliant people I've ever had the pleasure to work with, and great fun to share a drink with after a long day at a conference. Tarter has also been a leader in inspiring young people (girls in particular) to take up careers in science. At the Paris meeting, she reported on the first year of operations of the Allen Telescope Array, an ambitious new dedicated SETI observing telescope she helped to create in Northern California, which was projected to expand into by far the most powerful instrument ever created for SETI searches. Since then, it's had a challenging existence, with expansion plans interrupted by budget troubles, and temporary shutdowns forced by funding problems in 2011 and a nearby forest fire in 2014. Still, it remains in operation.

I maintain hope that the universe will show us a sign, that my (I tell myself) scientifically informed faith may be rewarded. Personally, I want this more than anything except perhaps world peace,* and it even seems possible that one could help bring about the other—that contact could catalyze the kind of sea change in human awareness and planetary identity that just might help put a damper on nationalistic and religious conflicts. Such a discovery and its unpredictable effects goes in the category of potential game changers for the human future.

I've been waiting for a signal all my life, and although our powers of searching are just beginning to ramp up, I have to admit that I've felt my expectations drop somewhat. At least my hopes of a message received in my lifetime have dimmed considerably. Perhaps this is irrational, but we're talking about hope here. If you try to look at the problem rationally and quantitatively, you know that we have not searched very far or wide, and there is still immense space within which a strong, unambiguous signal may be found. Yet, as the decades pass, I notice the persistent cosmic silence, and I have to deal with the increased likelihood of no signal while I'm here, of never knowing for sure.

As a teenager in the 1970s, I was, appropriately, full of what Shklovsky once called "adolescent optimism," the exaggerated hope of the early SETI theorists that detection would come almost immediately. I'm still full of many things, but no longer this. I may once have been slightly brainwashed by the radio optimists I grew up around, but I also suspect that this change

---

* It would also be nice to see the Denver Nuggets win an NBA championship some decade, but now I'm really dreaming...

has more to do with being middle-aged than with any rational consideration of the observations to date.

It seems common for "contact pessimism" to increase as one grows older. Shklovsky and Sagan both became less optimistic as they aged. Both were also influenced by their increasingly dim view of the human future, but for SETI enthusiasts, this change in expectations might have to do with coming to grips with mortality, and becoming resigned to the likelihood of personally missing out on contact. So we egocentrically extrapolate this to the possibility of no signal ever.

My colleague Seth Shostak, senior astronomer at the SETI Institute, reminds me that there is much reason for continued optimism. The power of our receivers is increasing rapidly, in a pattern similar to Moore's law, by which computer power increases exponentially, doubling every two years. Our electronic ears can listen much farther out into the cosmos with every passing year. Seth says that, given this, we'll know in twenty years. He's been saying this for a few years now, and I notice he hasn't decreased his rough estimate as the years tick by. Still, I agree with his overall point. There is every reason to believe that if there are radio signals, awaiting a slightly more thorough or sensitive search, then the next couple of decades are propitious.

Much that we have learned in over a half century of space exploration seems to tell us that life and complexity are bound to be anything but rare. The basic ingredients and conditions that facilitated the origin and evolution of life here seem to be widespread throughout the universe.

When we started doing SETI, one of the great unknowns was the number of life-friendly planets. We're finally nailing this down, and the truth is much more consistent with the dreams of the wild optimists who saw a universe teeming with life than with the galactic grumpycats who said we must surely be alone.

When Sagan, Drake, and the rest of the Order of the Dolphin tried to quantify the number of galactic civilizations in 1961, they guessed the number of stars with planets to be between 20 and 50 percent, and the number of habitable planets per system to be between one and five, which leads to a predicted number of habitable planets per star of 20 to 250 percent. (It can be over 100 percent if, on average, each star has more than one.) Their guess, it turns out, was pretty good. The big reveal from the Kepler mission is that almost all stars have planets. The first extrasolar planets found were all non-Earths—mostly freak-ishly huge and hellish "hot Jupiters." This was expected, since it's so much easier to detect the star-broiled giants calling atten-tion to themselves on hyperfast orbits. It's harder to tease out the signal of worlds like ours, smaller, more distant from their stars and cool enough that our kind of chemistry, our brand of self-replication, would stand a chance.

Yet now we've started to find the tinier, less flamboyant Earth-class worlds. Preliminary estimates are that something like one in five Sun-like stars has Earth-like planets, defined crudely in terms of size and distance from their star. That's more than eleven billion planets. More than one for every person alive today. So the planetary population of the galaxy turns out to be within the range guessed by the Dolphins many decades before we had any data. We don't yet know how many systems have more than one habitable planet—perhaps our own came close to having three.

If you include stars that are not as Sun-like, adding in the more abundant red dwarf stars, and a sprinkling of other types, then we're talking about perhaps more than fifty billion hab-itable planets in our galaxy alone. That's nearly one for every person who has ever lived. This provides an astounding number of places where things could happen.

Combine this with what we've learned from the extremo-phile organisms on Earth, displaying the tremendous resource-fulness of life to thrive in so many kinds of environments and to extract free energy from its surroundings wherever it exists. These "lovers of extremes" have given us, more than just an expanded appreciation of life's limits, a sense of the impressive diversity of life's skills. They've taught us that organisms can tap into a huge assortment of alternative energy sources. There are, for example, creatures who are living off the hydrogen com-ing from nuclear decay in rocks a mile underground. This is impressive, and quite encouraging for the prospect of life on other planets.

All these discoveries encourage us to think that we live in a biophilic universe, one that is widely productive of life. Our gal-axy is chock-full of potential Edens. So where are they?

The Meaning of Silence

A century is nothing in the life of a star, a planet, or a species. Yet the people who started SETI are growing old and dying with-out receiving so much as a whisper from the stars. We've heard plenty of noise but no signal.

What does this really tell us? Some will say not much, because we, and our search, are so young and green. Success would mean everything. Failure means very little. Or does it? Is there any message in all this silence? It may be true that we've only scratched the surface; still, we scratched it and there was nothing there.

It's become a truism, a mantra, that "absence of evidence is not evidence of absence." Yet a search that turns up nothing is different from no search at all. The question is how do you interpret it? A negative result could always mean you're just looking in the wrong way. Still, the silence does tell us something about the kind of universe we live in. Most Sun-like stars in our immediate vicinity (within a few tens of light-years), we can now say, do not have a certain type of aliens, behaving a certain way, on their surrounding planets. If space were full of noisy neighbors cranking their radios, we would have heard them by now. Our galaxy does not have civilizations occupying nearly every Sun-like star and broadcasting strong radio signals.

Beyond this narrow conclusion, we can only speculate and argue over the significance of the negative results. There is no shortage of speculation or argument. An extensive literature offers to interpret this "great silence" and the closely related Fermi paradox: If there are aliens out there in abundance, then surely some portion of them must be technically very advanced, and capable either of fantastic works of stellar engineering, sending messages, or ships to colonize the entire galaxy. If that is the case, then why haven't we heard from or seen them? This conundrum is named the *Fermi paradox* after physicist Enrico Fermi,[14] who, legend has it, posed the question to his Manhattan Project colleagues over lunch one day in Los Alamos in 1943: "Where are they?"

Fermi was known for his simple but deceptively instructive questions, which often concealed deeper lessons and have spawned a whole class of questions-as-thought-experiments that scientists call Fermi questions.[15] When we geeks get together for a hike or a beer, these kinds of questions come up: How long would it take a bullet train to get to Alpha Centauri? Which is larger, the number of atoms in a human body or the number of stars in the universe? Before you know it, we're reaching for our pens and scribbling "quick and dirty" calculations on napkins. Part of why Fermi is a geek god is that he was a master of this kind of thinking. When Fermi asked his colleagues, "Where are they?" he was implying more than a wistful look toward the heavens. He was suggesting a calculation. If you make a few reasonable assumptions about the behavior and capabilities of advanced aliens, it's easy to show that their presence in our universe should be obvious. The fact that it's not is then the paradox.

What are these assumptions? That a civilization slightly more technically advanced than ours should be able to travel between the stars, and that, having done so, they, or their off-shoots, would keep doing so. Even if such travel takes millions of years, they should have had more than enough time to explore or occupy the entire galaxy. Given the age of the universe and the number of places where life could get started, even if a minuscule number of these places produced advanced alien civilizations with this behavior, we should have seen them by now.

So where are they? The problem Fermi identified seems compounded now by the "Great Silence," the fact that in nearly sixty years of searching for radio signals or optical laser signals, we have not really detected anything promising.

Many answers have been suggested.[16] Here I'll summarize a few:

Fermi Answer 1: They are out there in abundance, but our assumptions about how to find them are wrong.

In 1959, Cocconi and Morrison made such a convincing case for radio SETI that it still dominates the way we think about detecting or contacting our sentient counterparts out there in the beyond. But it's possible that the aliens don't read *Nature* magazine, and never got the memo that this is the way to do it. Radio SETI presupposes that they are more advanced than we are, but not so much more advanced that they have abandoned radio in favor of some other communication technology that we have not yet invented or imagined. Maybe for a mature civilization using radio to communicate between the stars would be like you and I putting messages in a bottle to reach across the ocean when we could just make a phone call.

Fermi Answer 2: They have no interest in us.

The Great Silence may be neither. We imagine ET civilizations as a slightly more advanced version of ourselves, and picture establishing a common language and exchanging information, as if they will be curious about us beyond a level of zoology. There is a lot of frail human ego wrapped up in this rational fantasy. How could it be that our star kin won't want to talk to us? Won't they be proud of us and our wonderful equations and symphonies? Of course they will think we are exceptional and be excited to share technological information with us. Yet what if they couldn't care less? What if our most sophisticated messages seem to them like the cute preverbal babbling of a baby? Even in the face of the solid quantitative conclusion that they will be incredibly advanced beyond our imaginations, we cling to this idea that they are going to be interested in talking to us more or less as equals, not as pets or specimens or curiosities, but as electromagnetic pen pals. This may make as much sense to them as it would for you to discuss philosophy with a frog.

Many have speculated that advanced civilizations may in fact not be living organisms but machines. Perhaps, as Arthur C. Clarke put it, "We're in an early stage in the evolution of intelligence, but a late stage in the evolution of life. Real intelligence won't be living." This is but one in a universe of possible aliens that are much more different from us than we can imagine. Of course, even if they are far beyond us, they might decide to stoop down to our level to say hello, or probe us with primitive radio the way we might dose a laboratory insect with some pheromone, as a means to get our attention. But what if they're just not that into us?

Fermi Answer 3: The Zoo hypothesis

It's possible that other civilizations exist but for some reason are shielding us from knowledge of their existence. Maybe they've decided to leave us alone, perhaps for our own good. They could be protecting young civilizations like ours from some harm or shock or other effects of contact that they deem inappropriate, similar to *Star Trek*'s "prime directive," which prohibits spacecraft crews from any interference in the development of local cultures, and specifically forbids alerting them to the fact that there are other, space-faring civilizations.

In 1897, H. G. Wells was inspired by the dismal history of European contact with native Tasmanians (none of whom survived) to write *War of the Worlds*, the archetype of a certain kind of unhappy extraterrestrial contact story. We know what happens when the guys with the fancy machines show up. It's often the beginning of the end. Perhaps those who have explored the galaxy thoroughly have experience with this kind of thing, and they know that it is better for us not to know about them.[17] Maybe advanced aliens maintain radio silence specifically because they know that primitive societies like ours are likely to conduct radio searches. Maybe they've decided that it is best that we not learn of their existence. Have we been wrapped for our own protection in something that filters out signals from interstellar space? An intelligent cosmos might prefer to mimic one devoid of consciousness for the benefit of primitives like us.

<u>Fermi Answer 4</u>: There is really nobody out there. We are alone.

It is certainly possible that there are simply no other civilizations out there. Maybe advanced technology always leads to a quick accidental suicide. Or we could be the most advanced in all the galaxy, the first to build radios and start listening. If you imagine evolution unfolding on somewhat similar tracks on worlds around different stars, sometimes taking a longer or shorter path to the stage of interstellar communication or travel, then somebody has to be the first, and for them it will seem like a quiet, lonely cosmos. This solution requires that Earth be special in an almost biblical way that seems inconsistent with the ridiculous profusion of planets where evolution could run its course.

The astrophysicist most identified with this point of view has been Michael Hart, whose papers in the 1970s and '80s were so influential that you sometimes hear reference to the "Fermi-Hart paradox." The galaxy is observably empty of other technological species, he claimed, because if they existed, then Earth should long ago have been visited or colonized. The lack of alien visitors and the failure of our early radio searches both point toward a galaxy free of other thinking, communicating beings. Hart and others have made mathematical models of Fermi's question. You can consider the galaxy to be like a giant sand pile, with stars as grains of sand, and with life and intelligence as an initially rare but self-multiplying impurity spreading through the pile. Then you can make assumptions about the colonizing behavior of individual civilizations and calculate how the pattern will move through the galaxy. These models show pretty convincingly that if the process started off anywhere at almost any time in galactic history, then the wave of colonizing technological intelligence should long ago have passed through the entire galaxy.

Of course, you do have to start with some assumptions about the behavior of individual civilizations in order to model the macroscopic behavior of colonization through space. This is a lot like deriving laws of economics by assuming certain simple things about individual human behavior, and then showing how the mass effects of that behavior will manifest. Your assumptions can always be wrong, bringing your conclusions into doubt. There is a lot of room for finding the answer that you are seeking—a dangerous possibility in a science such as SETI, which obviously contains a great deal of subjectivity. Hart's calculations, and others that followed, have often assumed that some technological species will be motivated to expand as much as possible, to colonize any planet they can get their grasping tentacles on, to keep launching new colonization missions until they have left their mark everywhere, and to maintain this behavior through many, many successive migrations over millions of years. It's the same narrative I critiqued earlier in this chapter, as the "inevitable expansion fallacy," often expressed as belief in the inexorable progression through Kardashev's types of civilizations.

Defenders of Hart's argument consider it obvious that at least some civilizations will want to keep sending out "colony ships" that will establish new planetary civilizations and send out more colonizing ships until they have filled the entire galaxy. Is this something that happens out in the real galaxy, or only in human science-fiction stories, projecting, onto unknown others, a mentality that is already becoming obsolete on our own planet?

If you accept the premises of the uniqueness argument, then if a technological civilization got started anywhere, by now it should be everywhere. In that case, absence of evidence really *is* evidence of absence.

Fermi Answer 5: The Sustainability Solution

It may be, as I argue earlier, that truly intelligent civilizations do not grow without limit. Maybe advanced societies inevitably turn against an expansionist mentality, and focus on the survival and quality of life on their home worlds. If it is rare or aberrant behavior for a civilization to expand relentlessly, then the chance that this pattern is maintained and repeated throughout countless generations of colonizing and recolonizing, unceasingly over millions of years may be effectively zero. This would resolve the paradox. They are out there but are not driven to keep expanding and to colonize the entire galaxy. The answer to "Where are they?" may be: they're all over the universe, cultivating their own gardens and minding their own business.

This idea has been explored by some young scholars who, appropriately, relate it to the current survival challenges of humanity in the Anthropocene. In 2009, Jacob Haqq-Misra, whose studies of Earth's future human-altered climate cycles I discuss in chapter 4, and Seth Baum, an ethicist who studies global catastrophic risk, published "The Sustainability Solution to the Fermi Paradox"[18] which included a critique of the assumption, based on human history, that civilizations must grow exponentially. They pointed out that not all civilizations have been exponentially expansive, and those that have done so often proved unsustainable and suffered collapse. They concluded:

> The absence of ETI observations can be explained by the possibility that exponential or other faster-growth is not a sustainable development pattern for intelligent civilizations.

As we evolve our way toward Terra Sapiens, a wise Earth, we may shed this reflexive desire for physical expansion. The *Exo sapiens* born of other worlds may have done the same. They may be out there in reasonable abundance without running rough-shod over the entire galaxy. The Great Silence might actually be a very good sign.

Fermi Answer 6: The galaxy is silent because contact is very dangerous.

There are several reasons why contact could be much more dangerous than we usually imagine, and even represent an existential threat to humanity. This possibility has been raised recently in a newly heated debate within the SETI community, which I will explore in the next chapter, over the wisdom of sending out messages from Earth to provoke a response. Advanced aliens might be malevolent or inadvertently harmful. It has been proposed that the Great Silence could result because civilizations have all been destroyed or because they know better than to attract attention to themselves.

Certainly it is possible that, even if extraterrestrials are not in any way malevolent, contact would still have an extremely disruptive effect on us. Many people believe—I think with good reason—that it would change us profoundly. Imagined scenarios run the gamut from salvation to damnation. Many people view advanced extraterrestrials as powerful and hopeful creatures who might come and save us from ourselves. This idea was explored in the novels of Arthur C. Clarke such as *Childhood's End* (1953) and *2001: A Space Odyssey* (1968). Clarke gave us utopian, transformative visions where extraterrestrials inducted us into our next phase of evolution, allowing us to achieve some kind of cosmic maturity that eluded us while we were isolated from galactic culture. SETI pioneers such as Sagan and Drake have proposed that advanced civilizations might send information designed to teach us the keys to survival. This utopian view of contact is in stark contrast to the idea of the invading, conquering, or destroying aliens portrayed in *War of the Worlds* and countless subsequent alien invasion stories.

Some SETI theorists believe that the silence is conspicuous and could mean that there are dangers lurking in the galaxy. If there are many civilizations out there, and none of them are announcing their locations, do they know something we don't know? Or, if there are no civilizations, did something happen to them all?

What if there is something out there that seeks out and destroys young broadcasting civilizations? One way to do this would be to build a fleet of self-replicating machines that spread throughout the galaxy, listening for the first radio signals of fledgling technical races, ready to sweep down on these worlds and obliterate all life. Such killer probes are called "Berserkers," after a 1970s science-fiction series by Fred Saberhagen that invoked relentless machines programmed to destroy all organic life. Yet, as is often the case in this field, when you study an idea carefully, you discover that Shklovsky and Sagan were already there. As they phrase it in *Intelligent Life in the Universe*: "Might an extraterrestrial society want to be alone at the summit of Galactic power, and make a careful effort to crush prospective contenders?"

There is no certain rational rebuttal to the frightening and bleak possibility of Berserkers, and (it seems) no good way to assess the likelihood of such scenarios. These hypotheses may seem paranoid, but that doesn't mean they are not true. As solutions to the Fermi paradox, they represent perfectly logical possibilities. How seriously you take them probably reflects your own psychology. Personally I don't believe that advanced aliens will want to harm or destroy us. I can cloak this belief in logical arguments, but honestly I think it comes down to spiritual reasons, arising from my faith in the basic goodness of life, intelligence, and consciousness. This may be the best retort against the logical possibility of evil, paranoid Berserkers enforcing a quiet and barren galaxy. As a Kwakiutl chief once advised, "Do not fear the universe."

I know some of this sounds more like cheesy science fiction than sober science, but all these ideas can be found, discussed, and debated within the peer-reviewed scientific literature. When it comes to the paths that evolution of intelligence may take elsewhere, science-fiction writers have done a better job than scientists of exploring the wide landscape of possibilities. They have the advantage of not having to undergo peer review, and they are paid to push the envelope, while the rest of us furiously scribble equations on the back of it. Consequently, some fraction of their guesses should turn out to be right.

Some people feel they know the answer. They are certain that we are completely alone in the universe, or they know for sure that we are not alone. Yet at present the only foolish answer is an overconfident one. Fermi's question, like Drake's equation, serves to spark a conversation that leads us through the logical options. As we learn more about the universe, the reasonable options slowly narrow. As of now, however, the possibilities are still wide open. Is the chance of intelligence evolving on an appropriate planet nearly inevitable, as the Dolphins of Green Bank concluded, or so rare as to make the search fruitless, as others have stated? Will advanced aliens display a universal altruism and wisdom that characterizes all minds winnowed through the bottleneck of powerful technology? Or will they be ruthless predators who have ensured their own survival by destroying all potential competitors? Are intelligent creatures out there seeking conversation, broadcasting high-tech survivalist sermons, or hunting for competitors to quash or worlds to steal? Sophisticated arguments appealing to history, anthropology, or biology predict everything from a lonely universe to inevitable, ubiquitous technological civilizations. In the absence of data, these questions are as much philosophy as science. Intellectual honesty demands that we remain comfortable with uncertainty as we explore, experiment, and speculate. And so we search. And listen. And wait...

FINDING OUR VOICE

The Destiny of Earthseed is to take root among the stars.
—Octavia Butler

Shouting in the Dark

It has taken far too long to get through. Finally, an answer comes, but the signal is weak with static. We search for a common language, encouraged by a shared desire to communicate, by an assumption of goodwill.

"My understanding of spoken English not so good," explains Alexander Zaitsev from Moscow. Graciously awkward, he has an easier time talking than listening, and our conversation does not get very far. "Better we e-mail," he suggests. I had called him in September 2007 to discuss a growing dispute in the SETI community, a group of scholars I have long admired for their selfless dedication to a long-term endeavor, for keeping their eyes on the cosmic prize. Yet Zaitsev was now playing a starring role in a festering quarrel that had opened into an angry rift.

The spat is over the idea of sending messages from Earth toward other stars, known as "active SETI," in contrast to the traditional, more passive approach of just listening for messages from others. Should we try to provoke a response, shout as loudly as we can toward the stars, and then cup our ears and listen? This would be a break with the sixty-five-year SETI practice of patient searching, assuming that others are broadcasting. Some scientists think it's time to shake things up a little bit and announce our presence to the galaxy. Others are not so sure this is a good idea, at least not without inclusive worldwide discussion and debate. Others still are dead set against it, regarding it as an unconscionably irresponsible act that could open up our world to unprecedented existential risk.

We always knew SETI could take a long time, and that success was uncertain at best. It was always envisioned as a multigenerational effort. Still, you cannot participate unless you imagine the possibility of success, picture the adrenaline rush of discovery and the mechanics of confirmation, announcement, and reaction. SETI practitioners have rehearsed for that moment many times, in their heads and in actuality. They've developed confirmation protocols and announcement procedures to follow, just in case. And they are only human. Who wouldn't feel a little impatient? Some feel that we have been waiting long enough, and it's now time to take a more active approach.

Zaitsev, chief scientist at the Russian Academy of Sciences' Institute of Radio Engineering and Electronics, thinks we need to let the other denizens of the Milky Way galaxy know that we are here, that they are not alone. He has decided that we can no longer wait for an answer from the sky, and he has taken it upon himself to start broadcasting. Rather than use "active SETI," he prefers to call his broadcast program METI, for Messages to

Extraterrestrial Intelligence. The two terms are now used inter-changeably. Zaitsev is reaching out to the inhabitants of nearby planetary systems because he feels he must, because it is the right thing to do. He feels he's speaking for all mankind.

Yet there are some who feel strongly that he has been speaking out of turn. And they would like the rest of us to get involved. Some members of the SETI community were offended by Zaitsev's efforts to initiate galactic conversation without at least first attempting to consult with the people of Earth.

The conflict came to a boil in October 2006, at a SETI meeting in Valencia, Spain, where there was a debate over active SETI and a contentious vote over new guidelines for initiating broadcasts from Earth. Later that month, *Nature* published a scolding editorial criticizing the SETI community for a lack of openness. According to the *Nature* editors,

> the risk posed by active SETI is real. It is not obvious that all extraterrestrial civilizations will be benign—or that contact with even a benign one would not have serious repercussions for people here on Earth...yet the Valencia meeting voted against trying to set up any process for deliberating over the style or content of any spontaneous outgoing messages. In effect, anyone with a big enough dish can appoint themselves ambassador for Earth.
>
> The SETI community should assess [the risks] in a discussion that is open and transparent enough for outsiders to listen to and, if so moved, to actively participate.

As a lifelong SETI enthusiast, I found it disconcerting to see the field so publicly chewed out. I reached out to several of the people involved to see if I could understand what had happened.

The Barn Door

Zaitsev, and other proponents of active SETI, like to point out that Earth has already been sending out radio broadcasts. So, in effect, we've already outed ourselves. As Sagan and Shklovsky put it in their 1966 book,

> It is of no use to maintain an interstellar radio silence; the signal has already been sent. Forty light years out from Earth, the news of a new technical civilization is winging its way among the stars. If there are beings out there...they will know of it, whether for good or for ill.

This is often called the "barn door objection," as in: it doesn't do any good to close the door once the horse has already left the barn. We no longer have the choice to evade detection. The aliens are already learning all about us by watching *The Brady Bunch* and *Gilligan's Island*.

Yet this does not persuade those who urge caution. It is true that an unstoppable sphere of radio signals is spreading out from Earth, its radius growing at the speed of light. Since Shklovsky and Sagan wrote the words above, it has traveled another fifty light-years. So isn't it too late? That depends on how sensitive the alien radio detectors are. Our television signals are diffuse and not targeted at any star system. It would take a huge antenna, much larger than anything we've built or planned, to pick up on them. From a radio point of view our planet is not completely hidden, but it is hardly conspicuous. This could easily change. Targeted messages sent directly toward nearby stars would cause Earth suddenly to turn on like a spotlight, becoming an obvious beacon announcing, for better or worse, "We are here!" So, while it's true that we no longer have the choice of maintaining complete radio silence, it is also true that programs such as Zaitsev's, if continued and vastly expanded, would fundamentally change the radio visibility of our planet. We would go from barely detectable to highly conspicuous for any alien SETI programs.

In fact, several targeted radio messages have already been sent, and are racing through space, unstoppable, toward their selected destinations. The first, "the Arecibo message," was sent by Frank Drake from the massive radio telescope (Earth's largest) in Arecibo, Puerto Rico, on November 16, 1974. The message, designed by Drake and Sagan, contained digitally encoded two-dimensional pictographs of some simple math (to get the small talk out of the way), amino acids, DNA, stick figure humans, the solar system, and a radio telescope. In the extremely narrow direction of the beam, the Arecibo message was the most powerful signal ever sent from Earth. It was aimed at M13, a globular star cluster about twenty-five thousand light-years away—which means that, at the earliest, we can expect a reply in fifty thousand years.

The Arecibo broadcast lasted all of 169 seconds and was never repeated. So even if someday it fortuitously crossed the receivers of alien radio astronomers, they might dismiss it as a momentary fluke. We would. Our current detection and verification protocols acknowledge only repeating or long-term continuous messages. In order to avoid hoaxes, mistakes, and false alarms, if we see something fleeting that seems like a possible message, we immediately try to get a repeat observation. If we can't, we don't consider it to have been a bona fide detection, no matter how promising it first seemed. So, as clever and powerful as Drake's broadcast is, if we caught a whisper of something similar, coming at us from a distant star cluster, we might not declare a successful SETI discovery. Earth astronomers seeing such a brief and powerful pulse of seeming coherence might say, "Wow! What was that?!" but probably not "Get me the White House!"*

Drake's message may never get the attention of extraterrestrial astronomers, but it was certainly noticed by Earth's astronomy community and the wider public. It elicited a furious letter of protest from Nobel laureate and British astronomer royal, Sir Martin Ryle, who described the transmission as reckless because "any creatures out there might be malevolent, or hungry," and called for an international ban on any further broadcasts. The editors of the *New York Times* (November 22, 1976) responded with an editorial entitled "Should Mankind Hide?" arguing that, "On balance, the chances of gain from communication with alien intelligence greatly exceed the chances of harm"; that "the universe seems too rich to require an advanced race to look hungrily on Earth's meager patrimony"; and that "Despite Sir Martin's eminence there is no reason to assume that alien intelligence among the stars must be hostile or predatory."

In reality, Drake's Arecibo broadcast was more of a demonstration project than a real attempt to open a cosmic communication channel. As a symbolic stunt to show off the reality of SETI technology and message construction theory, it worked brilliantly. In recent years, however, Drake has said he regrets sending this message. He sees it as pointless and frivolous, likely to add to misconceptions about SETI and to distract from the important work, which to him means continued and expanded search programs.

---

* There is one famous example of such a tantalizing but nonrepeating pulse being observed, from the Big Ear telescope at Ohio State in 1977. It looked for all the world like a genuine alien radio pulse, and the telescope operator who first saw the printout recording the event circled it and wrote, "Wow!" in the margin. Despite years of searches of this same swath of sky, it has never been seen again. It is known, appropriately, as the "Wow! signal."

Danger

There are really three reasons that SETI might possibly be best advised to maintain a listen-only policy. The first is practical: the expected asymmetry between our capabilities and those of any other civilizations. As I discuss in the previous chapter, it is likely that young civilizations will have more success in listening than broadcasting. Second is the possible threat to SETI itself from a public perception of putting humanity at risk. SETI has always had to worry about respectability. Even back at Byurakan II, in 1971, there was discussion about downplaying any perception of risk so as not to threaten public support. One Soviet participant urged that:

> There is nothing more dangerous than to speak about the danger of communication... If we would like not to lose off the problem all together, forever, our obligation is, I feel, to stress that, in any sensible way, this problem has no danger for human society. I believe we can give a full guarantee of this.

Third, also as discussed in the previous chapter, there is the possibility that real danger may lurk.

Yet what are we really worried about? Alien invaders? Unfortunately, it is hard to raise this possibility without invoking associations with bad science fiction. The problem with movie aliens is that they are generally just minor variations on Earth life. Creatures from planets around other stars will be products of alien evolutionary convolutions. They are not going to come eat us or subject us to cruel sexual slavery. Still, there are some contact scenarios that cannot be entirely dismissed, in which meeting up with an alien civilization would be dangerous or disastrous. This ground has been explored in numerous scientific papers and in decades of good "hard science fiction," the branch of

SF literature whose writers and readers pride themselves on, even in their wild imaginings, being constrained by science and logic. In fact, one of the most adamant and vocal opponents of laissez-faire active SETI is a well-known science-fiction writer. David Brin has published both peer-reviewed astronomy papers and award-winning, best-selling science-fiction novels, including *The Postman*, which was made into a Hollywood movie starring Kevin Costner. Brin—that's Dr. Brin to you—has a PhD in planetary astronomy, and his fiction draws on a solid background in physics and astronomy coupled with a freewheeling imaginative mind. Nobody alive today better represents the convergence of hard science and fictional speculation.

I first met Brin at a space resources workshop at the Scripps Research Institute in summer 1984, when I was in grad school and he was collaborating with my adviser, John Lewis, on a study about mining asteroids.[1] Back then he was already a famous sci-fi writer, and in the decades since, I have enjoyed watching his star rising further. In 2007, when I learned he had become embroiled in the controversy over active SETI, I contacted him by e-mail, reminded him that we had met twenty-three years earlier, dropped John Lewis's name (to make sure he knew I was not just another fanboy), and asked if I could call him to discuss the issue.

When we first spoke, he had just returned home to California from science-fiction conferences in Japan and China. He was obviously multitasking when he answered the phone, judging from the sounds of rustling, typing, and padding about. Yet he answered my questions in thoughtful, measured, fully composed paragraphs. His ability to do this, speak in paragraphs that seemed as if they had been written and carefully edited, reminded me of Sagan's.

Brin pointed out to me that he is "familiar with literally hundreds of contact scenarios." This is why, he said, "if there ever is a delegation of people chosen to first meet with the aliens, there should be some science-fiction writers among them."

In Brin's widely referenced 1983 paper, "The Great Silence," he provided a thorough taxonomy of Fermi answers. These include the more disturbing possibilities that I described above as Fermi Answer 6: that nobody is on the air because there is something out there that seeks and destroys everyone who broadcasts, or everyone is being quiet because they are aware of, or concerned about, some threat. As long as we cannot dismiss these possibilities, shouldn't we think twice before we naïvely start transmitting loudly in search of conversation?

Brin points out, rightly, that those using the "barn door" defense of METI to justify new broadcasts are being inconsistent. On the one hand, they deny that they are creating any additional risk because they claim that Earth is already widely observable due to our "leakage." On the other hand, they want to start dedicated broadcasts in order to get the attention of ETIs who have not yet noticed us. If they believe they can change Earth's visibility (which is, after all the point), how can they also claim that Earth is already completely visible to anyone out there?

Another tactic used by METI proponents that really annoys Brin is to ridicule his concerns with references to all the silly sci-fi movies about evil aliens—or, as he puts it, by reducing all the subtle and sophisticated arguments that can be found in the literature expressing concern about METI dangers to "fear of slathering Cardassian invaders."

On the other side of the debate is Seth Shostak, from the SETI Institute. A handsome man with a graying Beatle haircut, Seth is quick-witted, gregarious, bright, and charismatic.* He

---

* True fact: Seth was once Bachelor Number One on the TV show *The Dating Game.*

has been involved in the science and policy of SETI for many years, but given his outstanding communication and media skills, he has increasingly taken on the role of public spokesman for the institute and, seemingly, for SETI in general. This is not always an easy task, given that "aliens" are such a cultural lightning rod and that the SETI Institute is dependent on the largesse of private donors. As comfortable answering questions from New Agers on Art Bell's *Coast to Coast AM* as he is speaking to the intelligentsia on NPR, Shostak skillfully uses humor to disarm potential critics and defuse tensions, while also retaining the gravitas of quantitative, skeptical science.

When I asked him about the uproar surrounding active SETI, Shostak insisted that *Nature* got it wrong in their editorial, that in Valencia there was no organized effort to encourage active SETI broadcasts or to discourage open and transparent debate about the wisdom of sending signals. Others clearly saw it differently, and felt that the mainstream SETI community was tacitly endorsing rogue broadcasters such as Zaitsev, and failing to consult the wider communities who have a stake in the way Earth represents itself (namely, everybody).

Back in the Former USSR

Meanwhile, Alexander Zaitsev was not waiting for any stinking international consultations. He had already begun sending greetings to the stars, at least to a few of them. Later that fall, in October 2007, I finally met Zaitsev in person, when the science team for a spacecraft I was working on met in Moscow. Our meeting was timed to coincide with the fiftieth anniversary of the launch of *Sputnik*, and the Russian space agency, hurting since the collapse of the Soviet Union but justifiably proud of their history, pulled out all the stops to celebrate the anniversary. Speeches and performances showcased the strangely mixed legacy of the space race, when humanity's beautiful dreams of reaching beyond Earth first soared on rockets fueled by threats of destruction and doomsday. In the surreal, larger-than-life, space deco environs of the Russian Academy of Sciences—I kept thinking, "Have I been transported to the Klingon embassy?"—we were treated to military marches, folk dance performances, and speeches interspersing Tsiolkovsky's futuristic utopian schemes about humankind's cosmic destiny with nostalgia about the Cold War.

Sasha Zaitsev is considered by some in the SETI field to be an irresponsible and irritating rogue operator. But—what can I say? I like the guy. Kind and bright, he was a good host. He walked me around town, showing me the sprawling Moscow House of Teen Activity youth center where, in 2001, he worked with a group of at-risk teenagers to create the "Teen Age Message to the Stars." This included greetings in Russian and English,* simple pictograms portraying Earth and its location in the galaxy, and idealized images of human family life, including a child holding hands with two parents. The Teen Age Message also included a fifteen-minute theremin concert for aliens, with classical and folk musical selections chosen by the teens. According to Zaitsev, the theremin is the ideal instrument for interstellar concerts because it consists of a simple sinusoidal wave easily converted from a radio signal back into sound. We might add that, conveniently for the only audience the Message is sure to get, it is instantly recognizable as the spacey instrument used in the music for the original *Star Trek* series. Teen Age Message was broadcast in August and September 2001, from the seventy-meter radio dish of Yevpatoria Deep Space

---

* Hedging bets, since nobody knows which language aliens will speak.

Center in Crimea, aimed toward six Sun-like stars (selected with the help of the teenagers) at distances between forty-five and seventy light-years.

Zaitsev invited me to give a seminar to his colleagues at the famed Sternberg Astronomical Institute, where Iosif

Alexander Zaitsev (*left*) with the author outside the Moscow youth center, where teens worked on a message from Earth.

Shklovsky worked throughout his career and where SETI research continues today. Walking there, across a city square with boom-box-toting teens doing skateboard tricks off a Lenin statue, past elegant old Russian Orthodox churches coexisting with McDonald's restaurants, I thought of the dampened but simmering historical conflicts between our societies. In an elegant wood-paneled auditorium that felt out of another time, I scrawled a few equations on the blackboard—yes, a real chalk-dusty, scratchy blackboard—and presented my thoughts about immortal civilizations and the consequences for Earth and SETI. The audience of mostly older scientists was attentive and full of great, challenging questions. Afterward, they showed me Shklovsky's modest, cozy, book-lined office, which has been preserved, shrinelike. The walls and shelves were unchanged in the intervening twenty-two years since his death, and it felt

Sagan smiling down over Shklovsky's desk (Sternberg Institute, Moscow, October 2007).

intimate, like slipping into his lively, luminous mind. They invited me to sit in the great man's chair. From his neat wooden desk, I gazed out through a dense tangle of poplar branches nodding rhythmically against a darkening fall sky. I imagined Shklovsky looking out on these same tall trees while writing his timeless *Universe, Life, Mind,* which grew into his classic collaboration with Sagan. On this space race anniversary, I felt the presence of futures past, the rushing ephemera of our lives against the patience of the cosmos. Looking up, I noticed a photograph taped to a glass cabinet filled with scientific reports: an eleven-by-fourteen glossy author photograph, slightly yellowed at the edges, of a radiantly smiling and handsome Carl Sagan.

As Shklovsky and Sagan stated in the early 1960s, and SETI has accepted ever since, our first and most obvious role is to listen, not transmit. The rationale for this did not stem originally from concerns about ethics or existential risk. It was more about knowing our station in life. As Seth Shostak puts it, "We've had radio for a hundred years. They've had it for at least a thousand years. Let them do the heavy lifting."

Zaitsev, however, believes that broadcasting to the stars is a good idea. The way he describes it, there is almost a moral imperative to transmit. At least, he feels, there is a logical necessity: if every civilization in the galaxy is listening but not sending, then there is no chance of success. He gave me a thick stack of papers he'd published on the rationale for METI. In one of them, he stated that "such an unselfish activity is natural for a developed civilization."

Throughout the early years of the twenty-first century, Zaitsev and his Russian colleagues continued their occasional series of messages from the Crimea to the stars. In addition to the Teen Age Message, they sent Cosmic Calls I and II, which were constructed along principles similar to those behind Drake's original Arecibo message, but contained significantly more information. Unlike Drake's broadcast, these were aimed toward relatively nearby Sun-like stars. So, if anyone wishes to extend the courtesy of a reply, we could expect it in the next century or two. Between May 24 and July 1, 1999, Cosmic Call I was transmitted four times to four different stars, each between fifty and seventy light-years away. Each transmission lasted for four hours. On July 6, 2003, Zaitsev's team broadcast Cosmic Call II to five Sun-like stars at distances between thirty and forty-five light-years, some of which have recently been shown to have planets orbiting them. Cosmic Call II was expanded to include a number of items that will probably be puzzling to any alien interceptors but are meaningful for present-day and future humans studying the broadcast, such as pictures by Russian schoolchildren and the song "Starman," by David Bowie.

Despite these efforts, nobody has really blown our cover yet. Zaitsev and Drake, and a handful of jokesters, stunt broadcasters, and self-appointed interstellar deejays, have, a few times, called out indecorously toward the stars in a stage whisper. Yet nobody is going to get Drake's original message. Even if we were actually worried about triggering a response that could come as soon as fifty thousand years, it turns out that, in reality, in twenty-five thousand years, when the radio waves reach that far, the targeted stars of M13 will no longer be where they were when the broadcast was made. For all practical purposes that message was aimed at nothing. Zaitsev, so far, hasn't given us away, either. His messages were so brief that residents of these planets would notice them only if they had already been continuously and intensively monitoring Earth. The only listeners who could conceivably pick up these targeted blasts are nearby, tech-savvy, and already regarding us with curiosity. They would already know about us and if they are there then the universe is so densely crawling with curious technological minds (a possibility that seems incredible but has not yet been ruled out) that it matters little what we do because our presence is already noted.

To date, all of these messages to aliens are really messages to Earth. They remind us of our tiny place in a big universe, demonstrate our latent ability to start reaching beyond our terrestrial nursery, and also subversively call attention to our common humanity in implicit contrast to whoever else might be out there.

Who Speaks for Earth?

After talking it over with all the main protagonists, I wrote an article for *SEED* magazine summing up the conflict. Entitled "Who Speaks for Earth?"* it was published in late 2007 and concluded:

> Even if no one else is out there and we are ultimately alone, the idea of communicating with truly alien cultures forces us to consider ourselves from an entirely new, and perhaps timely, perspective. Even if we never make contact, any attempt to act and speak as one planet is not a misguided endeavor: Our impulsive industrial transformation of our home planet is starting to catch up to us, and the nations of the world are struggling with existential threats like anthropogenic climate change and weapons of mass destruction. Whether or not we develop a mechanism for anticipating, discussing, and acting on long-term planetary dangers such as these before they become catastrophes remains to be seen. But the unified global outlook required to face them would certainly be a welcome development.

---

* Borrowed, of course, from Sagan's brilliant *Cosmos* episode with this title.

I began to see this as a series of dilemmas nested like Russian dolls. Can today's SETI community agree on a policy about active SETI? Even if collectively forged and broadly ratified, would such an agreement actually control or change global behavior, as perceived from the outside? How would you get everyone to go along? Can human society in some sense agree on active SETI? Should we, as a species, cautiously try to hide our presence, or hopefully announce ourselves to the universe? Beyond the scientific politics, there is a deeper story here, of humanity struggling to find its coherent voice, its place, and a sense of self strong enough to survive local and self-imposed threats, let alone those that come from other star systems.

The question of how we should present ourselves to the cosmos hinges on what kind of behavior we expect from our brothers from another planet. We filter this question through our ideas about human nature and our suspicions about the universal nature of sentient life: Do we have any basis from which to predict the ethics of advanced technological aliens? That depends on where you think moral and ethical principles come from. Can they be derived from universal logic? From the laws of thermodynamics? From evolutionary theory? Or do morals come from God? We may not have any answers, but in thinking about the possible morals of alien species we consider these questions from a wider perspective.

Is it incumbent upon us, as would-be participants in the interstellar chat room, to stop lurking and to reveal ourselves? At some point, some wise species must break the silence. Yet are we anywhere near that point? If not, what do we have to do to get there? The question forces us to consider where we are in the potential spectrum of advanced technological, intelligent life-forms.

I was gratified when, after my *SEED* piece was published, the partisans on either side seemed pleased with it, and remained friendly to me. When Brin, Shostak, Zaitsev, and others all praised my summation, it furthered my sense that common ground could, in fact, be found. In this piece, I expressed hope that the impasse could soon be resolved, and pointed out the many areas of agreement.

Unfortunately, people get entrenched and divisions tend to solidify. After Valencia, nothing much seemed to happen. No agreements were made, and the community was left in an unhappy standoff. Brin tried several times to create a forum for wider discussion of the issue, beyond the narrow boundaries of SETI insiders. "As newcomers in a strangely quiet Cosmos, shall we shout for attention?" he wrote,

> Or is it wiser to continue quiet listening? We propose an inter-disciplinary symposium, to be the most eclectic and inclusive forum, by far, to deliberate the METI issue. It is not too much to ask that METI people hold back until the world's open, scientific community can get a chance to examine their proposal.

He pushed for a plenary session at a meeting of the American Association for the Advancement of Science (AAAS). This is the world's largest general scientific society, and the huge annual gatherings always highlight, in addition to new scientific results, discussions of societal and policy issues related to science. Despite the name, it is increasingly an international meeting, and would make a good forum for starting global discussions. For several years Brin's proposal went nowhere. The organizers of the meeting did not feel that a discussion of sending messages to aliens was worthy of such an august gathering of serious scientists.

Planetary Protection

After several years of stasis, this conflict has recently flared up again. Doug Vakoch, a social scientist, has stated his intention to initiate a new program of active SETI broadcasts using the powerful Arecibo antenna. For sixteen years Vakoch was at the SETI Institute, with the job title of Director of Interstellar Message Composition. Recently, he left to direct METI International, a nonprofit dedicated to preparing and sending messages to extraterrestrials. It's not clear if Vakoch will get the support and permission he needs from the National Science Foundation, which controls the telescope, but he has kicked the hornet's nest, riling up the opposition and sparking a new round of intense debate. Vakoch argues that his plan is really just an expanded kind of SETI search, a way to find those civilizations that may be merely waiting for a signal from us to begin a dialogue. We might need to prompt a response from them, initiate contact ourselves. Unlike Zaitsev's moral arguments for the necessity of sending messages, Vakoch's are pragmatic. He regards the initiation of systematic broadcasts as a useful strategy to help achieve the goal of SETI. It is, after all, possible, that the Great Silence would end as soon as someone received a poke from us. It might be up to us to get this started.

This reminds me of something Art Bell described when I was on his radio show, *Coast to Coast AM*, a few years ago. Bell, an avid ham radio enthusiast, told me a frequent experience is to tune in and at first hear just long stretches of complete stillness. After listening for a while, though, it sometimes becomes apparent that there is a whole crowd there, just waiting for somebody to break the silence, at which point, suddenly, everybody chimes in at once. It is conceivable, I suppose, that interstellar radio could be in that state, full of lurkers who are shy or not convinced we are ready for conversation. In this case, someone needs to break the ice. Vakoch believes it is worth a try.

Given the resurgent controversy, Jill Tarter proposed that the AAAS hold a public debate on the question. With her gravitas in the scientific community, her proposal was successful. The symposium, Active SETI: Is It Time to Start Transmitting to the Cosmos? was held at the AAAS meeting in San Jose, California, on February 13, 2015. I was pleased to be invited to speak, along with Brin, Shostak, Vakoch, and U.S. District Court judge David Tatel, with Jill Tarter presiding. Holding the symposium at the AAAS was something of a victory for David Brin, who had been trying for years to get such a hearing at this or some equivalent highly visible professional venue.

The day before the symposium, we all spoke at a press conference organized by AAAS (and also attended by Frank Drake) that generated a large number of articles about the controversy. What science writer could resist a debate about the dangers of communicating with potentially hostile aliens? Taking advantage of the fact that we were all in town, a daylong workshop was arranged at the SETI Institute for the day after the symposium, where we could involve more people and hash things out less in public and more in detail.

When I was invited by Jill to participate in these events I was heavily immersed in this book, reading, thinking, and writing daily on the Anthropocene and its cosmic dimensions. I began to see the METI debate as emblematic of many of the dilemmas we are facing as part of this planetary transition in which cognition is starting to play a central role in the workings of the planet.

As the reluctant carriers of those inchoate global cognitive processes, we have some choices to make. Changing Earth's visibility in the galaxy is one of many actions that could affect every inhabitant of Earth (not just the human ones). These are discussions we can't really have without considering ourselves as long-term global actors responsible to the planet as a whole. We'll have to act without ever completely understanding the risks, so how do we decide whether it is worth it? Who is "we"? Who gets to decide? Even if we're able to choose what seems a wise course of collective action, what is to prevent some individual or group from acting alone? In all these dimensions, the active SETI debate is similar to questions about other technologies with global consequences that can open doors to dangers and/or opportunities. In my AAAS presentation, I tried to draw upon a few useful analogies.

The first of these is "planetary protection." This is the term we use at NASA for our efforts to guard against inadvertent interplanetary contamination. It seems like a very long shot that our little spacecraft could end up being the vector bringing some horrible infection between worlds. Yet, given our ignorance, and the very high stakes of being wrong, it is appropriate to exercise caution. From the very beginning of planetary exploration, NASA has taken this responsibility seriously.

Though the opportunity to explore space came on the coattails of a Cold War that threatened planetary devastation, the scientists designing planetary craft on both sides acted with great sensitivity to possible threats to Earth's biosphere—and other potential biospheres. Early in the space age, international committees were formed and guidelines adopted. NASA and the Soviet space agency both instituted policies of sterilizing spacecraft intended to impact, or even pass near, other planets. They did so even when it meant increasing costs or compromising mission goals.

In 1967, the United States, the Soviet Union, and the United Kingdom ratified the UN Outer Space Treaty. The legal basis for planetary protection lies in Article IX of this treaty:

States Parties to the Treaty shall pursue studies of outer space, including the Moon and other celestial bodies, and conduct exploration of them so as to avoid their harmful contamination and also adverse changes in the environment of the Earth resulting from the introduction of extraterrestrial matter and, where necessary, shall adopt appropriate measures for this purpose...

This treaty has since been signed by almost all nation-states, including all the current and aspiring space-faring countries. Additional recommendations adopted in 1997 require that when Mars samples are returned to Earth, they should be contained and treated as hazardous until proven safe. Further, it is agreed that if sample containment cannot be verified en route, the sample should not be returned.

Are these real threats that we need to take seriously? It certainly doesn't seem like it. Diseases and parasites on Earth have evolved in close concert with their hosts, and the same should be true of evolution anywhere. The chances that a streptococcus bacterium could infect a truly alien life form *seems* to be zero, and it's likely that our cells would be hostile environments to alien germs. Rather than a home, a host, or a food source, we would seem like poisonous monstrosities to be avoided by any creatures that evolved with their own unique biochemistry.

Yet it would be wrong to be overconfident here, for two reasons. One is the possibility that our learned opinions could be false. To the modern, scientifically informed mind, it seems incredibly unlikely that Martian organisms could infect humans or other terrestrial organisms. Yet how unlikely is "incredibly," and is this good enough when we are talking about the possibility, however remote, of destroying life on Earth? Our ideas about life are both highly informed and based on complete ignorance. Biology and its subdisciplines and related sciences have made wondrous progress in understanding the rich biota and natural history of Earth. Yet we are in the dark about life elsewhere and how separate biospheres may interact with one another. So, while it seems that there should be no danger to us from Mars bugs, do we really want to bet the entire farm on our being right about this? I, and many scientists, strongly doubt that it would ever be a problem, but we do not *know*.

The other reason to err on the side of caution is the issue of public trust. Policy is guided by, but not only by, expert opinion. We are exploring the solar system, extending our grasp beyond Earth "for all humankind," and we owe it to our planetary compatriots to do so responsibly and with appropriate humility.

We astrobiologists have had to face the possibility that, through our science and exploration, we could bring about low-probability events with horrible consequences. We look at our designs for curiosity-driven exploration and ask: what levels of risk are acceptable, and what is the right way for us to behave with respect to possible unknown life? We're not just wondering about these questions in an abstract sense, or discussing them in seminars, but implementing policies and making real decisions about how to build and operate hardware based on the answers we reach, balancing our curiosity with responsibility.

In the case of planetary protection, I think we've done things correctly, and I'm proud of my community for the way this has been handled. We have protocols in place to guard against both "forward contamination," the possibility of carrying some harmful biological agent to other worlds, and "back contamination," bringing back something dangerous from another planet that might contaminate Earth. To enact these, we have NASA's Office of Planetary Protection, and there is even someone with the wonderful job title "planetary protection officer." Since 2006, the job has been held by biologist Catharine "Cassie" Conley, who has worked diligently to guard Earth and the other planets from inadvertent contamination. Cassie comes across as serious and studious, but once, after dinner in Washington with a group of NASA people, I asked her if her job title invited a lot of *Men in Black* questions. With little provocation, she suddenly reached into her purse, donned a pair of sleek black sunglasses, frowned seriously, and flashed a very official-looking "Planetary Protection Officer" badge at me.

The fact that NASA has such a program makes me feel very good about working so closely with the agency for which I have sat on innumerable review panels and advisory boards, and that has funded most of my research throughout my career. I am well aware of the intertwined histories of warfare and space exploration, of how the moon rockets that got NASA off the ground and gripped me as a child were motivated by the dangerous nuclear arms race and largely designed by Nazi scientists who "aimed for the stars, but sometimes hit London." Yet, after the Cold War ended, we kept on going, investigating the solar system with smaller budgets and greater international cooperation. These continued efforts required ingenuity and clever technology to do more with less, but were driven more by curiosity and less by deadly competition, and so carried less of a moral deficit. Given this history, I love that NASA is now a scientific agency with an ethical wing, the Office of Planetary Protection, meant to ensure that we not only do good science and exploration but that we also "first, do no harm."

Planetary scientists in general take this responsibility very seriously, and I regard this as a credit to our profession, a hint of maturity, a sign that when push comes to shove, we humans can rise to the challenge of melding ethics with science and do the right thing. Of course, implementing these principles always involves making educated guesses about what is safe, which carries with it all the gaping uncertainties in our knowledge about life in the universe.

Recently I was visited in Washington by my old pal Dorion

Sagan, whom I have known since I was six years old. We spent an afternoon together looking through boxes of his father, Carl's, papers in the Library of Congress. In a batch of correspondence with Isaac Asimov, Dorion found a 1976 letter from Carl discussing planetary protection. He wrote:

> As with all questions of interplanetary quarantine or recombinant DNA we must ask not only what is the most likely theory but what is the probability that this theory is incorrect. The potential consequences of back contamination of Earth are so severe that I would require any theory to have a probability of being incorrect of 10^{-6} or less to be believable. Since it is unlikely that any existing theory can have anything approaching this level of reliability I believe the only responsible reaction in the face of our level of ignorance is caution.

Dorion found the use of 10^{-6} hilarious, asking me, "How the hell could you ever know if a scientific theory had a probability of 10^{-6} of being wrong?" For decades we've had a running joke about our very scientific fathers, how seriously they take their own ideas and their (we think, at times) excessive faith in quantitative solutions to intractable problems. So we spent the rest of that evening saying things like "This has got to be the best vodka tonic I've ever tasted. There is only a possibility of 10^{-6} that it is not."

All joking aside, though, the letter quoted here demonstrates not arrogance but the opposite: a kind of humility. Despite the (perhaps slightly laughable) quantitative precision, this was just Carl's nerdy way of saying, "We had better be very certain we're right, and there is no way we can ever have enough faith in our scientific theories to risk the whole game." I repeatedly find myself impressed and grateful for the job that Carl and his contemporaries, the first generation of planetary explorers, did putting in place the planetary protection protocols we still follow today. This seems relevant to the debate about METI broadcasts. We are trying to agree on policy for a proposed scientific investigation that carries some risk. The level of danger, impossible to estimate precisely, seems quite low. Yet if we're wrong, it could change everything.

Rise of the Machines

The next point I made in San Jose was:

> If we're really concerned about advanced intelligences of unknown motivation that might harm or destroy us, then we should take note of the fact that right up the street here in Silicon Valley there are people who are spending all their waking hours trying to develop such entities.

As I discuss in chapter 5, some people feel that we are actually close to developing machines with intellect that will surpass that of humans. As with killer aliens, the idea of killer robots is so intertwined with tacky science fiction that it has an aura of inanity. However, if machines do indeed become conscious and autonomous, we cannot really know what their motivations, or their attitudes toward us, will be. If they get to the point where they have cognitive abilities superior to ours, and they apply this to building still-smarter machines, it could cause a "singularity" or "intelligence explosion," and then all bets are off. As with aliens, we don't need to postulate that they will be evil in order to imagine that they could represent a grave or existential threat. They may simply have needs and motivations we cannot fathom and that differ from our own. Or they'll be so fantastically effective at misunderstanding our instructions that, as in all those fables about genies granting three wishes, we'll wish we could take them all back.

I tend to be among those who do not fear the development of superior machine intelligence. I don't doubt that artificial intelligence is going to change our world in surprising and unpredictable ways. Yet I don't think machines are going to suddenly become conscious and decide to kill us off or enslave us. Nevertheless, as with the Fermi paradox, an honest assessment must include the admission that nobody knows—because nobody understands consciousness.

Although I am not terribly worried about either possibility, smart machines or superior aliens, it does seem that the problem of dangerous machines is potentially more imminent than possible "spirits from the vasty deep" we may summon from hundreds of light-years away. Yet the questions are comparable. We can't prove they won't present an existential risk; so how should we proceed?

In January 2015, Max Tegmark, an MIT physicist and founder of the Future of Life Institute, organized a conference in Puerto Rico on the Future of AI: Opportunities and Challenges, at which an international, interdisciplinary group of experts gathered to address the direction of AI (artificial intelligence) research, and the prospects for enhancing promising outcomes and avoiding pitfalls of a future "intelligence explosion."

This conference was well timed, as there seems to be a recent sea change in the AI community. A critical number of researchers has realized that they may be close enough to their goal that dangers need to be addressed and, it is hoped, managed. Many who went into the field years ago with the simple goal of manifesting "human-level AI" as soon as possible are now turning to questions about how to do so safely. Rather than simply try to make machines as smart as they can as quickly as they can, they are considering how to push their newly created intelligences in certain desired directions. This brings us back to Asimov's laws of robotics, and some interesting technophilosophical questions such as how we might instill values in machines, what values really are and where they come from, and how you would code them algorithmically.[2]

At the Puerto Rico gathering, a consensus emerged that it was time to redefine the goal of AI research away from simply making artificial brains and toward making things that are going to be beneficial to society. All participants signed a letter that has since been signed by dozens more scientists, technologists, entrepreneurs, and scholars. It notes the rapid increase in machine capabilities and the likelihood that this increase will accelerate, leading to societal issues that must be anticipated and addressed proactively with new research directions. On the plus side, the letter notes,

> The potential benefits are huge, since everything that civilization has to offer is a product of human intelligence; we cannot predict what we might achieve when this intelligence is magnified by the tools AI may provide, but the eradication of disease and poverty are not unfathomable.

The letter doesn't dwell on the negative possibilities (for instance, destroying all of human civilization), but it does state that

> We recommend expanded research aimed at ensuring that increasingly capable AI systems are robust and beneficial: our AI systems must do what we want them to do.

The letter was accompanied by a list of research priorities to be funded initially by a $10 million donation from Elon Musk. Many of these are focused on the near-term problems arising from somewhat smarter machines: possible legal issues associated with self-driving cars and the economic disruption that will be caused by machine displacement of workers across many sectors.

Near the end, this letter finally mentions the possibility of an "intelligence explosion" and the range of opinions on how likely such an event really is. It then suggests, vaguely, that more research should be done to ensure that if it does happen, there might be some way for humans to maintain control and seek a positive outcome. When you finish reading this statement, you realize that all these bright and well-intentioned experts don't really have a clue as to whether or not this could happen and how to affect the outcome if it does.

I appreciate these efforts by the AI community to develop, in tandem with machine cleverness, the wisdom to guide these growing powers. Their efforts will no doubt lead to some interesting and worthwhile research results that will help to improve AI systems. Will they be sufficient to avoid the worst possible outcomes? Not in the worst-case scenarios. Even if we really knew enough to enact rigorous guidelines, I'm not sure to what extent those competing with other companies for profits, or those working for militaries trying to make sure their autonomous weapons outperform those of their enemies, would temper their efforts if they saw such constraints as limiting their opportunity for advantage.

I am encouraged, however, that many of the best minds in the field are attempting to guide their community toward responsibility. Yet, when I think it through, my lack of concern about a dangerous "intelligence explosion" does not come from confidence in the AI community's ability to control truly sentient machines, or infuse them with values and goals that make them benign. I just think they are overconfident about their ability to make such machines anytime soon.

Nobody is seriously proposing a ban on artificial intelligence research, perhaps because it is obvious that such a ban would be impossible to enforce. Nonetheless, at least leaders in the field are discussing the questions, and thinking about ways to foster a culture of responsibility and to encourage safe best practices among those coming closest to success.

No Choice but to Choose

What is our role in this universe? Do we know enough about planets, and about ourselves, to be the shapers of worlds? In chapter 4, I discuss the problem of geoengineering and the related question of terraforming. Are these always bad ideas even when they might prevent mass extinction or allow bio-spheres to flourish? With these choices, philosophical and spiritual questions about what we *should* do are deeply entwined with technical questions about what we can do with, or to, planets.

Climate change is here, and there are some who, unwisely, advocate that we should go for the quick techno fix. Fortunately, among those who study geoengineering and climate policy, it seems that a consensus view is developing that these are last-resort options. We are still too ignorant, lacking even the knowledge to assess the risks adequately, but we need to keep doing research, and acquire that knowledge, because sooner or later—and I hope it's much later—it will become necessary to do some high-tech geoengineering. There is no question that if we manage to survive and become long-term actors on this planet, someday we will want to deploy more active geoengineering as a defense against the dangerous natural capriciousness of planetary climate systems.

Picture the next several millennia. Intrusive geoengineering is a bad idea now but will someday be necessary. Intelligent machines are likely to become permanently integrated into human societies and Earth systems. Active SETI is also something that can be done well only by civilizations with a multimillennial outlook. With each of these issues (geoengineering, artificial intelligence, and METI), we'll soon know a lot more. We need to keep studying these problems, and we need to take a long view. We have a lot to learn about our climate and cognitive systems. We are on the threshold of being able to study exoplanets and learn whether inhabited worlds, and possibly technically altered worlds, are common. Our listening technology is growing in such a way that even in twenty years we'll have significantly more evidence that bears on whether there really is a Great Silence.

If we want to become a broadcasting civilization, we can only really do so effectively if we can move far beyond brief one-off stunts and begin a project that will have global buy-in and will last for millennia. Yes, it is hard for us to imagine that now. That is reflective of one of our biggest problems, but we will need to think and act this way both to manage ourselves well on this world and to start to reach out effectively to others.

It so happened that when we gathered at the AAAS for the symposium on active SETI, the subject of geoengineering was in the news because just that same week a much-anticipated report on climate intervention had been released by the National Academy of Sciences.[3] The analogy between METI and geoengineering was stressed by our fifth panelist, and the only nonscientist, David Tatel, a federal judge on the U.S. Court of Appeals for the District of Columbia. Tatel was appointed in 1994 by President Clinton to fill a vacancy left by Ruth Bader Ginsburg.

How did a judge and lawyer with a background in civil rights law get involved in the METI debate? Well, as I discovered when he invited me to lunch in his expansive offices in the Federal Court Building in Washington, DC, he is a man of wide-ranging and eclectic interests. He has also been blind since 1972, which is neither here nor there, except it was fascinating to notice his use of alternate and innovative communications technology to read, write, correspond, and deliver public lectures. He certainly navigates this world more effectively than most of us do, and I did wonder, though I was too shy to ask, if his experience of doing so with a fundamentally different sensory palette might give him any ideas about or insight into the challenges of communicating with extraterrestrials who will likely have their own, differently evolved range of senses. Judge Tatel's entree into the world of SETI, it turns out, came about in a roundabout way, through his father's service in World War II. At the end of the war, Howard Tatel and some colleagues managed to capture a large German radio antenna and used it to help establish American radio astronomy. Tatel Senior then developed some innovative engineering concepts that were used to build the large telescopes of the U.S. National Radio Astronomy Observatory in Green Bank. So the telescope Frank Drake used for project Ozma was called the Tatel telescope, and it was at a celebration for the fiftieth anniversary of this facility that Judge Tatel met several of the scientists involved in the current heated debate over active SETI. He kept in touch with them and, through his intellectual gregariousness, got involved in enough discussions on the subject to generate an invitation from Jill Tarter to represent a nonscientist, policy voice at the symposium.

In addition to geoengineering, Judge Tatel suggested, as analogies for the policy challenges of active SETI, both recombinant DNA research and laboratory studies involving dangerous pathogens, areas where scientists have struggled to balance the value of unfettered curiosity-driven research and the free exchange of ideas, with the threat of conjuring something truly destructive and potentially unstoppable. Especially given the self-reproducing, mutating, and adapting quality of organisms, in biotechnology there is a possibility of unleashing a genie that cannot be put back in the bottle.

There is no way to entirely enforce a worldwide ban on any research, even if that were seen as desirable, but there are ways to encourage best practices. In each of these areas the potential dangers have been recognized and widely discussed by international groups of concerned professionals. These processes were accompanied by temporary suspensions of government funding to slow down the research while guidelines were developed and adopted. Through collaborative conversations, the scientific community focused its intellect and devised agreements for how to proceed.

Judge Tatel suggested that geoengineering is the closest analogy to the problem of METI broadcasts. Although biotechnology is moving rapidly in the direction of DIY genetic laboratories, where any "maker" in a garage can cook up their own organisms, it still takes a pretty sophisticated government laboratory to work on a pathogen like H5NI (Asian-origin avian influenza, or "bird flu"), which means there is still some possibility of enforcing meaningful guidelines for research. But geoengineering is more like METI in that a "rogue" experiment can be attempted by any rich person or group.

Tatel suggested that the National Academy geoengineering report could serve as an initial model for an approach to METI, and he finished his talk by quoting from the conclusions of the report:

> Planning for any deployment of albedo modification would bring unique legal, ethical, social, political and economic considerations. Open conversations about the governance of albedo modification research could help build civil society trust in research in this area. If new governance is needed, it

should be developed in a deliberative process with input from a
broad set of stockholders.

It is true that anyone can buy or build a radio telescope
and start broadcasting, but it remains the case that the more
powerful facilities, those with the power to significantly change
Earth's visibility, are controlled by larger governmental and
academic organizations that could be expected to conform to
widely accepted guidelines. As radio technology progresses, it
will, as has biotechnology, move in the direction of empowering
individuals who wish to do their own powerful experiments.

For this reason, there may be a window of time during which
we can seek consensus. As the march of technology moves in a
direction that would facilitate lawlessness and rogue actors, there
will be perhaps two or three decades when the major broadcasting power is still locked up in larger institutions. Whether
or not Vakoch is able to start his desired, controversial Arecibo
transmissions, there is nothing on the drawing board that would
immediately and drastically change Earth's galactic visibility.
Nobody on Earth is yet capable of broadcasting the powerful,
continuous beacons that successful METI would likely require.

Depending on whom you talk to, that is a reason why we
really don't need to worry about this now—or why right now is
a crucial time, an important window of opportunity for trying
to get the global conversation going before new technological
capabilities inevitably complicate the situation.

This sense of a window that may soon be closing is part of
what motivates David Brin in his activism on this issue. Brin
used his time at the AAAS podium to urge our community to
start a process of inclusive international discussions about the
wisdom of active SETI. He held up the "Asilomar process" as a
model. This refers to the time, in the 1970s, when there was a lot

of fear of recombinant DNA research and its growing power to create and possibly release dangerous, unstoppable agents. At that time, the DNA research community agreed on a voluntary hiatus, and a gathering was held at a conference center at Asilomar State Beach, in Northern California, to discuss potential hazards and regulations. Out of this came a set of guidelines for containment of some experiments, whereby those judged to have higher risk would require additional levels of containment. There was also a class of experiment (those involving deadly pathogens that could not be effectively contained) judged so risky that by consensus they were banned. These guidelines were incorporated into the culture, laboratory practices, and facilities of the burgeoning field and industry of biotechnology. It is widely believed that by effectively self-policing, the biotechnology community avoided government regulations that would have slowed or halted progress in research.

For active SETI, Brin is urging some comparable communal process of risk assessment updated for the twenty-first century. He envisions a global effort that would use modern telecommunications tools and include public input and professional guidance.

At the AAAS symposium, the opposing view that "We should launch active SETI as an ongoing complement to our traditional passive SETI projects" was represented by Doug Vakoch. He defended his plan to start broadcasting from Arecibo with what he described as "a more aggressive, ongoing series of transmissions." Unlike previous broadcasts, his would repeat messages to the same target stars over periods ranging from weeks to years. He justified this as a next, more mature phase of SETI. Though he expressed support for broad-based discussions of active SETI, he also declared an intention to begin his program of transmitting without waiting for these discussions to occur, saying,

We need to get over this idea of either promote consultation or go ahead with active SETI. We should be doing both. We are growing up, I would say, as a civilization. SETI has been going on for fifty years. In this first half century, we've focused on what we can gain from it. But I think it's a sign of maturation of SETI that we're now thinking about what we might give to future generations of humans and even other civilizations.

His SETI Institute colleague, Seth Shostak, however, in his presentation, belittled the idea of international, public input. Seth asked,

How do you do that? If you have this World Wide Web inquiry, and you say, "Look! Push this button if you think we should and push that button if you think we shouldn't." And, say, 55 percent of the people say, "Yes, we should," and 45 percent of the people say, "No, we shouldn't," what do you do? You say, "Well, okay! Now we know whether it's safe or not." Do you really think that? I don't know how that tells you whether it's safe or not.

Seth has a point that a broad public process would be difficult and perhaps unsatisfactory. Yet I think he is missing the larger point here, perhaps willfully. There are ways to develop tools and educational products that are more sophisticated than a simple yes-or-no, push-one-button-or-the-other vote. And, really, the benefit of broad public input would not be to "learn whether or not it is safe." Obviously, this would not tell us that. But, if we are going to act, and speak, on behalf of Earth, a good-faith effort at gaining some sense of buy-in is necessary. If it turned out that a majority of public input reflected that people thought it was a really bad idea, well, that would be useful data to which we ought to pay attention. It could mean that those of us in favor of transmitting someday would have our work cut out for us. Maybe, if it really is such a good idea, then after a generation or two of public education, people would come around. After all, what is the hurry?

A woman from the audience asked, "What are the benefits of doing it now rather than thirty or fifty years from now? You are talking about a round-trip communication time of hundreds or maybe thousands of years. Is there an advantage of getting that time shaved by a few years?"

My answer to her was:

> There's obviously no real urgency to commence broadcasting, because this is inherently a project of decades, centuries, millennia even. There's perhaps impatience among individuals who are champing at the bit, in feeling that one wants to do something while one is still alive and working. We shouldn't allow that to be the motivation that decides how we proceed.
>
> But there is an urgency in the need to learn how to have a global conversation and decision-making process, in imagining how we could speak with one coherent global voice, and learning how to act on behalf of the planet and its biosphere. We've mentioned climate change, geoengineering, and the perils and promise of biotechnology and artificial intelligence. We have increasing existential threats right now that are very much related to our difficulty in making global decisions, with a long-term outlook, about deploying powerful technology, and our inability to speak with one voice and act, in a sense, with one mind. So, that part of this discussion, I think, is actually very much related to things we need to be doing immediately. The question of METI can at least help encourage us in the direction toward becoming the kind of global civilization that will be able to hang around for many more millennia and perhaps ultimately be able to engage in interstellar conversation.

A final bombshell was dropped at this session: a new statement was circulated, and released to the press, arguing strongly against unsanctioned METI broadcasts. It had come out of the well-respected Berkeley SETI Research Center and was signed by Elon Musk and a crew of heavyweights from the worlds of astronomy, SETI, and astrobiology. This document left no doubt that, regardless of how it played in the wider world, any new, powerful METI broadcast would be extremely divisive and disruptive in our professional community. The statement read, in part:

> We feel the decision whether or not to transmit must be based upon a worldwide consensus, and not a decision based upon the wishes of a few individuals with access to powerful communications equipment.

It ended with this declaration:

> Intentionally signaling other civilizations in the Milky Way Galaxy raises concerns from all the people of Earth, about both the message and the consequences of contact. A worldwide scientific, political and humanitarian discussion must occur before any message is sent.

At the SETI Institute workshop the following day, Jill Tarter presided with her calm, inclusive intellect and things stayed for the most part very civil and friendly. Vikki Meadows from the University of Washington, whom I have known since we were both little postdocs learning at the knees of our common mentors, and who is now a formidable leader in studies of exoplanet habitability, treated us to a masterful talk on the exciting prospects over the next few decades for identifying signs of life on distant planets. It was good to be reminded that this landscape is not static, but evolving rapidly and in promising ways. This seems a strong counterargument to any sense of urgency in commencing a broadcasting program. If we wait just a few decades, which amounts to nothing at all in the necessary time-scales of interstellar discourse, we should know an awful lot more about what kind of universe we're dealing with.

The most passionate and challenging presentation was given by John Gertz, an important figure in the SETI world, having served three terms as chairman of the board of the SETI Institute. He was by far the most aggressive opponent of METI in general, and in particular of Vakoch's plans to initiate powerful METI broadcasts. As he described it, "Some of us who have thought a lot about this actually are deeply worried that aliens could present a serious danger," because "Our inadvertent electromagnetic emissions may have the unintended consequence of inviting death and destruction from malevolent aliens." His talk, first of the morning, woke us all up with its intensity of emotion. He became visibly angry while describing how "a handful of misguided individuals propose we scream out our coordinates to attract the attention of ET intentionally." This, he said, is not science; it's unauthorized diplomacy, and it should be forbidden.

In his view, not only should active SETI be banned, but we should take other measures to reduce our visibility to potential threatening aliens. For example, we should curtail our use of planetary radar. These are the powerful radar blasts that we send occasionally toward asteroids or other planets in order to study the reflection we get back, and so learn something about the targets. Several observatories are engaged in this, including Arecibo, and it has provided our best images of numerous asteroids and important information about the surfaces of other planets.[4] It's a completely harmless activity—unless you are worried about the fact that these concentrated radiation bursts will be highly visible from many light-years away to anyone who happened to be observing Earth from the right place during the brief instant when one of those flashes occurred. But planetary radar bursts are not aimed at any star. If you really want to be paranoid, you could fret over the fact that sometimes stars are randomly lined up behind whatever planetary object you are targeting. This is exactly what Gertz is worried about. His suggestion is that we should adopt best practices so we don't turn on our radar systems when they are pointed toward nearby stars, or intersecting the plane of the Milky Way galaxy, where there are, unavoidably, many stars in the background. It ought to be mentioned here that planetary radar is one of our main tools for learning about the properties of asteroids that may someday threaten life on Earth, and how we might mitigate against them. So any serious curtailment of this technology to avoid one *suspected* existential risk might cause increased vulnerability to a *known* one.

Gertz's radical anti-METI view put David Brin in an interesting position at the workshop. I think it helped him out by revealing his position to actually be quite measured. Brin has never said he supports an outright ban on METI, only that it is irresponsible, arrogant, and callous to proceed with such broadcasts without first conducting a good-faith effort at inclusive global consultation. After Gertz's talk, Brin went out of his way to differentiate himself from the extreme of blanket opposition to all future broadcasts. Suddenly he was the moderate in the room.

Changing My Mind

Up to this point, in my public discussions of this conflict, I've remained neutral. Now, however, after these events of 2015, I'm ready to get off the fence and join with those who urge a moratorium on broadcasting while we sort this out.

During these three days of intensive deliberations over

METI broadcasts (press conference, symposium, and work-shop), I found myself, as I often do, in somewhat of a diplomatic role, thinking and saying, "You're both right!" On one side are those who warn of some possible grave danger if we start screaming into the dark. They urge, at the very least, some glob-ally inclusive evaluative process before we start. On the other side are those who feel that it makes no sense just to listen and not to send, and that a mature technological civilization must reach out for others. I find both arguments worthwhile and the discussion itself valuable, as it forces us to think hard about what might be out there, what role we might play in the uni-verse, how we evaluate and handle existential risk, and how we might try to speak for all of Earth. At the time I wrote my *SEED* article in 2007, I was careful not to take sides. This wasn't hard: I am quite fond of all these brilliant people and felt that they all made excellent points, albeit with certain blind spots.

Still, at that time, in private, my sentiments were more with the broadcasters. Although I found the call for a global process of consultation intrinsically worthwhile, I thought, "What's the harm in sending a few messages while we work toward a global consensus?" I am sympathetic to Sasha Zaitsev's argument that if we are going to listen, we should also send. Maybe SETI karma requires broadcasting in order for listening to work. If we want there to be a signal, it seems the least we could do is send one. Sure, let's talk it over, I thought, but don't waste time worrying about a moratorium. Let's become a broadcasting planet: start sending and see what happens. In my heart of hearts, I don't really believe there are grave dangers involved in reaching out to our ET cousins. Yet, my views on the subject have evolved. I am increasingly swayed by those who urge caution, who say we must at least talk it through before we sanction more aggressive broadcasts.

It is still my personal belief that technologically superior aliens would be very unlikely to present any danger, and I can tell you plenty of reasons why. It's quite unlikely that aliens so advanced they *could* come here would also be so primitive as to want to harm us or so inept as to harm us inadvertently. It seems extremely unlikely that such an advanced entity would perceive of us as any possible threat. I can describe my sense that true intelligence will not be aggressive or careless or clumsy. I can rationalize that technically advanced civilizations are likely to have survived their own adolescence by becoming "morally advanced" in a way that kept pace with their technical achievements. I could offer my opinion that, if anything, such civilizations might be more inclined to have an ethic of wanting to help species such as ours make our way through the bottleneck of technological risk. Still, I must also admit that these are just my opinions, semi-informed at best. We absolutely can't know any of this. Maybe it's all wishful thinking. There certainly are logically valid arguments for the possibility of great dangers. So how do we proceed, if the risks seem absurdly low, but the cost of being wrong is everything we have, everything we love?

Ultimately I do favor active SETI. For me, the rationale is similar to my enthusiastic support for returning samples from Mars and other planets, but only with appropriate precautions. Even though I personally don't think there is any significant risk to these missions, I agree that we must take seriously the problem of containment. I could list other scientific experiments that I support even though they carry nonzero existential risks. (Biotechnology comes to mind.) During the first atomic bomb tests, the brilliant Manhattan Project physicists/designers were pretty damn sure (almost 100 percent certain) that this would not initiate a chain reaction that would destroy Earth. Although I wish now they had never built those damn things, at the time, I would have done the same. And what about those giant particle accelerators that some physicists admit may have a teensy, weensy chance of destroying the known universe? Some level of risk is inevitable. The only way to be completely certain of safety would be to not explore at all. Even then, there is no guarantee of safety. If the universe does have dangerous elements, what is to stop them from coming and looking for us? You can't be too careful—or, rather, you can. You could never leave the house in the morning so as not to expose yourself to random events. Even so, something could still fall on your house. At some point in the future, we'll want to reach out to other minds across the void, even though we'll never know that it is entirely free of danger.

One could make an argument that even just listening with passive SETI is not without some risk. What if we discover something horrible or depressing or shocking or otherwise dangerous? What if a message is designed to trick us into behaving in some way that does us harm? Few would argue that because of these dangers (which seem absurdly remote), we should not proceed with listening.

Yet if we really think there could be something dangerous out there that might come to harm us, to what lengths should we go to prevent it from learning of our existence? If we really wanted to do everything to guard against a possible existential risk from evil aliens, we should not only ban METI broadcasts but also make ourselves invisible. Should we turn out all the lights, treat the planet like London during the Blitz and enforce a global blackout? Ban all radio transmissions? No, of course not. At some level we live with risk because it would compromise us too much to worry about it. Right? So we agree we are operating on a continuum bounded on one end by excessive caution and paranoia. Then it's just a question of where to draw the line.

With planetary exploration, after assessing the risks of interplanetary contamination, and a process of international consultation, we elected to proceed. Ultimately, I think this will be the right decision with METI as well—but only when we're ready, after we have some kind of inclusive global process and agreement about how and whether it should be done.

The Berkeley statement does not argue that METI broadcasts should be forbidden. It says, "We strongly encourage vigorous international debate by a broadly representative body prior to engaging further in this activity."

I've come around to agreeing, and believe that a voluntary moratorium while such a process is worked out would be the responsible approach. Arguably, if some humans, on behalf of our race, speaking for our planet, plan to announce our appearance on the galactic scene, we all ought to have a say in the matter. I think that having such a dialogue is more important than asserting anyone's near-term right to broadcast.

Stop, Look, and Listen

A cost-benefit analysis is particularly hard when you are dealing with very remote risks that could possibly be completely catastrophic. One of the points I made at the AAAS press conference was that, in addition to remote but real risks, we should also consider possible existential benefits of alien contact. By "existential benefit," I meant something that could fundamentally help us or change us in a way that allowed us to persist, to survive, to thrive.

A message *could* provoke a response that has very positive effects. After all, we are currently living with some existential risks, as the famous Doomsday Clock of the *Bulletin of the Atomic Scientists* reminds us. You may quibble about how many minutes from midnight we are, but you'd have to be extremely naïve to think that with our current nuclear arsenals, changing climate, and a host of other issues there is not some risk in maintaining the status quo. We don't know how to construct a sustainable technological civilization with assured longevity. Any advanced civilizations out there will have solved this problem. Perhaps they can give us useful information. This is no more far-fetched than the notion that they would want to harm us. We sure could use some help in solving this global civilization puzzle. If we are depending on alien intervention to solve our problems—well, that's a pretty thin reed of hope. Yet, even the discovery that they do really exist could help us achieve a more united outlook, and could also be seen as an existence proof that there are solutions to our seemingly intractable problems.

Right now I would submit that lack of self-knowledge is an existential risk. An inability to act with global intent and consideration of multigenerational timescales is an existential risk. It may well be that the greatest value of METI will come not from anything we learn in response to a message we send, but from what we learn about ourselves in the process of attempting to reach some common ground and find our global voice. If we decide to send a message to possible extraterrestrials, we are also sending a message to our descendants. We are gifting them with possibilities of both benefit and harm. Such an endeavor requires us to form an alliance with future generations, to enter into a common project with them. That is clearly something we need to learn how to do. So, then, starting the conversation about whether to broadcast, the effort to have a globally inclusive process, becomes a worthwhile goal in itself.

As with biotechnology or artificial intelligence, you won't be able to stop someone from going rogue. Yet what we could do, by declaring a temporary moratorium and getting a lot of weight behind it, is to nurture a culture of responsibility in the field, as Max Tegmark and his colleagues are attempting to do with AI. Nobody is legally forbidden from doing anything, but the culture in their field is shifting. Similarly, we can create guidelines, urge our community to abide by them, and let it be known that it will be considered rude and egotistical to broadcast during this consultation phase. Say we declared a voluntary moratorium for ten years. If anyone unilaterally decided to start a broadcast during that time, it would amount to a stunt, comparable to those that have already been conducted, and would not obviate the need for a coordinated global approach. If anything, it would highlight the need.

What might such a process look like? There are two components. We would need a wide professional consultation, perhaps modeled after the Asilomar process for biotechnology, or the Future of AI conference in Puerto Rico. It should be broadly international and interdisciplinary, including experts in ethics, international law, and the study of existential risk. The goal would be to come up with some kind of consensus road map and best practices.[5] We would also need some kind of mediated global public discussion.

At the very least we could take some time to collectively assess the potential risks and benefits. We could also discuss what kind of message we would like to send, how to inclusively construct such a message from Earth, and what we will and won't reveal about ourselves. Some have argued that we should avoid drawing attention to some of our less exemplary characteristics, such as our history of deadly internecine conflict and our current inability to stabilize our global environmental impact. It has been suggested that exposing these warts might even be dangerous, revealing us to lack promise or seem like a threat. Others have promoted honesty. Lewis Thomas once wrote that we could send only Bach, "but that would be boasting." True. Or we could send only Hendrix, but that would be showing off.

As I've described, projected advances in radio technology suggest that the time when any lone yahoo can claim to represent Earth, and effectively announce our whereabouts, is coming, but not for a couple of decades, during which time a moratorium could more or less be expected to hold. If this adds a slight sense of urgency to the conversation, maybe that is all for the good. During that twenty years, things would not remain static. We are going to be developing the tools and building the instruments to scour the atmospheres of exoplanets, perhaps finding biosignatures or even technosignatures. Such a discovery would obviously force the issue of active SETI. In this same time frame, we can increase the power of our radio and optical searches by many orders of magnitude, and see if, under more systematic scrutiny, the silence holds up. Also we can attempt to involve the people of Earth in an extensive exploration and discussion of what our role in the cosmos should be.

How would we do that? What might such a public process look like? This question was not the main focus of the 2015 workshop, but after the symposium, I had lunch with Brin and Tarter and we batted some ideas around. Jill, with her mix of pragmatism and idealism, spoke of a wish to find the right tools that might be used for a Web-based project where broad global input would be sought and somehow organized and facilitated. I love that idea because if we could do this right, the process might also be helpful for a wide range of issues that humanity is facing right now involving tough global choices on how to wisely deploy transformative technology.

Several groups have initiated various online processes to try to get broad cross-cultural input and acceptance of content for possible messages from humanity.[6] Curating content is easier than gaining buy-in, but these efforts could form a template for the tougher question of how to get some sense of global approval for the idea of messaging and some wider appreciation of the issues.

Brin's idea is that this might be started with an innovative, widely aired television series. He suggests that such a deliberation would actually make for great drama, airing the debate for the global masses, and potentially reaching an audience of millions or billions. It's an enticing vision. Done well, it could serve a massive educational mission, covering astronomy, planetary science, biological and cosmic evolution, human history, communications technology, risk assessment, and ethics— all in a fun and compelling "reality" show in which articulate hosts hold a friendly debate over whether humanity should decide to purposefully reveal ourselves to any other civilizations out there. And then what? Hold a vote? Why not?*

In his AAAS talk, Shostak belittled the value of attempting a global process. Sure, it will never be perfectly inclusive or democratic (or substitute any value that you would like to see a global conversation have), but that is no reason not to give it a try. In fact, it is quite clear that we need the practice. I think that Shostak and Vakoch are right that achieving global consensus will be very difficult. That difficulty speaks directly to our most challenging problem as a species right now. That is why I've decided that Brin and the signers of the Berkeley statement are right. If we really can't hold some semblance of a global discussion about this, then perhaps we're not really ready to start shouting into the cosmos.

Doing It Right

I hadn't heard from Zaitsev in a couple of years, but as I was finishing up this book I reached out to him by e-mail and heard back almost instantly. He is no longer pursuing his provocative messaging activity. His last interstellar transmission, called "A

---

* Any such effort would be rightfully critiqued as not reaching or representing all the people of Earth. It would be necessary to counter this creatively and aggressively, with effective worldwide outreach.

Message from Earth," was sent in October 2008 from Yevpatoria toward the star Gliese 581, a red dwarf about twenty light-years away that is known to have several planets, one of which may be habitable. As of this writing, the signal is nearly halfway there. If a reply is sent instantly, it could be received on Earth by 2048. After this project, Sasha suffered some health problems and retired from the Institute of Radio Engineering, where he had worked for forty-four years, since 1968. Today he is in good health, living as a pensioner in Moscow, and still writing about, and advocating for, METI. He is not planning any more broadcasts.

I don't think Zaitsev has really done any harm. I'm not that worried about consequences from his few targeted broadcasts. Maybe in a roundabout way he did us a favor by forcing this issue to the fore of our awareness.

Seth Shostak addressed the active SETI controversy in a 2015 *New York Times* op-ed in which he argued against a moratorium, ending with the statement:

I, for one, would hesitate to let a paranoia based on nothing more than conjecture shackle the activities of our children and our children's children. The universe beckons, and we can do better than to declare that future generations should endlessly tremble at the sight of the stars.

I am with him there. I would like to see humanity proactively answer the call of the universe that beckons like a liberation. I would also add that there is no hurry, and much to gain by doing this the right way. I don't think we can properly "speak for Earth" if we're in too much of a rush.

Many scientists are uncomfortable with the idea of taking the public pulse before embarking on any research project. We turn on our computers and TVs and are confronted with a cacophony of ignorance and superstition, and we don't want to hold our ambitions for scientific or policy progress hostage to the illiterate mob. In some areas this is understandable. What if we were required to take a vote about programs involving vaccination or the fluoridation of water? (Well, actually, in a democracy, we sort of are.) Should we wait for consensus before implementing every solution that science says is clearly needed?

No, but this is different. Science itself is divided. One cannot look at the list of signatories to the Berkeley petition and credibly claim that expert opinion does not swing both ways on this debate. We have the obligation to seek some semblance of a global consensus. We could try to do it right, and in some small way demonstrate that this can be done.

Making Earth drastically more visible to extraterrestrials is not the most worrisome technological change facing our planet. Unwanted and dangerous alien attention is far from our most pressing problem. We've got more immediate worries. Yet this is something that our community, my community, needs to address responsibly, and in a small way it could serve as a model for how we could approach such issues. Because it is not pressing, and is even kind of fun to talk about, perhaps we can use it to widely engage people in a discussion of humanity's role on Earth and in the universe, with an inclusive process by which a policy, an approach, a set of guidelines, is hammered out. No, we would have no authority to impose or enforce our rules on anyone. Yet if we do our job well, perhaps we can gain some measure of moral authority, of respectability, and help create a culture of responsibility to guide our behavior so that when we do reach out to the universe, we will seem like a civilization that truly intelligent minds might actually want to engage with.

We can't avoid these kinds of questions. Like it or not, we humans are changing our world, and it would behoove us to learn how to make decisions about how we should do so. Clearly in order to survive to become one of those advanced species, capable of sustained galactic discourse, we need to develop mechanisms for collectively anticipating, discussing, and acting on long-term threats before they become crises. As I'll discuss in the next chapter, we are developing such mechanisms.

Vakoch and Zaitsev are right when they say that becoming a broadcasting civilization is part of growing up. Even if nobody is listening, in becoming the kind of species capable of making purposeful broadcasts, we'd be changing our role on the cosmic stage and acknowledging our changed role on this Earth. But you can't will yourself to be an adult by starting to do grown-up things without the maturity to handle them. You don't hand a child the keys to your car in the hope that driving it will force them into maturity. They are (hopefully) not allowed to operate heavy machinery until they are ready.

We do need to transition to the kind of species, the kind of civilization, that can initiate projects that will run for generations, for millennia. Yet we can't really do that by having some of us just decide to start broadcasting. We need buy-in. We need patience. We need a thousand-year plan. In my discussion of "the continuity criterion," I explained why for intelligence to be an observable phenomenon it cannot be ephemeral. Given the vastness of time, in order to be picked up, a signal or beacon needs to be on air for millennia at least. Can we do that? We have the technical knowledge but not yet the self-knowledge.

Philip Morrison once suggested that perhaps after one hundred years of just listening it might be time for us to send some messages. By then we may know a lot more. Maybe we could decide now that we want to start broadcasting in another forty years if we haven't learned anything before then that would indicate that this would be a mistake. Such a way of thinking, making such a long-term plan, seems foreign to us, alien, if you will—which is why it seems to me like a good idea. Maybe it's okay to admit that sending is a project for the next generation. By making a multigenerational plan, we begin to become that which we seek. I think we should plan to broadcast for one thousand years and see if we get anything in return, but it's fine with me if we wait forty or fifty years to begin.

We are indeed still so young. Sixty years after *Sputnik* and Ozma, the *Voyagers* are still just beginning their journey out of the solar system, now about twenty billion kilometers from the Sun, about 0.05 percent of the distance to the nearest star, and the *New Horizons* spacecraft will soon join them. Having passed by Pluto, it, too, will now coast out toward the stars— five tiny pieces of Earth stuff returning to the galaxy. A sphere of weak radio and television babble also expands outward. But don't worry: our embarrassing belches of sitcoms and infomercials are probably too weak to be picked up. We have not really announced ourselves to any galactic community that may be waiting.

All meaningful communication involves risk. Every new relationship starts with taking a chance. If it weren't the unknown, why would we need to bother exploring it? As a society, and as a species, we have to become more comfortable acting without definitive knowledge. We have to be smart and play the odds. To take the most pressing current example, to wait until we are absolutely sure about the risk of climate change is to wait too long to do anything about it. Someday we will want to shout our names out there, and we will have to do this without fully knowing what we're getting into.

Ultimately there may be no completely rational basis on which to decide. We can discuss probabilities and scenarios and continue to gather evidence, but the decision whether or not to make ourselves known may come down to what kind of universe we think we're living in. I still feel that we cannot be frightened of the universe. I believe that we should start pursuing active

SETI, reaching out to our space brethren and sistren, letting them know they are not alone and seeing if we can spark some cosmic conversation. There is no way to defend ourselves from, or hide from, some superadvanced entity that means to do us harm. So let's not cower, but let's take a good long look before we leap.

We already know a lot more than we did when we started doing SETI, and we will know so much more in the next twenty years. At some point we'll have to decide we're ready and go all in. The Universe or nothing. When will we be ready? When we've examined a lot of exoplanets. When we've pursued SETI (the listening kind) for another couple of decades. When we've been able to have some kind of satisfactory global conversation about it.

In the meantime, let's figure out how to manage ourselves as a planet, and speak with one voice. Let's grow up, and then introduce ourselves.

8

EMBRACING THE HUMAN PLANET

We travel together, passengers on a little space ship, dependent on its vulnerable reserves of air and soil; all committed for our safety to its security and peace; preserved from annihilation only by the care, the work, and, I will say, the love we give our fragile craft. We cannot maintain it half fortunate, half miserable, half confident, half despairing, half slave—to the ancient enemies of man—half free in a liberation of resources undreamed of until this day. No craft, no crew can travel safely with such vast contradictions. On their resolution depends the survival of us all.

—Adlai Stevenson

We need to overcome the habit of considering outcomes of human activity as more imperfect than those of nature's activity—understandable as such a habit may be at the current stage of development—if we are to talk about what is going to happen in a faraway future.

—Stanislaw Lem, *Summa Technologiae*

Originally you were clay. From being mineral, you became
vegetable. From vegetable, you became animal, and from
animal, man. During these periods man did not know where
he was going, but he was being taken on a long journey
nonetheless. And you have to go through a hundred different
worlds yet. There are a thousand forms of mind.

—Rumi

Usufruct

It has taken us a long time to realize that the world is not infinite, and not infinitely resilient. This discovery is still ongoing and is at the heart of our angst about the Anthropocene.

Here in DC, working close to the pulse of power there is much that remains inscrutable to me. During the time I've been working here, the U.S. Congress has been widely described as a place where nothing gets done, an institution impervious to data or insight. How strange to be ensconced right across the street, in the Library of Congress, arguably the greatest collection of knowledge in the world. The John W. Kluge Center, where the scholars work, is in the Jefferson Building, the original and most impressive of the Library's structures. I've heard it described as the most beautiful building in the United States, and that's hard to argue with. The architecture and interior details seem to be relics of an American government with different values. It is a lavish temple to scholarship, indulgently and lovingly crafted, built to last, full of artwork, symbolism, and inscriptions meant to provoke and inspire. Looking out my window across First Street to the Capitol Dome, you can feel the drift of standards and ideals over the generations. You cannot escape the shadow of Thomas Jefferson here at the Library, which he founded. After the original burned in 1814, he sold his own books to Congress so they could start afresh. Today these are on display, and I've enjoyed browsing through Jefferson's eclectic original collection. His six thousand volumes seem to cover everything: poetry, gardening, architecture, literature, law, humor, philosophy, and all the natural sciences. At the time, it was possible for an educated person seemingly to master all areas of scholarship.

One of my favorite places to work is the elegant, Art Deco–style Science Reading Room, in the Adams Building. The buildings, each named after one of our founding fathers (Jefferson, Adams, and Madison), are connected by underground tunnels, hidden veins, coursing with people and books, running beneath the streets of our capital. In order to get to the science collections, I walk beneath Second Street from Jefferson to Adams, take the elegant paneled elevator to the fifth floor, and emerge into the expansive reading room, its heavy wooden tables lit with old-timey lamps. Facing me across this soothing, capacious expanse is a wall engraved with a quote from Jefferson over a mural displaying brave and noble-looking people subduing the wilderness with horses, tools, and guns. It reads,

> The Earth belongs always to the living generation. They may manage it then and what proceeds from it as they please during their usufruct. They are Masters too of their own persons and consequently may govern them as they please.[1]

When I first saw this, I wondered at that strange word. *Usufruct?* I assumed it was some fancy elaboration on the root revealed in the first syllable, so in my head it translated as "use." It bothered me that Jefferson seemed to be saying that the world is there only for the use of the current generation. Future generations can worry about themselves. This seemed emblematic of the mind-set that got us into the Anthropocene dilemma.

We are all products, and prisoners, of our times, no matter how enlightened we fancy ourselves, and in Jefferson's time, the notion that Earth is infinitely durable could have seemed reasonable and not in conflict with even a very learned person's knowledge and experience. Today we know that the solar system and the universe are not infinite but so vast that we cannot really imagine a day when they would be filled up. So, to us, they are functionally infinite. In Jefferson's day, even though they knew the finite geometry of Earth, with the western frontier still so vast and open the world must have seemed functionally infinite. The notion of a finite planet that must be cared for, and stewarded for future generations, might have seemed strange. At least to a European man of means in North America, the West seemed unlimited, unconquered, unowned. Of course it was easier to think of the world as infinite and yours for the taking if you did not see that those lands were already occupied by human beings with their own civilization. Pondering this inscription enhanced so many of my contradictory feelings about the founding fathers, and Jefferson in particular.

The second half of the quote carries an admirable message of personal freedom. He was also saying, it seems, that people are masters of their own persons and that the living generation must be free from the shackles of their ancestors. Yet who was enabling the leisure time for all his reading and writing, design and philanthropy? I remember as a student visiting Monticello and marveling at the ingenious inventions and architectural innovations. At one point our guide proudly pointed out that all the furniture in the larger rooms was cleverly designed to be moved around and set up differently for different occasions. Well, that is nifty, but it sounds like a lot of work. Who was moving all that furniture around so Jefferson and his family and visitors could dine and plan and read and invent? Oh, yeah, right. It was easier to live a life of leisure, creativity, and scholarship if someone else was doing the lifting, building the libraries, and cooking your meals. Embodying the morally complex origins of the United States, Jefferson, the silver-tongued champion of freedom and architect of our Constitution, was also a slave owner. That has always been hard to square with the Jefferson of "all men are created equal."

When we think about the moral and intellectual failings of wise and enlightened people in the past, it is always worthwhile to try to imagine how our current culture and ideas will look in the future. Which of our mores will seem ghastly or shortsighted? I've played this game with many people, and answers that arise include eating meat, eating any food that comes from living beings, keeping animals as pets, the outrageous economic inequities that we accept, the fact that we go about our daily business while some people are starving, militarism, nationalism, the fact that we still accept to the extent we do sexism, homophobia, racism, and ableism. Lately I've been wondering if someday soon the insistence that everyone have a job in order to have a decent livelihood will seem a strange and primitive relic of the time before machines did most of the grunt work. In the context of the subject of this book, I've been wondering about our neglect of global thinking and the fact that we so often do not act out of concern for our descendants. In some more enlightened future, these habits may seem as unpalatable and unacceptable as cannibalism or child abuse does to us today. If the human race is to thrive, if our civilization or anything that we might want to have evolve out of it is to have a bright future, then it is these patterns of thought (the shortsighted, the tribal, the less than global) that we will need to move beyond.

One of the great benefits of working at the Library of Congress was contact with scholars from other fields who were thinking about some of the same topics but were approaching them from very different perspectives. Among the first people

I encountered was a young French environmental historian named Jean-François Mouhot, who was collaborating with John McNeill, the historian at Georgetown who has been a leading scholar of the Anthropocene. I was lucky that Jean-François was studying there when I arrived. He immediately started turning me on to books and suggesting sources. We ended up collaborating on a program called the Evolving Moral Landscape: Perspectives on the Environment, Literary, Historical and Interplanetary.[2] At this event, Jean-François made a very provocative comparison between the institution of slavery in antebellum America and our use of fossil fuels today. Many people at the time, he said, acknowledged that slavery was evil but kept slaves anyway because it was how they powered their society and they could not conceive of life without them. So they lived with a moral deficit they could never pay. Clearly our society has still not settled that debt. Similarly, today, Jean-François has written,

> Our abundant energy gives us an extraordinary power, and this is why it is so hard to do away with all the luxuries provided by our modern machines—even when we are convinced that using them is morally wrong.[3]

We power our society in a way that we know is doing harm, but still we persist. So we are running up a moral debt. Yet can we really compare driving cars and heating our homes with coal to the great evil of holding human slaves? I found his comparison offensive, but in a good way, the way that shocks you into realization. The slave trade resulted in millions of deaths, untold more ruined lives, unimaginable suffering, and an enduring legacy of poverty, disenfranchisement, and trauma. Jean-François argued that if we include the extinction of many species, and the mass displacement, hunger, and warfare that could result from the scarier climate change scenarios, then these crimes are not so obviously beyond comparison. He later elaborated on this in several published papers, including an essay called, "Thomas Jefferson and I,"[4] where he specifically compared his own fossil fuel use with Jefferson's holding of slaves. I am still not altogether comfortable with this comparison. I think this discomfort is exactly his point.

Having spent two years looking up at this mural and ruing Jefferson's limited outlook on the future, I recently learned that I had misconstrued his words. I discovered that Jennifer Harbster, one of the wonderfully resourceful reference librarians at the Library of Congress,[5] has studied the original letter from which this quote came, and from her I learned that I had gotten the meaning wrong. Jefferson was not saying that the living generation has the right to do anything with the world and leave it in any kind of shape. The word *usufruct* does not mean what I had assumed. It's actually a concept that comes from Roman law and means the right to use something temporarily, for a limited period of time, only if it is left unharmed for future users. So if you rent a car, you can drive it anywhere, get it dirty, and put it through its paces, as long as you don't dent or scrape it. That is your usufruct, and to violate it will cost you. Jennifer, having studied the full text and the context for this quote, has convinced me that Jefferson was actually saying that the living generation had the right to Earth only insofar as they left it in decent shape for their descendants. Well, what do you know?

Searching for Terrestrial Intelligence

We've shed so many illusions, but some stubbornly remain. We are still getting over the illusion that our world is infinite. We teach in school that it is not, but have not yet integrated this fact into our lifestyles and global economy. Another illusion is that without our help or hindrance, "Nature" will take care of us and our planet will always be a benign place for human life. It's easy to understand how we could have gotten that impression because we've been so lucky. We were born in a garden, and we've become dependent upon a very steady climate that is not a long-standing feature of Earth. It's shocking to realize that we are disturbing the pristine balance of nature in a way that threatens our future wellbeing, along with that of so many other species. It's perhaps even more shocking to learn that this balance is an illusion. Now we're not just altering the planet; we're populating it to the point where "normal" climate changes of the kind that, over the ages, have occurred routinely would be calamitous. Fortunately, now we can begin to make our own luck.

We can't return to that garden of ignorance and bliss, letting the planet take care of itself and of us. So we'll have to take a more active role in our planet's path, but not in the random, unconscious way we have been doing. We've tasted the fruit of science and technology, and now our best chance for survival lies in cultivating planetary knowledge and a planetary identity, in awakening to and embracing our part in this world.

So let's become planetary gardeners. Once we get over the strangeness and fear of this responsibility, we can also enjoy the privileges and freedoms that come with this role. This will include freedom from planetary changes that are "all natural" but massively lethal. It's tempting to romanticize the past, but cowering in a frozen cave while deadly predators growled and licked their chops just outside was probably not all that much fun. We've survived a couple of ice ages, barely, and in the process we became modern humans. With our current population we'd have a harder time making it through another glacial period. Fortunately, we won't have to. In avoiding this fate, we will continue the journey we began in those Pleistocene caves.

What really distinguishes humans, or the human age, from

the rest of life, the rest of Earth history, or the evolution of matter over cosmic history? When we talk of seeking intelligence, consciousness, or self-awareness elsewhere in the galaxy, we refer to something that we possess, seemingly headquartered inside our skulls, and that a head of lettuce does not. We wonder how much dolphins and monkeys have, and marvel at how much cats and dogs seem to have at times (and, at other times, how little). Yet what is it? It is not something that resides fully in an individual. We are cultural animals, and we cannot separate the evolution of individual human intelligence from our novel capacities to pass on information in structures that outlive individuals, in artifacts, songs, rituals, and stories.

The question of what, if anything, makes human beings exceptional is historically and scientifically vexed. Attempts to distinguish human behavior from that of the rest of the animal kingdom have focused, always contentiously, on such achievements as tool use or language. All these efforts have failed or been fraught with ambiguity. In the mid-twentieth century we had "man, the tool maker," but in Tanzania in October 1960 primatologist Jane Goodall, observing a chimpanzee she had named David Greybeard, saw him, and later other chimps, stripping the leafs off twigs to fashion an ideal tool for fishing termites out of a mound. When she informed her mentor, Louis Leakey, of this by telegram, he replied, "Now we must redefine tool, redefine Man, or accept chimpanzees as humans."

Although not everyone has gotten the memo, we've discarded "man" so as not to implicitly exclude brilliant scientists like Jane Goodall. So we needed to redefine "human." Since Goodall's breakthrough observation, many other species have been observed making tools, including elephants, dolphins, octopuses, and several kinds of birds. None of these manipulate objects as intensively, constantly, and innovatively as humans, but these are differences in degree not in kind.

Other candidate attributes for human uniqueness include language and the ability to recognize that other individuals have their own thoughts and feelings, also known as "theory of mind." Elements of language can be seen in birdsong and the infrasonic rumbles of elephants. Yet these generally lack the syntactical complexity and flexibility of human language, which gives us the open-ended ability to express an infinite number of new ideas. With the still-mysterious songs of the cetaceans being a possible exception, no other species writes poetry, constructs complex narratives, and produces an endless supply of gossip. Observations of dolphins and other animals interacting with mirrors suggest that humans are not alone in possessing theory of mind. Perhaps uniquely we have "metacognition," the ability to think about thinking, to reflect upon and make judgments on our own thoughts, to second-guess ourselves, questioning our memories and our decisions.

One human capacity that is very powerful, and apparently unique, is abstract thought: our ability to imagine alternative scenarios, to invent stories and characters, to think beyond the here and now. We can picture distant places and hold different points of view in our minds, project ourselves into a wider frame, learn from the past and imagine the future. This equipped our species, as none before, to survive new challenges.

Abstract thinking allows us to consider hypothetical and unlikely future possibilities. If we can model scenarios in our heads, we don't have to learn by trial and error. We can play out some disasters only in our minds, learning from (and avoiding) potential bad outcomes without needing to let them unfold in the external world. It is actually a form of trial-and-error learning, but we can make some of our biggest errors in our imaginations only, so that we don't have to live through them in real life.

Our powers of symbolic representation and abstract imagination facilitate two uniquely human ways of engaging with the world, one ancient and one brand new: art and science. The making of art may be, at present, the best candidate for a unique activity common to all humans. Every culture has made art, and archeologists use this to distinguish sites of early humans from those of other primates. Today we are unique in doing science, producing ever-evolving technology and, through these activities, changing the world.

Though each term is fraught and debatable, no other species on Earth has the combination of intelligence, consciousness, foresight, culture, tool use, and artistry possessed by *Homo sapiens sapiens.* I don't mean to exalt humanity or present us as some end state of evolution, as the rightful rulers of Earth. I do not regard us as the apotheosis of intelligent life on Earth. As I've mentioned, I'm not even sure that humanity qualifies as an intelligent species, and I've tried to describe how we're in a difficult situation for which we are ill-prepared. Yet denying "human exceptionalism" can actually be a way of shirking our responsibility. So the Anthropocene does put us on a pedestal, but also on the witness stand, and on trial.

These "human qualities" have uniquely enabled us to create a technological civilization, which is the behavior that concerns us here. The Anthropocene transformation of Earth marks us as distinct from other animals in being able to induce, for better or worse, this particular kind of change to our planet. We are equipped to build a global civilization, but can we keep it? Can we maintain it for one hundred thousand years or longer?

We humans are hard-wired for collective action. It goes back to our hunter-gatherer days. A human being alone is not an impressive hunter, but a band of humans is well equipped to survive. We were good not because we were the fastest runners on the savannah, or had the sharpest teeth in the jungle, but because we developed language and strategies and tools. When we invented agriculture and settled into towns, we developed larger social structures and new customs that made use of these innate proclivities to communicate and cooperate, and invented new tools to solve the survival challenges we created along with our new environs. For example, urbanization created horrible public health problems, many of which were solved by the invention and wide adoption of sewage systems. Now we are in need of new tools to handle our global effluence.

Perhaps our core problem today is that, though we are well evolved to cooperate in groups the size of hunter-gatherer clans (dozens or hundreds of people), we struggle to organize ourselves well at larger scales. However, we have evolved, and continue to develop, communication and organizational tools that have repeatedly changed our large-scale social dynamics: written languages; cultural, economic, and political institutions; books, artworks, and recordings; broadcast media, telephones, and the Internet.

At one talk, where I described the list of human qualities (awareness, collective memory, and foresight), my friend the Rev. Dr. Jeff Moore, agent provocateur and planetary scientist extraordinaire, noted that, ironically, this was also a pretty good list of those qualities we might be said to lack in sufficient quantity to navigate our current global situation. He's got a good point. Certainly, in lacking these qualities, previous species of planetary home wreckers cannot be said to have been responsible for their actions. They were innocent in a way we can never be.

We, however, have tasted the fruit of science, bulldozed the walls of the Garden, driven the serpent to near extinction, and genetically reengineered the tree of knowledge. We have a lot to answer for, and unlike the cyanobacteria, we can answer. Unlike them, we have consciousness, intentionality, and choice, and therefore responsibility.

But do we really? Do we, humanity, have consciousness and

intentionality? Does humanity or human civilization as a whole know what it wants, what it is doing, even know that it exists or possesses free will? It's a different question from asking whether human individuals possess these properties, just as it is different to ask if your individual cells have them.

Who Is This "We"?

A friend of mine recently wrote online, "We definitely have the ability to run a clean planet. It's annoying beyond words to not be doing it." To which I responded, "Depends on what we mean by the words 'we' and 'ability,' doesn't it?" After all, if we have the ability, why *aren't* we doing it? If we are talking about viable technical solutions for our energy and climate problems, I believe that humans can easily devise them. In this sense we have the ability. Yet does global humanity have the ability to implement them? It's not simply a question of what choices we should make. There is also the question: can we make choices?

This book and all our conversations about the Anthropocene, climate change, and the human future are rife with prescriptions for what "we" must do to solve "our" problems, to survive and prosper. *We* must end *our* addiction to fossil fuels. *We* must change *our* behavior or expand *our* ways of thinking. But just who is it that is supposed to be making these changes and learning how to run this planet well? "We," the people reading this book, the voting public, the rich people in developed nations with more choices, the entire human race, the biosphere, or the planet itself? Who is this "we"?

When I say that humanity is gradually becoming more aware of our global role, I am fully aware that not every person on Earth is, and that knowledge and conversations are diverse and culturally specific. When I say that we are responsible for getting ourselves in this mess, I acknowledge that many people have not had the information, freedom, or power to make choices that affect our global environment. There is a convoluted and shifting geography of responsibility. Yet it is not meaningless to speak of humanity as a whole, something that is new to Earth and having an unprecedented effect. There is an important sense in which we (the human race) are facing challenges that can only be addressed on a global level. It is essential that we search for, and cultivate, a planetary identity with which to respond to planet-level threats.

This kind of talk rubs some people the wrong way. Some object strenuously to the notion of humanity as a collective entity, in any sense sharing agency, responsibility, and destiny, facing common choices and needing to make collective changes. With its new prominence in scientific circles, the concept of the Anthropocene has diffused far beyond the scientific community and into numerous fields of the humanities. It is also escaping the walls of academia, spreading to humanity at large, where it might really do some good. Various academics from such fields as critical theory, philosophy, environmental ethics, postcolonial studies, and economics are taking potshots at the Anthropocene, subjecting it to critical scrutiny—as well they should. One popular view is that by representing humanity as a single, monolithic entity, we are glossing over various histories of oppression and inequity, and letting the rich peoples and nations of the world off the hook.

In the humanities, compared to the sciences, there is more awareness of the fraught history of claims of the universal made by Europeans, sometimes used to cover over racist assumptions. As literary scholar Ursula Heise explained it, at the Symposium on Longevity of Human Civilization that I organized at the Library of Congress in September 2013,

> The assumption that humans are somehow one entity, because of their biological species, tends to come very easily to scientists, but it's an extremely difficult assumption for a humanist... [A] lot of humanists are very cautious about using terms such as "humanism," "humanity," "mankind," because of the lousy track record these terms have in Western history. A lot of us look at histories of imperialism and colonialism where it was very common to say, "Oh, we stand for the universal human," and the universal human is French or the universal human is British.[6]

Several writers have suggested that the "species view" or "species talk" is inherently oppressive. Listening to Ursula, I better understand the source of this concern, but I worry that this viewpoint sets regional, national, or identity politics against the global view, and I'm concerned that this could itself have an oppressive effect, hampering our ability to care for future generations.

This is not to belittle questions of justice and responsibility and blame. There is a legitimate critique of simplistic species-level thinking. Is climate change a threat to the whole species? Certainly, yes. But certain places will bear the brunt, and the global poor are most at risk. Has climate change been caused by the whole species? Not equally. A privileged few are wreaking disproportionate havoc, and so those of us with any influence whatsoever—and that probably includes most people reading this—are especially obligated to try to mitigate this. An obvious and deserved target is the fossil fuel industry. Some with vested financial interest in perpetuating our current energy infrastructure are doing great damage and endangering all our futures. These same industries have supported a successful effort to sow doubt and confusion about climate change. In the end, this smokescreen will not succeed in obscuring the truth. It is already wearing thin, but it has been a successful delaying tactic, preserving profits for one generation at the expense of others, and has set us on a more perilous course. Is it all the fault of the CEO of ExxonMobil? No. Most of us drive cars, most of us vote. So blame is distributed, but in this analysis, it does not extend to those in poor countries who have neither a car nor a vote.

However, it is sometimes economic disadvantage that leads people to harm their environments. For example, in some places, deforestation and desertification advance as a consequence of the fact that people have no choice but to gather wood for cooking. A huge amount of deadly particulate air pollution comes from the cooking fires of those too poor to use anything but the wood and dung and peat that is available to them. They damage the land, the air, and their lungs because they have to.

If you're in a boat and you realize that the current is pulling you dangerously close to deadly rocks, it is a good idea for everyone to row like hell in the opposite direction. Maybe some of the passengers are more responsible for getting you in that situation. Some are sitting in drier, safer, more comfortable parts of the boat. Some have been well fed while others are too hungry to do any rowing. Nonetheless, you all had better pull together now. Sure, the strongest rowers or those holding the biggest oars should do the most, and those too hungry or tired won't be of much help. A discussion over fairness is vital for running a better boat and avoiding the rocks the next time. If everyone had been consulted and well cared for, and if there were not so many people crammed on the damn boat, you might not be in this situation. But with the sound of the surf crashing on the rocks and waves breaking over the bow, it's best to put your energy into pulling toward safety. This may be promoting a "boat view" that distracts from the issues of fairness and responsibility, but still, it's best to row while you work things out.

Or let's grow this analogic boat into a luxury liner and imagine it as the *Titanic*. It's true those in steerage got a raw deal, but everyone went down. Well, most did. While the rich, and the women, survived disproportionately, I don't think any of the crew or passengers had a great night. From a "ship view," everyone would have been much better off avoiding the iceberg.

A species view makes more sense when we consider a deep-time, evolutionary perspective. A million years hence, our time will have been a moment at which life redefined its relationship with the planet, a juncture in the evolution of the biosphere, one caused by a single species. This doesn't mean that these changes happened all at once, everywhere, caused equally by every human being, but from this distant view, you would not be aware of these complexities. You would see human civilization as something that suddenly came along and had a major effect on the planet.

Surely during other important evolutionary developments there was inequity and injustice. In chapter 5, I describe how we narrowly avoided extinction and survived a genetic bottleneck back when we all were living in southern Africa. At the time, there were surely cultural complexities, inequalities and tragic unfairness. Yet we can describe this now as part of our collective story, as an evolutionary event that the human race endured and survived. We. Humanity. We did not get together and vote that in a certain year we would all give up hunting and gathering, start agriculture, and move into villages. We can describe these historical events meaningfully with a "species view" even though, in detail, they must have been much more convoluted. When we say that humans left Africa and populated the rest of the continents, we are imposing a species narrative on something that was undoubtedly culturally complex. Only some humans left. Maybe only the elites got to go. Or maybe the elites got to stay behind and others were forced to leave. There's so much we'll never know. There are details glossed over, responsibilities not assigned, and injustices left unaddressed when we simply say, "Humans left Africa and peopled the rest of the planet." Yet this is a true statement, just as it is now a true statement that the human species is altering planet Earth in an unprecedented way. The granular does not invalidate the general, or vice versa.

Humanity is one thing right now, a differentiated and conflicted thing that is having an unprecedented effect on our planet. We are not a harmonious, coherent entity. That's the problem, but that does not obviate the value—no, the *necessity* of taking a species-level view of ourselves. In fact, you could argue that even the species view is still too narrow and self-centered. We can draw back farther and ask, "What is the biosphere doing to itself right now with this strange cephalization and self-torture, and why is the planet behaving like this?" It's fine to assert the importance of difference, of complexity within the human superorganism, but to deny the validity of a parallel and conjoined species level global narrative is regressive, and ultimately runs counter to the needs and aspirations of all peoples.

This is one of those false dichotomies, amplified by binary political thinking, that interfere with sensible approaches to our global problems.[7] People love to find a villain and to map the climate crisis into their preconceived political notions. They want to blame it on capitalism, colonialism, racism, elitism, or other -ism schisms. All these have some grains of truth. The error is to assume we can't be concerned about humanity at multiple levels simultaneously. We need the ten-thousand-year view alongside awareness of the daily struggle. We must see both the local and the global if we are going to get anywhere. Those who refuse to do so are asking us to pit survival of global humanity against the survival of, and justice for, local cultures. That is a lousy choice, and one that need not be made if we allow room in our minds for both views. One hallmark of maturity is to be able to see two seemingly contradictory viewpoints simultaneously. I'm not advocating that we deny or downplay the need for regional climate justice. It's clear to me that in addition to cutting back on fossil fuel use and hastening alternatives in developed nations, the other essential component in ensuring a sustainable future is to work for a more equitable world. People, women in particular, who are educated and not in poverty have smaller families and more freedom of choice as to how and what they consume. As awareness and opportunity spread, population growth slows and poverty declines. As nondestructive choices are made more available, local concerns align with global needs.

The global view has historically been a source of compassion for the less fortunate. Identification with everyone on Earth includes our brothers and sisters in poor and struggling regions. Thirty-six years before the word *Anthropocene* became a subject of scientific debate, economist Barbara Ward, who coined the term *Spaceship Earth*, explicitly connected care of the planet with justice for the poor, writing, "the careful husbandry of the Earth is *sine qua non* for the survival of the human species, and for the creation of decent ways of life for all the people of the world."

Don't dull your appreciation of the evolutionary significance of our time and the global nature of our struggles, responsibilities, and tasks by focusing exclusively on our internecine struggles. In the absence of alien invaders or some other outside threat, we have to find it within ourselves to recognize that we are all in the same boat. If we really want to address climate and the other problems that threaten our future, what we need is a human identity politics. I don't offer this in opposition to any particular identity politics, but in addition to these. Wise women and men have always understood that fighting for the rights of their people is not counter to, and indeed ultimately requires, a vision of the dignity and common purpose of all people.

So I reject the notion that *species talk* is inherently dismissive of the needs of oppressed people, but there is indeed a problem with our uncritical use of collective terminology. When we, as we always do, use that term *we* to talk about what we humans need to do, we are often invoking a sense of ourselves that is quite different from the way we are. We are individuals acting with certain values and desires. Yet, collectively, we behave in ways that do not reflect the desires and values of some of us, or perhaps even any of us. Most of us are not consciously choosing to endanger the elephants or melt the polar ice. But we get down on ourselves. We talk about what horrible, selfish, destructive creatures we are. Yet we are not. We are just confused. There is a great divide between the clear desires of talented and compassionate individuals and the perplexed actions of the collective. We conceive of ourselves, humanity, as a whole, and talk as if we were, collectively, a decision-making entity. This comes naturally when we discuss threats that are global in nature, and we don't always realize the disconnect. So we ask, "If we have the ability to run a clean planet, why aren't we doing it?"—not recognizing that we are talking about two very different *we*s in the same sentence. This identity crisis is at the core of the Anthropocene dilemma.

Global problems require global responses. When we use this language "In order to survive and thrive, we must..." change our habits, cut down our emissions, intensify agriculture, and so on, we are actually describing planetary changes of the fourth kind, the desire to take action globally, collectively. Yet have we, as a species, as a global civilization, ever made a decision or deliberately undertaken any kind of action? Is this even possible? Yes, indeed we have, and it is. In chapters 4 and 5, I give several examples. While excessive nationalism and tribalism are a hindrance to planetary changes of the fourth kind, and thus a threat to our survival, our world is, over time, being knit more closely together. Globally, we are not as isolated from one another as we were as recently as a century ago, and our communications with, and connections to, one another are greatly enhanced. This will be even truer for our kids, who are digital natives. Online discourse does not stop at borders. Even those societies that try to stamp it out and control it are fighting a losing battle. They provoke a reaction, a desire to join. History and changing technology are with those seeking wider connection.

Worldwide Mind

Referring to the global "we," do we make choices? Are we conscious? Seriously, what are the properties of consciousness, and can you say that we have them? We have economic models of our behavior, assuming that each of us acts and responds in some predictable way. We have models that tell us world human population will peak at about ten billion later this century. It's almost as if we don't have any say in the matter, but are just fated to play out the projections. What if someone told you a model predicted you and your family would move to Texas? You'd say, "No, we are going to stay right here. We don't even like barbeque and line dancing." Yet you can't argue in the same way with a model that predicts that a certain number of people from California are going to move to Texas. It's as if in groups above a certain size we are not conscious actors, or even if we are, we're as predictable as inanimate objects.[8]

I'm always struck by the language we use to describe markets, or "the market," as if it were a metaphysical creature with a mind of its own. Just to pick one typical article at random from the *New York Times*, we learn that "If the market really believed there was a high probability of default, you would see a much more negative reaction," "The markets are sending this complacent message," and "markets have been relatively sanguine." The market has beliefs, moods, and feelings. It exhibits behavior and sends us signals that we try to interpret. We read about the market being spooked, the market feeling confident, the market reacting emotionally to this or that event; we speculate endlessly on what the market will decide to do.[9] These seem to be slightly more than metaphors. The market really seems to have a mind of its own, but is this perception just a bit of magical thinking? Maybe not. Perhaps financial markets really are some sort of primitive, collective global mind that is evolving.

I notice something similar in the way we talk about "humanity." Within our family or our team or department or company we feel some agency, but when we discuss the actions of humanity, it is often as if we're describing some large external agent beyond our control or responsibility. Are those disasters caused by humanity caused by us? Or are they like tornadoes and asteroid impacts? Is humanity something we are in charge of? Or is it like the market? Is our global human civilization capable of acting with conscious intent? As individuals or in small groups, we're capable of taking in new information, making decisions, changing plans, and acting accordingly. We can adapt our behavior to unforeseen and unprecedented circumstances. Yet, globally, we seem more akin to what we would consider a primitive organism, acting unconsciously with a much simpler, less adaptable nervous system.

To consider the source of planetary changes of the fourth kind, of global-scale intention, we need to have some understanding of consciousness and intentionality on an individual level, but the question of where and how intentionality arises is deeply mysterious.[10] How does it happen that one group of atoms (you), built into cells, organized into a human brain, can anticipate the future, desire to change things in your surroundings, and then actually do so? We are so used to this we don't see how profoundly weird it is. We think that telekinesis, the ability to move external objects just with one's mind, is impossible, and that to believe in it is some kind of flaky woo that breaks the rules of science. Yet every time you do something as mundane as raising your cup of coffee to your lips, something nearly as strange happens. How can one assemblage of matter, your brain, decide to move another assemblage of matter, your arm and hand, and cause yet another object to move? Other matter doesn't behave this way. I don't recommend thinking about it too deeply before you've had any coffee or after more than four cups.*

Nobody knows where intentionality comes from, but it is likely an emergent property, not wholly manifested in any location within our cognitive apparatus but rather emerging from the pattern of interactions between the parts. Does the whole shifting, neurotic circus of consciousness arise from feedbacks between neuronal circuits constantly switching and firing? This mystery is related to our Anthropocene dilemma, because there are interesting parallels between our political and deliberative processes and those interior workings of a brain that seem to give rise to conscious behavior. These processes may even represent forms of cognitive activity. Our collective mechanisms for recognizing and responding to dangers, for coming to conclusions and choosing actions, may seem, from here on the ground, like a hopeless cacophony of competing voices, ideas, and values. But maybe this is just how our global brain operates, how it makes up its mind.

What if you could reduce yourself to the size of a neuron

---

* That stereotypical stoned college student asking, "Hey, have you ever *really* looked at your hands?" is actually on to something.

and descend into your own human brain to observe it while you were working something out. It might be disconcerting. Groups of neurons firing everywhere, some stimulating their neighbors, and some inhibiting others, perhaps holding different thoughts in different locales, all battling, trying to win the day. Even a coherent thought, if you could see the process up close, might seem to arise out of a mess of confusion—the chaos and conflict of an active mind making itself up. Perhaps consciousness is always some messy kind of a Darwinian or democratic process with a balance of competitive and collaborative forces winnowing patterns and ideas and weighing actions. You make up your mind, but you still harbor doubts. Any smart person pondering any complex situation always will. Consciousness need not be neat, orderly, and unanimous.

I suspect that markets and democracies, political discourses, competing and interacting institutions, and the Internet itself are all emerging and evolving forms of global-scale cognition, made up of components evaluating information and urging different actions, feeding back upon one another, and producing some sort of imperfect, weighted consensus to achieve thoughts and take actions. Collectively, these arguably have the properties of some kind of mind.

Who is to say that some such pattern of globally connected nodes cannot develop some sense of identity, a purpose, and an instinct for self-preservation? As this global mentation arises, we (individual human beings) may not be aware of it. Is a neuron in your brain aware of your thoughts? We may be participating in a form of large-scale consciousness, and intelligence, that we don't individually perceive, and then observing the behavior of that emergent being, sometimes being surprised by its actions. In this view, with democracies, market regulation, environmental activism, international diplomacy, and other activities, this globally embodied mind is learning to exert some self control to avert its most self-destructive behaviors. With these activities we are helping it to perceive new threats and respond constructively. Through us, and our technology, the global biosphere is developing a mind. This way lies survival.

Can Earth really be said, through us, to have, or be gaining, consciousness? Suggesting that our planet is gaining the quality of intelligence or consciousness is not the same as some mystical notion that every rock, tree, and slug is humming with an aura of mind (although I kind of like that...). When you say that a person or a whale is intelligent, you probably don't mean that its every cell, its big toe or dorsal fin, is full of thought and intention. Yet we say humans are smart, not just that brains are smart, because our brains are tightly wired to the rest of us, and the parts are closely coordinated through a dense mesh of signals, feedbacks, and physical connections. Whether or not our planet is in some way literally becoming intelligent, I think it can help us to think of our role as bringing consciousness and awareness into the cycles of the planet.

Just as the Gaia hypothesis redefined life as an inherently planetary-scale property, we can envision planetary intelligence as a global property that emerges from the interactions between our collective thoughts and actions and other global systems, and one that could become a long-term stabilizing influence on the planet. The rapidly coalescing, technologically interconnected global noösphere is developing some of the qualities of a mind, and may even already have some limited ability to act with intention. We don't need some moment of singularity or sudden awakening of machine intelligence to manifest this increasingly effective global cognitive system. It's already happening. In myriad accelerating ways, technology is profoundly changing our modes of interaction with others in distant places, and enabling the birth of a new global community. Now when I hear about something big happening in, say, Egypt or Zimbabwe, I can take a gizmo out of my pocket and directly ask a friend there on the ground, "What's up?" This instant connectedness is new, powerful, and in its infancy. Machine-enhanced communication is drawing the world together, making us one people. Strangely, it is also pulling us apart, enabling disenfranchised and violent people to find one another and coordinate their actions. Global civilization breeds global resistance. In partnering with interconnected computers, we have unleashed powerful new tools for inclusion, democracy, and distributed community, as well as vectors for oppression and destruction.

The 2009 Green Movement in Iran, though it faltered and failed to bring down the theocracy, provided a glimpse of how social media and the Internet are making people, even in totalitarian regimes, less isolated and more hopeful of joining the modern world. The Arab Spring has also faltered. Yet it revealed a powerful new dynamic at work on the world stage. Especially in its earliest stages, the movement was largely begun and fostered on the Internet, on social media sites. At crucial times it became a battle for the hearts and minds of the world, fought by firing off tweets and posts. Unfortunately, these same tools became the loci for rumors, recriminations, and hate speech, helping to hasten the demise of that brief flowering of hope and democracy. This failed revolution revealed the promise and pitfalls of online community at this early stage.[11] We have a lot to learn about how to use these tools wisely and productively, but change is in the air and, especially, online.

If our global-scale connections and interactions are forming some new kind of still-primitive brain, how does it measure up to an actual, biological brain made of neurons? Your brain is not simply composed of one hundred billion neurons (about one for every star in the Milky Way). Its most crucial physical property, the one that enables it to do what brains do (think, sense, feel, and act) is the huge number of "synapses," or connections between neurons. These number about one hundred *trillion,* a thousand times as numerous as the neurons.

The human population of Earth will never approach the number of neurons in the human brain, and the number of connections among us will never approach that number of synapses. However these comparisons don't tell us much. A human being is very different from a neuron, and its behavior is much more complex. When we consider that each of the billions of elements being connected in some fashion to construct this putative emergent brain is itself already a brain, then we realize that the quantitative analogy doesn't tell us much about the potential for such a global brain. Even if we understood how brains worked, it wouldn't tell us much. Still, it does make sense that the abilities of any emerging global intelligence will likely depend crucially on the number and speed of connections between the disparate nodes spread around the globe.

We hear different opinions on whether machines themselves will actually become conscious actors in our near future. Even without such a transformation, there may be a way in which interconnected and powerful processors, working in concert with human beings, radically change the role of intelligence on our planet by continuing to increase the speed and level of connectivity of our human networks. Perhaps there will not be machine intelligence as such—the rise of the sentient machines—but rather, much faster and more connected machines augmenting and acting in concert with much more widely connected humans.

Such a worldwide mind may already be on its way to attaining cognitive abilities equal to the survival challenges before us. Compared to the pace of most geological change, this is all happening lightning fast, but compared to individual cognition, this global mind still seems pretty dumb and so damn slow to learn and respond. Can it get better?

Born of Climate Change

When we consider the challenges facing us, most obviously climate change, we are aware of serious new global problems caused by humanity. Yet if you look at our evolutionary story, the plot thickens. You could also regard this as something the planet has done to itself. Humanity is creating climate change, but it is also true that climate change created humanity. We were born of climate change.

One of my colleagues who has taken a great interest in the origins and implications of the Anthropocene is Dr. Rick Potts, a jovial and erudite paleontologist who directs the Center for the Study of Human Origins at the Smithsonian Museum of Natural History in Washington, DC. Rick is also a member of our Washington Anthropocene Group, and we've had many enjoyable times shooting the breeze in conference rooms and bars, comparing perspectives on this infinitely interesting topic. For the last twenty years, Rick has spent his summers in East Africa, digging up clues to our origins. He has found that human evolution is surprisingly coupled to the ups and downs of climate. Indeed, Rick has concluded that many of the major evolutionary innovations that have most defined and differentiated humanity occurred during brief periods of tumultuous climate instability, separated by long periods of more stable climate during which our evolution was also more static. To me this is fascinating, and hopeful, as it suggests that in a real sense our unique brains and their powerful social capacities evolved as climate change survival machines.

African climate has always been strongly influenced by the Milankovič cycles I describe in chapter 4. The wobbling of Earth's axis and the vibrating eccentricity of its orbit, forced by the gravitational pull of the other planets, leads to phases of high and low climate variability. At times, Africa has experienced extreme and rapid fluctuation between dry and wet conditions, forcing extinctions and rapid adaptations. What Rick and his colleagues have found is that all the major genera in our family tree, including Australopithecus (around 4 million years ago), Paranthropus (2.7 million years ago), and our own genus, *Homo* (about 2.8 million years ago), first appeared during these periods of erratic climate change.

Periods of intense climate havoc seem to have provoked the origin of the most important anatomical, behavioral, and technological transitions in human evolutionary history. The earliest-known appearance of bipedalism, about 6 million years ago, came during a time when Earth entered an extended period of climate change. Upright walking could have come in handy (so to speak) for surviving in diverse and changing habitats. The most extreme climate swings of the last 3 million years occurred between 800,000 and 200,000 years ago. During this phase, we experienced our most rapid increase in brain size, which enabled our ancestors to better survive in unpredictable surroundings. Near the beginning of this prolonged stage of crazy climate, we first controlled fire, a powerful new tool that provided warmth during icy periods and a gathering place fostering our accelerating social evolution. Fire also allowed for a diet richer in meat protein, which likely contributed to the development of larger brains. It also began our long relationship with combustion and carbon, which has recently become so fraught. In chapter 5, I describe the genetic bottleneck that occurred around two hundred thousand years ago as an ice age engulfed Africa, and how modern *Homo sapiens* may have been forged from the trials of survival during that period, developing symbolic language, new cognitive abilities, and sophisticated technology as means of coping with that harsh climate.

Later, during the most recent ice age, in the late Pleisto-cene, low sea level facilitated new global migrations, for exam-ple, opening up the Bering Strait land bridge that allowed the migration of Asians into Beringia, and later into North Amer-ica. When the ice receded, we emerged from the caves as fully modern humans. Having completely domesticated fire, we began to clear land for agriculture and, without knowing it, started the process of remaking planet Earth. Around six thou-sand to seven thousand years ago, sea level stabilized after a multithousand-year period of rapid rise. The first large coastal settlements on several continents all date to this period. The high-protein fish diets made possible by stable sea level and con-sequent coastal settlement contributed to the rise of complex societies around the world.

In all these ways, climate change made us into humans. Now we've succeeded so well that we've blundered into causing the next episode of dangerous climate change, one that threatens to outstrip our ability to cope. These big heads of ours have, at least for the moment, got us in big trouble. Can they also, used appropriately, come to our rescue? We do have a history of evolutionary innovation in response to climate change. We've survived existential challenges by cooperating, communicat-ing, and innovating, by becoming, in a sense, successively *more human*.

Becoming Fully Human

Maybe the source and the solution to our Anthropocene dilemma are one and the same. What makes us human, more than anything else, is our ability to work together to modify our environments in creative ways, powered by abstract think-ing. When we examine our species in an evolutionary mirror, we discover that our best hope actually lies in enhancing those same qualities that got us into this mess. Our biggest obstacle is that we're not particularly good at seeing ourselves as a global entity, or focusing on the long-term consequences of our activities. In other words, we must improve and enlarge the things that we're supposedly already so good at: anticipating, cooperative planning, passing on collective memory, and acting on accumulated knowledge. Fortunately, we don't need to invent new capabilities, just refine those that have been the secret to our success so far. When we look for the essential obstacle to surviving the Anthropocene, we see that we are just not equipped with quite enough of these human qualities.

Or not *yet* so equipped. Our capacities are, in fact, changing rapidly. As we have in the past when faced with extinction, we need to widen our frame, this time in such a way that a global, long-term identity becomes a reflexive part of how we see ourselves, and how we act.

In chapter 6, I propose a new, aspirational definition of technological intelligence, "true intelligence" or "planetary intelligence," which can be said to have evolved on a planet only when it can pass the test that now faces us, to act with intention on a global scale. I mean, really, what good is all this so-called intelligence if we can't get our act together to ensure our civilization's survival against our own technical cleverness? If our kind of intelligence is ultimately self-destructive, well, that would seem pretty stupid. What kind of intelligence is that?

What is required of us shouldn't be that hard. Cognitively it is on the level of a child becoming toilet trained. Just don't shit on yourself: that's the lesson humanity needs to learn. When we increased in numbers and built cities, we had to learn not to throw our waste in the streets. Sewage systems, a marvelous invention we usually take for granted, allowed us to live collectively in much larger numbers. Now we've again increased our numbers and energy use to the point where our waste is piling up in an unhealthy way. Only, this time it's not city streets but the air itself that is getting dangerously soiled. This forces us to confront the fact that we didn't evolve for global self-management. Fortunately, we're not stuck with only the tools that biology gave us. We now must seek to cultivate in our world culture the powers of foresight and preemption we have as individuals. We don't need some utopian, perfect world completely free of all conflict and inequality. Our new global energy system will be the sewage system of the twenty-first century. By the twenty-second century, people will take it for granted, and marvel that there was a time when we were so backward as to mess up our atmosphere driving around in clunky cars run on fossil fuels.

Although, as I've described, humans are wired for cooperation, today we are well aware of the difficulty we have in trying to extend this sense of cooperative community to a global scale. There seems to be a limit to the size of groups within which we are equipped to play nicely together. Our current cognitive and social skills served us well through the challenges of the Pleistocene and the Holocene, but they may not be equal to the task of long-term survival on the Anthropocene Earth. However, within our evolving global culture, we can see the seeds of something else that might do the trick. The signs are everywhere that, along with our penchant for conflict, we have proclivities toward global cooperation. Trade is rapidly pulling the world together into a functional whole. Whatever we think of the social effects of globalization—and any astute assessment gives it mixed reviews—we should be encouraged by its inherent tendency to foster far-flung connection. Devastating all-out global conflict, and equally devastating lack of cooperation on global environmental issues, become less likely as we become more obviously interdependent and perhaps, as I've described, begin to develop a global cognitive apparatus, a worldwide mind.

Science itself is perhaps the best example of the inexorable march toward interconnected global mental activity. The Earth systems we study know no borders. Migrating birds, evolving weather systems, and globe-spanning ocean currents don't stop at national boundaries. Scientists, while we are driven by all the petty and egocentric concerns of all people, also have a shared, universal value system that motivates a common, global, inter-generational project. We share a genuine desire to ferret out physical truth, and this leads us, perhaps in spite of ourselves sometimes, to a culture of global cooperation.

The history of my own field of planetary exploration illus-trates this. Even at the height of the Cold War, when the separate Soviet and U.S. space programs were locked in furious competi-tion to be "first" at every goal, the scientists who started sending out interplanetary spacecraft in the 1960s would share infor-mation with their "adversaries" whenever they could—because, really, what sense is there in having two separate, disconnected efforts to explore Mars and Venus? When the Russians were designing their next Venus probe, American scientists would find a way to share the results from our last probe (and vice versa), to help improve the design and maximize the probabil-ity of success. In chapter 6, I describe how the most important book in the history of SETI was written by Sagan and Shklovsky as a product of such defiant cooperation, hampered but not stopped by the mutual paranoia of their governments. Science is inherently global. And today, even in our factious, conflicted world, we routinely have international conferences that function more or less as meritocracies, where contributions are welcome from anyone who can further our understanding. [A global map showing four years of the geography of scientific collaboration is shown on page 8 of the photo insert.] The World Health Organi-zation and the Intergovernmental Panel on Climate Change are modern examples of effective global organizations that spring from the reality that urgent, collective problems such as epidemics and climate change do not restrict themselves to artificial national borders. Except for local and (one hopes) temporary cases of collective insanity, such as the Taliban attacking workers trying to distribute vaccinations in Northern Pakistan, enlightened self-interest pulls all parties toward cooperation.

The noösphere we have been building, and becoming, is now extending beyond Earth and into the surrounding space. In chapter 4, I describe the start, in 2014, of the European Space Agency's Copernicus program, which will provide continuous scientific observations of the health of our planet. We've had uninterrupted weather monitoring for decades, but scientific measurements have been more ad hoc and spotty: one satellite is up for a while, and returns some data until it fails. Our data and our records have been discontinuous, but the plan now is to keep things going. It is assumed that satellites will fail, and when they do, a replacement will be launched. Continuity of observations is built into the program. Is this planetary self-monitoring activity now a permanent feature of Earth? If it ever ceases completely, it will be because something has gone terribly wrong with our civilization. Such a termination would be one viable definition of the end of the Anthropocene. Yet I believe our orbital self-examination is here to stay. It's a new and hugely constructive development, perhaps even an early physical manifestation of the coming of Earth's Sapiezoic Eon.

Electronic connectivity and the increasingly obvious commonality of our environmental challenges are binding us together. We don't need world government, just effective global governance, something that is, out of shared necessity, slowly manifesting. A gradual change to a planetary worldview is aided by the proliferation of views of Earth from orbit, and the experiences that some human beings have had of actually physically going into space, gazing down upon our world, and reporting back. Many who have been there have reported a common and profoundly transformative experience. The number of people who have had this experience is as yet very small, but their influence is outsize. Stimulated by the sight of Earth looking alive, fragile, and achingly beautiful, framed by the blackness of space, they report a powerful sense of identity with the entire human race, the entire biosphere, and the entire planet.

The first to fully articulate this, and still the best, in my view, was American astronaut Rusty Schweickart. In March 1969 he got much, much higher than a kite, and had a much trippier time than anyone at the Woodstock festival of five months later when, on the Apollo 9 mission, he was one of the first to float freely, outside a space capsule, with only an umbilical cable connecting him to any other human creatures and artifacts, and only his helmet window separating him from the vacuum of space. As he floated in orbit and gazed down on Earth, he experienced an overwhelming moment of awareness of not only identity with the entirety of Earth, but also his role as a sensor for humanity.

When humans go into space, the biosphere is extending a fragile eye and looking down on itself. At that moment, Rusty felt acutely aware of experiencing this not just for himself but for all of us, and a sense of responsibility to communicate his experience. In the five decades since that moment, he has dedicated himself to speaking out about the space perspective, and in particular he has become a prime mover in the movement to get humanity to prepare for planetary defense against dangerous Earth-crossing asteroids. There is no way to talk about that problem without promoting a global, long-term view of ourselves as citizens of the planet who, through our special skills, have an obligation to care for it.

We're early in the space age, and the space perspective is still slowly infiltrating our consciousness. Two generations have grown up with images of the whole Earth. Now, with the recent launch of NASA's Deep Space Climate Observatory and the Japanese *Himawari 8* satellite, stunning high-resolution views of our always-changing planet, from multiple hemispheres, are continuously available to the growing percentage of our planet's people with connectivity. I recognize that awareness of and excitement about this imagery is at its highest in my own community of space geeks, but I also see these images of Earth's shifting beautiful wholeness steadily diffusing outward, seeding messages of unity.

The world is stitched together more tightly than could have been imagined two hundred years ago. Compared to sailing ships and wagon trains, we now travel at warp speed. With electronic communication becoming an established form of social interaction, our virtual relationships and experiences are getting both more real and quotidian. We are evolving new kinds and categories of relationships, and new protocols and modes of behavior to match. I'm not in any hurry for the boundaries between the real and virtual to disappear, and I don't think we're in danger of that, but as virtual experience improves and we find new ways to share, it becomes an augmentation to awareness. As bandwidth increases and machines are more closely integrated into our communications and our very thought processes, and as immersive visualization and engagement of other senses proceeds, we gain new access to shared experiences and forms of communion.

One of the downsides of moving to Washington from Denver has been less contact with big clear western skies. During the "blood moon" lunar eclipse of April 2014, I had a new and different kind of experience—I had been hoping to watch it from a nearby park, but the sky was hopelessly overcast. Yet, late that evening, I found myself vicariously enjoying the eclipse in real time through a multitude of pictures, videos, explanations, clever simulations, and reactions from my virtual friends viewing it across North and South America and Hawaii.

The previous lunar eclipse I had found memorable was a very different experience. On winter solstice 2010, I lay on a rooftop in North Denver, on a couple of layers of blanket and a bunch of pillows, huddling against the winter breeze in a pile of friends and one very large and heavy dog who didn't quite know what we were up to but who very much wanted to be part of the pile. We lay there for hours, watching Earth's shadow slowly pass over the Moon, rendering it a deep, smoky red. It was a mesmerizing, slowly shifting spectacle that none of us will ever forget. Lunar eclipses are not as rare or spectacular as solar eclipses. You don't have to travel to see one. Just wait a year or so, and one will come to you. Still, they are lovely and moving, and each one is different in unpredictable ways.* Up late into the night, out in the elements, seeing it with your own eyes with friends close at hand—there is nothing like that.

For this one in April 2014, I didn't have that option, but even though I was stuck in DC with thick clouds and rain, it ended up being thrilling in a different way. It would not have been the same if I had never seen one with my own eyes, but having done so many times, I felt connected to all those tweeters and posters sharing their views and their joy. It became a spontaneous global online celebration of the celestial. We experienced it together.

Now I've learned I can expect something similar anytime I get clouded out of a sky event, or even when I don't. On February 20, 2015, there was a close triangular convergence of the Moon, Mars, and Venus, and even though it was again overcast here, I "saw" it along with everyone, in a wave that followed the dusk, starting in Europe, crossing the Atlantic, and sweeping

---

* Because the red light you are seeing comes from all the sunsets circling the Earth at that moment, the exact shade and color depend on the weather around the planet at the time of the eclipse.

westward across this continent. I viewed this magnificent trio peeking through a stand of trees in Frankfurt; over a lake in Orlando, Florida; over a space shuttle replica at Kennedy Space Center; and over Ottawa, Canada. I tweeted, "So many stunning pics of young crescent moon consorting with Venus and Mars as evening sweeps across. At its best this thing is wonderful." Then my virtual/real friend astronomer Natalie Batalha tweeted a video loop shot from the International Space Station by astronaut Terry Virts, showing the moon settling down and softly disappearing into the cloudy curving edge of Earth, with the words "As if the #moon could be any more beautiful, here she is setting over Hokkaido and Vladivostok." Something new is happening here.

We planet people have been doing a version of this for a while, out of a need to share the experiences of our robot craft on other planets, but the recent *New Horizons* flyby of Pluto felt like a breakthrough. At the time, I called it the "first post-human spacecraft encounter," because of the way so many people spread around the globe were able to experience it together. The sense of real-time global participation and multiway communication was very real.

Even though Pluto is by far the most distant planet we have visited, and the pictures and data took more than four hours to reach Earth at the speed of light, there was still something wonderfully immediate about the experience and our ability to share it. It had been twenty-five years since the last flyby of a never-before-seen planet. The *Voyager 2* encounter with Neptune in 1989 had been a formative experience for me and for many of my friends on the *New Horizons* team. We were all students or postdocs then, and now are . . . well, slightly less young. There was a familiar feeling about the accelerating approach, over several days seeing a planet expand from a dot to a disk with barely discernable features. Then, all of a sudden, you're there and the details are revealed with startling quickness and clarity. And then—zoom—it's over—a last few shots of the bright crescent fading into the distance, and we are left with images and data to treasure and pore over for years.

Yet this Pluto flyby was very different due to changes that had happened in the intervening quarter century. With Voyager there was a certain inevitable elitism, a feeling that to participate fully you had to be in the right place at the right time. The few best pictures went out in press releases and made it into the *New York Times* or were flashed briefly on the evening news, and eventually showed up in their full glory months later in *National Geographic.* But, if you wanted to see all the pictures in real time, you had to be there, in the small room with the imaging team at the Jet Propulsion Laboratory in Pasadena, looking at the big fat, curved screen monitors.[12]

For New Horizons you didn't have to be there. It was cool to be present at Applied Physics Laboratory (APL) in Maryland, where the mission was run, with scientists and engineers who had put decades into the mission, the gathered crowd of enthusiastic space nerds, the press circus, celebrities, and politicians. Pluto didn't disappoint. There was the genuine relief from anxiety that the damn thing worked, didn't hit anything and die. And there was the sheer joy at how mysterious, strange, complex, and lovely a world Pluto turned out to be—worth the trip. Yet even there at APL during the heat of the encounter, people were spending a lot of time online, looking at their screens, sharing images, information, and impressions in real time with a worldwide community that was seeing it all, and chiming in, as it happened.

During the moment of the actual flyby, I participated, from there in Maryland, in a multisite live broadcast organized by Carter Emmart, the astrovisualization guru at the American Museum of Natural History in New York. We were all connected with cameras and microphones, watching what was happening at Pluto and sharing the experience simultaneously with crowds gathered in Buenos Aires, Bolzano, Hamburg, Ghana, Tokyo, Singapore, Brisbane, and several American cities. We could see and hear the people in all these places and respond to their questions. Strangely, even though the events were taking place five billion miles away, much farther than any planetary encounter had ever been, it seemed as though we were able to cheat the speed of light and expansive scope of our planet. It really felt as if we were all there together, riding along with *New Horizons* as it swung close past the icy dwarf planet and slid into its shadow. Times had changed, and you could watch and participate from everywhere. You no longer have to be there. If you are connected, you are there.

Our evolutionary history shows that we have reinvented ourselves several times. Often in response to climate turmoil, we've found completely new ways to live, and have redefined our relationship with the rest of the world. We came down from the trees; invented language and story; learned to make tools and plans and to hunt in groups; discovered fire; invented agriculture, cities, books, the Internet, and Earth-observing satellites. The rest is history, but it doesn't lock us into any mode of being.

During the last ice age, we huddled around campfires and told stories, solidifying our identity as social, collective problem-solving beings. Today the world is being woven together rapidly, and we're building distributed electronic campfires and gathering around them, struggling to find a newly enlarged sense of identity and purpose. We are a global force now. We are not going to relinquish this capacity, so we need to finish what we started in the East African savannahs and Pleistocene caves. We have new tools at our disposal that may allow us to change again to meet new challenges. Don't write off our potential to wake up, to grow up, to "human up" to our responsibilities and our capabilities. If we are going to live up to the *sapiens* in our name that Linnaeus optimistically gifted us with, then we need to become fully human. We have to remake our world and ourselves. We have to create *Terra Sapiens*.

The Power of Negative Thinking

We've all heard perhaps a bit too much about the power of positive thinking, but I am concerned about the power of negative thinking. Pessimism is easy, but certainty is uninformed. Our situation is shocking but not hopeless. We don't know enough to draw that conclusion. Yet, with the shock of the new, the shock of the now, a pessimistic outlook seems to be almost reflexive. Sometimes I think this knee-jerk pessimism, so prevalent in discussions of our future, is its own kind of irrational "magical thinking." It may be more dangerous and destructive than the rose-colored New Age magical thinking so often decried by skeptics. This reflexive pessimism can be contagious and corrosive.

People love to talk about how human beings suck. There are many who insist that there is no positive outcome to be imagined from the human presence on Earth, that any other thought is delusional. A persistent current of misanthropy has crept into many present-day environmental narratives. There is an awful lot of this human bashing, a repeated message about how awful the human race is, how the world would be better off without us.[13] It seems to push some satisfying self-righteous button to say this kind of thing. We can get stuck in this chorus of self-deprecation that, oddly, sometimes contains a strange note of gleeful exoneration. In a perverse way, this trash talking of humanity seems to make people feel good. Do they feel that by making it clear how much they personally hate humanity they can dissociate themselves from these crimes and somehow be excused or forgiven? It's as if we can become exempt from judgment if only we repeat loudly and insistently that we know how truly horrible we are.

Look at the metaphors we use most often to describe our global role, negative and violent images of disease and crime. Humanity is a scourge upon Earth, a cancer, a virus, a rapist, a mass murderer, a killer asteroid. Now, these comparisons are not without value. Clearly there are some dark truths here about our nature, our collective behavior. When we view ourselves from above, in satellite or aerial images, our patterns of construction and destruction do often appear ugly, metastatic, unhealthy, or stressed. As Edward Abbey said, "Growth for the sake of growth is the ideology of the cancer cell." The implicit warning is that we could kill off our host or force ourselves into rapid decline.

And yes, we have committed horrific crimes. When I hear of some of our accomplishments, such as the current near extinction of the northern white rhinoceros to supply traditional doctors in China and Vietnam with medicines of dubious efficacy, it pisses me off.[14] It fills me with shame that *Homo sapiens* is doing this to such a magnificent species that, as of this writing, is believed to be extinct in the wild, lost to habitat destruction, hunting, and poaching.[15]

Yes, these are crimes. Yes, we have been a cancer. But what shall we be now? Once we experience the shock of waking up and finding ourselves in the midst of committing these horrifying acts, well, then what? Cancers do not wake up and decide to stop being cancerous. I worry that repeating these self-descriptions can concretize them and limit our options, hampering our ability to move beyond the behaviors they emphasize.

There are many flavors of this pessimism. There's the sometimes gleeful misanthropy just mentioned: we are the species we love to hate. There is the self-righteous pessimism that makes us feel morally superior to the rest of the harmful entity we see ourselves as the better part of.[16] Closely related is guilty pessimism: by confessing our sins, are we exonerated? There is well-informed quantitative pessimism, by far the hardest to counter. If you take a look at the various indicators of the Great Acceleration, you can certainly muster a credible case for doom. Yet it's not the whole story, as we are also the only exponentially changing phenomenon (at least on this planet) actively engaged in studying its own patterns and reimagining its future.

Then there is dutiful, alarm-sounding pessimism. Many concerned, altruistic, compassionate people seem to feel that voicing visions of apocalypse is their civic duty, to ring alarm bells and motivate action. If you are concerned that not enough people are paying attention, you may feel morally obligated to spread a bleak message. Wanting to be a force for change leads some to repeat the gloomiest scenarios. Here I think people are conflating the need to be responsible with the need to be pessimistic. They want to shock their fellow humans awake, but they may be numbing them asleep instead.

Persistent messages of doom can also spread nihilism and defeatism. In March 2014, a young journalist named Clive Martin wrote a piece in *Vice* entitled "What the Fuck Are We Meant to Do with Our Lives When We're Told the World Is Ending?"[17] It described how NASA had concluded that our society is heading for inevitable collapse due to climate change, and bemoaned the fact that this left young people no room for hope or cause for action. In reality, NASA has made no such conclusion. This writer made the common mistake of assuming that an author listing a NASA affiliation equates to "NASA saying" whatever the paper concludes. The study in question is full of guesswork, assumptions, and simplifications. It is more of an interesting cartoon than an accurate simulation of the world.

Still, the message "Scientists say there is nothing we can do" is spread over the interwebs by concerned activists. And we have this poor young writer stating,

> The problem is that ignorance is bliss when the truth means knowing that you and all of your friends are staring down the barrel of fate. If nothing can be done, then it seems better to just live our lives as we always have: networking, hobnobbing, chitchatting until the sun goes black and the birds start to fall out of the sky.

I see similar statements every day on social media. If this human bashing and doom prophesizing is tactical, I think it's backfiring. It's more likely to become a self-fulfilling prophecy than to rouse people to action.

Currently I feel that spewing misanthropy and random anti-human sentiment is just as dangerous as emitting CO_2 into the air. It is the opposite of activism. I know that people spreading these messages mean well. They want to shock others into realizing the effect we're having on the planet, but there is a real danger of unintended consequences, of encouraging people to give up. Spreading messages of doom is a form of inactivism.

There is a widespread propagation of apocalyptic images in our culture right now. At some dosage these are valuable as a warning, but we should be wary of this kind of future becoming an expectation. If you conclude that we are invariably, inevitably incapable of finding a long-term balance with our planet, then you are giving up on Earth, and selling out the rest of the biosphere. Far from spreading a sense of responsibility, it spreads "disaster fatigue" and resignation. Pessimism, if it becomes a habit, can reinforce a narrative of unstoppable decline. If there is nothing we can do, that releases us from our obligations.

There's no future in despising humanity. Self-flagellation may feel good to some, but how does it help move us toward solutions? Surely we can find a way to love Earth without hating ourselves.

Earth is a stunningly lovely planet for so many reasons. Among these is the wondrous presence of curious, artful, inventive humanity. Whenever I see a nighttime picture of Earth from space, with its glowing lights, I am stirred by its beauty. It's a different sort of awe than we get from the opposite view, looking up at the numberless stars—the *mysterium tremendum* we feel facing all that infinity. Looking down, we know these lights scattered around the world reveal a pattern of human lives. Not only that, but these pictures were taken by curiosity machines sent into orbit. What kind of a planet does that? We are leaping beyond this sphere and looking back down at ourselves. To me that is spectacular, encouraging, and inspiring.

Often, when I post such a picture online, someone will comment about how ugly it is, because of light pollution, showing what a hopeless cancerous influence we humans are. I don't see it that way. I, too, am concerned about light pollution, and I support efforts to control and mitigate unnecessary scattered light. Yet I am overwhelmed by the grandeur of our world, and that includes the interconnected glowing networks we have woven around it. Ours has long been a beautiful and rare planet, but it is especially so now because of the human presence. Earth—the planet that wonders and sings.

There's more to my argument than just "put on a happy face." This negativity is suspect, tactically (it doesn't work) and philosophically (it reinforces Earth alienation rather than identity). It is also logically suspect, feeding a false narrative about climate change and our responsibilities. Many people see the fight to halt global warming as an impending either/or situation. We're going to stop it by a certain date or we're not—and it looks like we're not. Therefore, humans and the Anthropocene must be bad.

This reasoning is simplistic and damaging. For well over a decade we've been seeing websites and articles and what-not promoting the idea of an approaching deadline by which time we will either have definitively acted to solve the problem or it will be too late. If we don't act in five years, we're dead. Every time one of these deadlines passes, this narrative becomes less convincing.

Some have likened climate change to an asteroid that is clearly heading for Earth. It's a good analogy in some ways. Scientists are certain that it is heading our way, and we can probably avert it if we apply the right resources toward a solution. Even so, people are ignoring the warning and going about their business while the preventable nightmare hurtles toward us. In some important respects, though, this is a bad analogy. Unlike the path of an asteroid, whose motion is determined by the relatively simple laws of gravitational mechanics, climate is horribly complex. We can't predict the exact trajectory with perfect reliability. With an asteroid, we can tell you exactly when it's going to hit, down to the day and even the minute, and roughly what will happen when it does. Once we've averted it, we can relax for a few more centuries until another one is spotted coming our way.

The asteroid will either hit us or it won't. With climate, it's not an either/or proposition. It's a constantly shifting trajectory that will require sustained attention and concern over years, decades, and centuries. Clearly we are not going to shut down all the coal plants in the next ten years. I believe, just as clearly, that they will all be shut down by this century's end. Between those two boundary conditions lies a huge range of possibilities. Yes, we are putting ourselves at risk, and yes, we should do whatever we can to move ourselves as quickly as we can toward new energy systems. However, there is no critical moment at which the disaster hits or will be averted. There are, rather, accelerating and decelerating trends, and an infinite number of possible paths. It's a slowly unfolding emergency that requires not a quick pulse of activity, not a fight-or-flight response, but long-term engagement, enduring changes in the way we live, through a long series of thoughtful readjustments. We don't need to panic or despair. We are in this for the long haul.

Can we at least envision a behavioral mode, a way of working with the rest of the world, a version of ourselves, that we could celebrate? If not, then what path is there but nihilism, fatalism, and resignation? If we really believed we could not change course, then concern about our future would be as useless as agonizing over an approaching comet in a world where we had no space program, no way to stop it. Yet we are the species with a unique ability to envision futures and sometimes work together to manifest them. As long as we can imagine a better path, of course we are obligated to seek it. This is why unwarranted pessimism about our future is actually irresponsible. The naysayers, prophesiers of certain doom, are giving us a way to avoid responsibility. Don't listen to them. If we don't know enough to know that we're doomed, then the drumbeat of gloom is not helping, it's hurting. Let's replace it.

Remixing Our Metaphors

We're not a cancer or a disease. We are organisms doing what all organisms do, surviving and reproducing as best we can. We are, however, a kind of organism that has never existed before, and we've gotten ourselves in a situation. Fortunately, we may be equipped to get ourselves out of it. A plague does not think. A cancer does not decide to change course. A weed does not weed itself. We could. So these metaphors may describe our past, but they needn't proscribe our future. When we look for solutions, all these negative metaphors do not serve us well. Why not convey some possibility of adaptation, survival, of hope and constructive engagement?

We're aware of the damage and loss associated with the proto-Anthropocene, but we also need to focus on gain and opportunity. We're at least partway through the transition from being the species that bumbles through its world-changing ways with no awareness whatsoever. We've figured out how to deal with the worst of acid rain and ozone destruction. (That was a close one.) We're starting to come to grips with the biodiversity and climate crises. Whatever happens to climate now, it would have been a lot worse without these recent waves of awareness and concern reverberating around the globe. If we get our act together, we still have the potential to consciously save many more species than we have inadvertently destroyed.

If you look from an evolutionary perspective, you might see us a little more sympathetically. You might realize that we're not inherently evil, destructive, or malevolent, but that we're unprepared and ill-equipped for the task we have stumbled into. We are uniquely outfitted with the power of imagination, and it seems clear that any robust solution to our Anthropocene dilemma will involve reimagining ourselves and our interactions with the world. Throughout this book, I've attempted to seed some alternative allegories. I've described us as being like sleepwalkers waking up in the middle of performing some task. There is a sense of discovering we're in a difficult situation that some version of us has gotten ourselves into, but that we have not been fully conscious of until forced to realize what we're doing. We now find ourselves in the unenviable role of sort of running a planet—a job we didn't ask for, don't deserve, and don't know how to do. Still, we have to find a way forward. We're like an unfortunate soul who has just woken up at the wheel of a big rig, a racing, out-of-control truck. We have absolutely no idea how to drive it, but everything we love is on board. We're heading furiously down a twisty road. We're starting to figure out how some of the controls work, but nobody's ever given us a driving lesson. We'd better learn in a hurry.

Like an orphaned baby left on a doorstep, with no training in life's hardest tasks, we need to figure out how to survive on our own. Assuming there is no SETI success around the corner, there is no adult coming to clean up, feed us, or set us straight. In some ways, we're like an awkward, naïve, and reckless adolescent experiencing a difficult and painful transition. We are self-conscious, aware of our new powers, but not fully in control of ourselves. We're learning that our actions have consequences, but unwilling to take responsibility, hooked on immediate gratification and not willing to clean up our room. We still have little awareness of limits, and we love to watch things blow up. We're confused by these new abilities, physical changes, and strange desires. We're not criminals committing premeditated planetary desecration. We're more like juvenile delinquents in need of rehabilitation.

There are parallels between the transition from the infinite resiliency of youth to the realized limitations of maturity and the journey of a young technological species beginning to feel the limits of planetary capacity.[18] As philosopher Frederick Ferré wrote in 1993,

> The Earth is no longer, and never will be again, naively bountiful, after three hundred years of avid exploitation by the ever-expansive appetites of the modern world. Much of the easily mined copper and other essentials for high technology, most of the easily extracted oil and other energy sources

required for recognizably modern living—these have been used up already in the centuries-long process of building the world we now inhabit.[19]

We'll need to adjust our habits and expectations because we will never again be those young creatures wild and free on a seemingly infinite world.

Several other biological or medical metaphors suggest themselves. Jim Lovelock has called the study of Gaia "geophysiology." It's an apt concept to capture his sense that the planet has its own innate homeostasis, an ability to care for and heal itself when not pushed too far. In this case, though, the physiologists live inside, or comprise part of, their only patient. Geophysicians, heal thyselves. This brings to mind the image of poor or neglectful self-care. If Earth is our body, are we damaging ourselves in expanding so fast and thoughtlessly? If so, is this a kind of self-mutilation? Do you want to punish or help this troubled creature?

There is also the image of the noösphere, and especially the Internet, as a developing nervous system of Earth, with us as the thinking, connecting nodes and all our satellite sensors as the ears and eyes of the world. Perhaps this baby brain is starting to realize that it needs to care for its body.

We can update one of the most oft-used negative biological metaphors, giving it a new spin that takes advantage of some cool new science. As I've noted, the human presence on Earth is often likened to a virulent disease threatening to destroy its host. Yet our understanding of the nature of disease has been rapidly evolving with growing awareness of the human microbiome. This term was coined by Nobel Prize–winning biologist Joshua Lederberg[20] to describe the vast microbial community living within our bodies.

I heard a narrator on NPR describe it by saying, "Ninety

percent of you is not human!"* I would say, rather, that 100 percent of you is human, but a human is not what we once thought it was.† Or, in a sense, an individual human being is an illusion. The nature and boundary of the self has changed. This requires an update to the germ theory of disease. Germ theory, the eighteenth-century discovery that diseases are caused by microscopic organisms invading our bodies, was a powerful advance in our understanding of the world, leading to huge breakthroughs in public health. However, the resulting picture of the human body, as a citadel of health occasionally invaded by dangerous, dirty microbes, wasn't quite right. Most of these invaders (or should we say visitors? residents?) are not bad for us. Sometimes they just get out of balance, or unwelcome guests show up and start to damage things. In fact, many parts of this community within us play vital roles necessary to keeping us healthy and alive. These beneficial organisms, referred to as the "commensal microbiome," are essential parts of our bodies that are actually comprised of other creatures. They are parts of us that are not coded in the DNA of our parents but acquired from our environments (including the bodies of our parents and those of other people). In the shock of realizing that our imbalanced growth threatens the health of the global systems we depend on, we have described ourselves as pathogens, invaders in the body of Earth. But now, in analogy to our new understanding of human health, rather than assuming we're the disease, can we seek to play the balanced and mutually beneficial role that would make us part of the commensal microbiome of Gaia?

Just as Earth is in some important respects an organism, it is now in some respects a garden or a park. We don't want to pave

---

* More recently revised downward to perhaps 50 percent.

† That is, a large fraction of your cells is not "you" in the way we thought. By mass, your microbiota is a much smaller fraction of you, about 1 percent.

paradise and put up a parking lot, so we need a different image. Humans aren't going away, but we can concentrate our activities into certain areas and have the wisdom to retain wide zones with more of their own wild character. The mere fact that these are our decisions to make does put us in some kind of a managerial role over the whole ball of dirt. Yet, if it's a garden, it's a strange one, without walls. It's finite yet unbounded, like the geometry of Einsteinian space-time itself—and we, the gardeners, must live within the garden and also cultivate and tend to ourselves lest we become weeds. We have been weeds and must learn to be gardeners.

Our large parks and nature reserves can feel pretty wild when you hike into the back country, but they are all heavily managed as well. They must be, to uphold anyone's idea of responsible conservation.[21] Our planetary park is one with no exit, and we ourselves are one of the species that must be managed within it. I don't mean to suggest that the way we have been spreading our constructions and our crap around Earth so far resembles good park management. Right now, we are as much the vandals as the wardens chasing after them. We have much to learn. Good management starts with deep knowledge of the way your landscape tends itself, of the innate patterns and flows of the life and landscapes within your park.

The idea of a domesticated planet makes us squirm,[22] but what is the alternative to some level of conscious global management? We are not going to disappear. Even if we do eventually reduce our numbers and our impact substantially, these would be choices. We would still be managing, cultivating. Our global garden or planetary park must include large wild places, but, inescapably, they have boundaries and thus they are managed to some degree.

Planetary park is a place we can manage but never fully control. Like any vast preserve, it has wild elements that will remain beyond our influence. We'll never be fully in control of Earth, nor should we seek to be. The convection of the mantle, the changing output of the Sun, the orbital forcing of the other planets—these are Earth forces we can learn to comprehend, movements and rhythms with which we can learn to dance. Yet we will never be leading.

This Generation Ship

There is something potentially mythical about this evolutionary moment we're in, something of the flawed hero's journey in this time of waking to realize that our world is not what we imagined. It's that story where you've been training for something all your life, and you thought you knew who you were and what you were preparing for, but then your true, larger task is revealed to you. We have an essential role to play, one that will test us to the limit and determine the fate of our people and our world.

In searching for the right metaphor to describe the strange situation of humanity in the twenty-first century, I find myself reminded of a "generation ship," a popular recurring motif in classic science fiction. It is conceived as an engineering solution to the problem of how to enable people to travel to exoplanets. Our lives are so short and the stars so very far away. If we're right about the laws of physics, the exoplanets are forever out of reach for a human being—but not for a society of humans. A generation ship is a large interstellar craft built for a journey that will take multiple human lifetimes to cross the fearsome distance to another planetary system. Those who arrive will be the descendants of those who originally set out from Earth.

Like so many ideas in space science and fiction, this one can be traced to Tsiolkovsky, who in his 1928 essay "The Future of Earth and Mankind" described a space colony, which he called

Noah's Ark, that travels thousands of years to reach other stars.[23] It's been explored in some classic sci-fi novels such as Robert Heinlein's *Orphans of the Sky*, and *Non-Stop*, by Brian Aldiss. They envisioned self-repairing, self-steering "world ships" with populations large enough to ensure genetic diversity and social contentment, built with enough room, comfort, and distraction to make life decent for the generations that will live out their lives on the journey. On board are engineers, pilots, and stellar navigators to oversee the machines and keep everything on course and running smoothly.

Yet in the story, something always goes wrong. There has been a mutiny, an accident or societal collapse, and the people have reverted to a pre-technological, superstitious state. The inhabitants have no idea they're on an artificial ship. To them, it is the entire universe, and they know nothing of planets and stars. Then the plot thickens when our heroes find a control room, or a porthole, see the stars for the first time, and realize that their world is not at all what they thought—and they discover that things have gone badly, dangerously off course, and the lives of all on board are threatened. Their only option is to try to convince everyone else to accept their radical revision of reality. "Our world is completely different from what we all assumed. We're on a ship. There's a whole universe beyond. We're all traveling somewhere together on a multigenerational journey. We have to learn how to drive this thing and figure out where we're going, or else we're doomed. We have to wake up, everybody!" Talk about an inconvenient truth.

To me, this seems like our current situation. Now we can see the stars for what they are and our home for what it really is—and we see that, unbeknownst to us, and whether we like it or not, we've been tasked with operating a world that we don't know how to run. Just as, once, with the Copernican Revolution, we had to completely reevaluate our place in the scheme of things, awakening now to our role as world shapers requires another painful shift in worldview. We're essentially on a generation ship, and we have to figure out how it works.

This is similar to *Spaceship Earth*, the term coined by pioneering economist Barbara Ward and popularized by Buckminster Fuller, who stated that we are all crew and there are no passengers. What I like about the generation ship metaphor is that it hinges on that dramatic moment of awakening to realize we have inherited a perilous situation, and that our lives and those of all our descendants depend on convincing our fellow travelers of the reality we have discovered. It also carries the hope that humans have the capacity to overcome the challenge. In the generation ship story, hope comes from discovering that humans were once great inventors and must have the capability to be so again. In our world, we must draw upon our imagination and ingenuity, the innate resourcefulness and resilience of our planet, and the knowledge that our species has a long history of responding to existential challenges through innovation and self-reinvention.

We are hurtling through space on the only place we know we could live, and we've discovered that it is indeed, in part, a kind of construct. We are piecing together its history, coming to understand our situation, and realizing that we have inherited a role for which we are not trained. Our current world, inhabited by seven billion, soon to be ten billion people, was created, in part, by the actions of our predecessors and will require smart engineering to return to a safe course. Our immediate task is to switch to auxiliary power and turn off the carbon generators that are overheating the ship. Our longer-term challenge is to shore up our world ship for the generations who inherit it.

Beyond its value as a font of science fiction and a metaphor for our Anthropocene dilemma, is a generation ship something we might actually build some day? I think so. We are a wandering species, and it's hard to imagine we will never, ever set out for the stars. When we do, eventually, I believe this is how we'll go. You can find many confident pronouncements by scientifically knowledgeable people declaring that interstellar travel is impossible. Yet the speed of light is no secure barrier against our moving out into the galaxy. Mature humankind would be perfectly capable of visiting exoplanets. No hidden physics is needed. All we need is a long-term view and a collective spirit. In short, the same attributes that will help us (or anyone) survive technological adolescence are identical to those that will, if we wish, bring us to the stars.

To do this, we might have to in a sense become trans-human, and that is not necessarily something to fear. I'm not referring to the abandonment of the physical body in some techno-Rapture, when we become uploaded machine versions of ourselves. I just mean the extreme commitment to the collective, to the long-term success of the group, that would allow us to feel right about joining such a venture. We would have to feel okay about committing our unborn children and their children to a journey they did not choose to be a part of. This introduces challenging problems of ethics and societal design. A world ship journey would require us to envision harmony and continuity of purpose over many generations in an unfamiliar and contrived setting. The members of later generations could not be given the choice of whether to make the journey. Their entire lives are committed in advance to a project they did not choose.

Perhaps, though, this is not all that different from past human journeys, when future generations were banished to challenging new circumstances by the decisions of their forbearers—for example, ancient humans originally migrating from our ancestral home in Africa, the first Americans crossing the Bering Strait, the first humans to reach Australia, Pilgrims shipping out on the *Mayflower*, or Mormons first moving to Utah.

There are similar examples for every continent but Antarctica. None of these pioneers could give their great-great-grandkids a vote in their plans. We have long been making commitments that later generations have had to live with. Here on Earth, the decisions we are making today about how to power our civilization will inalterably circumscribe the lives of our descendants for many generations.

If we go to the stars, we will go as collectives, as communities dedicating their lives to a voyage they will not complete. We will make these ship/worlds fun and interesting places to be, with forests and privacy and infinite diversions, with places to run and swim and be alone. They have to be places where you would actually want to live.

A world ship is a project for another time, but there is something we can do in the coming generation or two that will be a baby step in this direction. We can indeed send something to the stars, something positive and hopeful for the future. With current trends in miniaturization and propulsion, it will soon be possible to launch tiny interstellar probes that could reach some nearby exoplanets on a timescale of centuries. The builders of these probes will not live to see the pictures and data radioed home from our nearest interstellar neighbors. We should do this as a gift to our near descendants.

Going interstellar means going long. We cannot imagine ourselves as interstellar actors without also conceiving of ourselves as intergenerational actors. We cannot reach the stars without a sense of identity and goals that span generations. This is true for interstellar communication as well as for travel. Neither makes sense unless we see ourselves as collaborating with descendants. To travel, or even send messages, to the stars, we will have to start conversations, projects, and journeys for our progeny to finish. This cements the essential bonds between generations. We won't be the first to attempt such projects.

The builders of pyramids and cathedrals mostly never lived to see them completed. Sometimes they worked under duress or coercion, but sometimes they were moved by spiritual commitment to something beyond their individual lives. I think of science itself as such an effort, with individual researchers fashioning bricks in an edifice each of us can see only partly constructed, knowing that our students and theirs will continue to build.

In the last century, we became aware of ourselves as short-lived creatures on a small planet in a long-lived galaxy of evolving stars and countless planets. In this new century, we must act upon that knowledge. What could be a better project for us to commit to scientifically and spiritually than creating a sustainable civilization that can thrive on this planet and begin to reach for the stars? Regardless of how we got here and whom we want to blame, we have to take stock of where we are and figure out how to proceed. We are obligated to understand our world ship and get it back in balance, back on a course where our descendants will be safe and free to pursue knowledge and goals that we can scarcely imagine. We need to have a vision of where we're going in the future. With this in mind, let's rename our ship. I would suggest we call it *Terra Sapiens*.

A Brief History of the Future

So, can we do it? Successfully navigate this proto-Anthropocene bottleneck? Survive our current global changes of the third kind and emerge into the stability and promise of a mature Anthropocene? Do we have what it takes? We don't have a crystal ball to tell us what's coming. We have climate models and other projections, all dependent on assumptions and limited understanding. We have growing knowledge of the history of Earth and other planets. And we have visions of humanity's future from literature, pop culture, and science.

As a lifelong student of science fiction and the history of science, I've enjoyed reading predictions of the future from different time periods. During my two years researching this book at the Library of Congress, I've read numerous expired forecasts about the human future. These give us perspective on our skill at prophecy. What we often find is a mixture of uncanny accuracy and wild error, of clear foresight and utter blindness. Reading the futuristic visions of past sages is an interesting combination of "Wow! How did they know *that*?" and "How could they have *missed* that?"

It is our ingrained habit, probably one built into our subconscious cognitive apparatus, to imagine the future as a linear extrapolation of current trends. Yet history shows us that the world changes most in surprising leaps, and the future will be no different. What will the world look like in the twenty-second century? The one thing I'm sure of is that nobody's predictions will be right. Stanislaw Lem, discussing the futility of technological prediction, wrote, "Nothing ages faster than the future." Nobody anticipates the game changers, the social or technological breakthroughs that rapidly alter the rules and assumptions of a society. Science fiction from the 1950s is full of sleek, futuristic buildings, space planes, and rocket ships carrying people throughout the solar system, but nobody foresaw the transistor, the microchip, the communications satellite,* or the Internet.

British biologist J. B. S. Haldane, one of the founders of population genetics, in a 1923 essay entitled "Daedalus; or Science and the Future," wrote,

---

* Arthur C. Clarke, famously, did describe the communication satellite in the 1940s, but did not foresee it becoming a ubiquitous reality in his lifetime, and in later decades joked that he wished he had patented the idea.

As for the supplies of mechanical power, it is axiomatic that the exhaustion of our coal and oil fields is a matter of centuries only. As it has often been assumed that their exhaustion would lead to the collapse of industrial civilization, I may perhaps be pardoned if I give some of the reasons which lead me to doubt this proposition.

He then goes on to predict that

Ultimately we will have to tap those intermittent but inexhaustible sources of power, the wind and the sunlight. The problem is simply one of storing their energy in a form as convenient as coal or petrol.

and continues:

Personally, I think that four hundred years hence the power question in England may be solved somewhat as follows: the country will be covered with rows of metallic windmills working electric motors which in their turn supply current at a very high voltage to great electric mains.

He also describes how this countrywide network of windmills will be used to chemically break down water into hydrogen and oxygen, and these will be used to drive a hydrogen economy. Haldane concludes that "Among its more obvious advantages will be the fact that...no smoke or ash will be produced."

Haldane's timescale was off, but he presciently foresaw the limitations of a fossil fuel–based economy, and even though he didn't exactly see global warming coming, he proposed a solution for widely available clean energy that still seems viable. However, in the same essay, he offers predictions that today seem ridiculous (as when he suggests that science may soon achieve communication with the spirits of dead people) or grossly offensive (as when he opines that one of the great hopes for the human future is the improvement of the species through eugenics).

Yes, Haldane and many of his learned prewar contemporaries seemed to think that eugenics was a great idea—that by purifying and improving the human gene pool we would rid ourselves of the undesirable traits of those they considered lesser people. In the long, dark shadow of the Third Reich, this scheme is repugnant and makes us deeply uncomfortable. This illustrates something else about the limits of prediction: morals and acceptable norms can change in highly nonlinear ways. We are capable of learning lessons that become embedded in our collective character.

Can we invent our way out of our current dilemmas? And can we change our "modes of thinking" sufficiently to win the quickening race between education and catastrophe? Currently there is a wide dispersion of views about our future, held by informed and reasonable people. We hear abundant warnings of impending doom and decay. There is ample ammunition for those promoting this outlook. The Intergovernmental Panel on Climate Change reminds us, with their graphs and figures, that if we continue on a "business as usual" path of carbon emissions, our world will be forced into an unknown, hard-to-predict, and scary climate regime. The Worldwatch Institute, with their regular comprehensive reports on global challenges, warns us of the multiple and converging threats of food scarcity, conflict, and drought, all of which may be worsened by unpredictable climate change.

Yet, juxtaposed against the hard and true math about climate change and limits to global carrying capacity, we hear a different set of truths: predictions based on the numerous positive long-term trends in global health, education, and eradication of poverty, and on the nonlinear, unpredictable potential of human creativity, which has historically come through for us when our species faced crisis. Can we, through adaptation and innovation, save ourselves from the strange fruits of our past cleverness? Many creative and credible thinkers claim that we can.

Will we harness the resources of our technology and intellect, raise up the living standards of the developing world, rein in the unsustainable consumption of the developed world, feed the world's people, stabilize climate, and develop a peaceful global civilization, evolving beyond war? Or are we headed for a world of scarcity, with an out-of-control climate causing crop failures, water shortages, and endless streams of refugees from newly infertile or flooded lands, fueling conflict, terror, and societal decay?

The outcome of this unprecedented experiment in accidental planetary (mis)management is unknown. There is widespread understanding that this is a time of risk and uncertainty, that there are dangerous possibilities, and that we need to meet the future by both inventing new technological solutions and changing our assumptions and behaviors. Reasonable people disagree about how dire the danger of various scenarios is, and how radically we need to change our society to fashion it into one where we can ensure that future generations thrive. Perhaps depending on our individual psychological makeup, we tend to embrace the good future or the bad. Today we live in a world with ample evidence for either conclusion, so it's very easy to cherry-pick your examples to shore up your view. We are addicted to certainty, but the Anthropocene raises questions for which we don't have answers.

However, we can take heart from the abundance of genuine examples of the kind of thinking and actions that, spread widely, could steer us to safety. The best involve innovative science and technology combined with multinational cooperative efforts to attack problems that themselves transcend borders. Some come from the realm of global health, where many infectious diseases have been fully or nearly eradicated. Our world is on track to be completely free of polio by the year 2018. In the domain of conservation, there is good news as well as bad. An abundance of heartening success stories counters the widespread disaster narratives. Consider the beautiful and elusive Asian snow leopard, which is threatened with extinction due to poaching and habitat loss. Remarkably, a growing effort involving local communities and nongovernmental organizations in Pakistan, Afghanistan, Tajikistan, and China is making new progress in protecting these magnificent creatures and bringing them back from the brink.[24] The diverse peoples of this area all value these legendary animals and are driven to protect them. We think of this zone as rife with sectarian strife, which it is. Yet people in communities across these many borders are coming to understand that only by working together can they get what they all want. Even in this war-torn region, the collective recognition of a crisis is revealing shared interests and values, and science and technology are providing new means to rectify past mistakes. The combination of satellite-enhanced tracking, sophisticated ecological modeling, and on-the-ground coordination among communities that might, on the surface, not be primed to cooperate, is a hopeful microcosm. It shows how the potent mix of innovative technology and innate human capacity to come together against shared obstacles can, at least sometimes, carry the day.

Likewise, for a wide range of global environmental problems, recognition, at least, is on the rise, if only sometimes out of shocking necessity. Chinese leaders might like to keep on burning their vast coal reserves, damn the consequences for climate, but the air in Beijing is now often frighteningly thick with smog, and they are working hard on alternatives. Yet the need to stop burning fossil fuels collides with some of the largest economic interests in the world. Our current global economy runs on coal and oil. How are we going to change that? We're not going to end capitalism or instantly end coal, but we can bend capitalism to make coal a losing bet. There are paths forward that we can see, some combination of solar, wind, geothermal, tidal, and possibly nuclear energy can replace fossil fuels as the international community slowly musters the will to address the growing threat of climate change. And there is now no technical problem with a bigger incentive for innovation than the need for breakthroughs in energy technology. There will be breakthroughs. It might be successful nuclear fusion or some fancy biotech invention that makes solar power hugely cheaper and more efficient, but revolutionary change in this area is coming.

In the meantime, our "modes of thinking" really are changing. There is a new global conversation going on now about climate change and energy choices. Our sense of ourselves as global actors evolves with the changing cognitive capacities of humanity augmented by satellites, woven in the Web. As I discuss in chapter 5, human social and political systems also go through tipping points every bit as dramatic and unpredictable as those of planetary climate and other complex physical systems.

If we make it through the next few centuries it will be because we've honed our survival skills and adapted them to work on a planetary scale. Once we achieve that, we'll have done much more than ensure our own persistence against near-term self-induced challenges. We will have unleashed the power of reason and foresight in permanent defense of Earth's biosphere.

In Our Hands

Vladimir Vernadsky wrote his prescient papers about the noösphere in the 1940s, as World War II raged. He was describing his vision of the long-term transformation of Earth toward reason and intelligence, but all around him the world was burning. Still, he saw that there would be a world left after the war and that, as horrible as that time was to live through, the cosmic evolutionary trends he was describing would be only momentarily interrupted. He wrote:

> Now we live in the period of a new geological evolutionary change in the biosphere. We are entering the noösphere. This new elemental geological process is taking place at a stormy time, in the epoch of a destructive world war. But the important fact is that our democratic ideals are in tune with the elemental geological processes, with the law of nature, and with the noösphere. Therefore we may face the future with confidence. It is in our hands. We will not let it go.

To perceive our place in cosmic and terrestrial evolution, we need to look beyond the immediate problems of this year and this decade—not to avoid them, but to try to see where we're headed as we engage them. We've steered our civilization into unsafe waters, and now climate change is coming at us like a giant wave. We can't avoid it, but we are going to buckle down and steer through it. We'll get to the other side—not unscathed, but still sailing.

Harlow Shapley, writing in 1952—before plate tectonics, before global warming, before space exploration or Gaia—said, "Climate and continental changes usefully force us inhabitants of this sun-controlled planet to evolve in adaptability." As it has in the past, climate change must catalyze a new phase in human history. The difference is that this time we see it coming. In the past, we adapted to climate change as it was happening or after the fact, by inventing new tools, by moving to new lands. This time we know that it is headed our way. To whatever degree we can't prevent it, we'll again have to adapt. I believe we will be able to forestall and even reverse much of the damage.

We are already behaving differently from that bacterial colony in a petri dish, deviating from the fatal S-curve, using our limited but growing global cognitive capacities to anticipate and soften or avoid the crash. We are waking up, and we can see the trends starting to turn. We are slowly rounding the corner on the related problems of poverty and overpopulation. The Bill and Melinda Gates Foundation, perhaps a trifle optimistically, now claims there will be no countries left mired in poverty by the year 2035. It is undoubtedly true that the fraction of desperately poor people is declining globally. As standards of living are raised, and in particular as education and options for women are increased, fertility rates decline. Population is going to level off by the late twenty-first century. Birth rates already have. If we continue trends in intensification of agriculture, we may well reduce the amount of land needed to feed the burgeoning population, taking pressure off stressed biomes. Some studies indicate that the rate of deforestation has reached its peak, and the reforestation of Earth can soon commence.

There is plenty of energy available, in the form of sunlight reaching Earth, to power our civilization many times over. We are going to wean ourselves off fossil fuels—not as fast as we should, but by century's end our energy production will be dominated by cheaper and cleaner alternatives. Our carbon emissions will slow and then basically cease, and we will probably be pulling CO_2 out of the atmosphere, bringing it back within a safer range.

The best science now seems to say that it is perhaps fifty-fifty as to whether the global average temperature will be three degrees hotter by the year 2100. This is not a comfortable place to be. If we end up in the high end of the range of uncertainty, there may indeed be terrible times ahead for a few generations.

Climate disruptions could make the twenty-first century as bad as the twentieth century, with its tragic famines and world wars that uprooted massive regions and cost hundreds of millions of lives, but it will not be the end of our civilization. It may be the beginning.

Much less likely are worst-case scenarios where unforeseen positive climate feedbacks induce more severe global changes. If that comes to pass, the coming centuries could be as bad as the great bottleneck our species suffered through over one hundred thousand years ago, when most of us perished and the only survivors were those who found a fundamentally new way to be human. Either way, we will survive. We will get by, and the real question is: how do we best equip ourselves to meet the challenge and, within the wide range of possibilities, to realize the most benign future? As I said above, we humans learn from both disaster and foresight. The more we can, collectively, anticipate and meet the future, the less we'll have to learn the hard way.

The temperature will not rise in a smooth fashion. Things will shift gradually, then shift back, then shift farther in the original direction. There will likely be, in the next fifty years, a year or several or a decade that seemingly bucks the global warming trend as long-term cycles of heat and motion fluctuate. It's to be expected even as the world warms. We could even get lucky. A variation in solar output or a few major volcanic eruptions could cool things down for years or decades, giving us time to catch up with our energy transition. We certainly shouldn't count on it.

By century's end, arctic sea ice may be completely gone. This is worrisome because we don't know what other effects this will have on global circulation of the atmosphere and ocean or what ripple effects it may have on weather, precipitation, and agriculture. There may be several feet of sea level rise, in which case, some large coastal areas and low-lying island countries may need to be abandoned. Climate systems and ecologies are both inherently hard to predict, and our knowledge of some key factors is still sketchy, so it can go better or worse than our predictions, maybe much better or much worse. By century's end, coral reefs may be disappearing. Or, with luck and ingenuity, we may yet be spared a world without reefs. Recent results suggest that some species of coral are more adaptable, so with some help from us, they may survive the worst of it. The cascading effects on marine ecosystems cannot really be predicted. I think we need to get lucky here—or not unlucky. Gaia is tough, and the ecosystems will work it out; they will recover from these insults, rebuild their webs, and carry on. Yet what we don't know is how these readjustments will affect the marine and terrestrial webs that we depend upon for food. These unpredictable changes in the oceans are the aspect that, frankly, worry me the most.

Many species will have gone extinct, but the worst of the potential mass extinction can be averted. As we stabilize population, we can also stabilize and then decrease the amount of land we use for agriculture. Our cities will be smarter, greener places. Over the centuries, we've made a lot of progress in learning how to urbanize. We invented plumbing and sanitation systems, learned not to stain our cities brown with coal ash, realized we don't want polluted urban rivers. We are still learning how to live well in cities. I bet twenty-second-century cities will be nice places to live.

Our civilization will be running mostly on solar power and/or nuclear fusion. The ozone layer will be largely restored and on a path to full recovery. The pacific garbage patch will be gone. We'll look back on the wave of destruction we caused in the same way today we look back on World War II and say, "Never again."

Science will rush forward, illuminating the world, including our role in it, with increasing depth and clarity. The exoplanets will start to come into view, revealing something of where we, our biosphere, and our noösphere stand. Climate modeling of that day will make our current efforts seem quaint. By century's end, we'll be undertaking the gentlest form of geoengineering: finding ways to reverse our injection of CO_2, largely by various forms of enhanced photosynthesis. We'll have realized that we need to set the thermostat—we can't avoid that responsibility—and found ways to do so with subtle tweaks to existing Earth processes, those "high precision negative feedbacks" that Sagan and Pollack wrote about in 1993.

A planetary defense system will be in operation. People will be living, still in small numbers, multiple places in the solar system. Our first robot probes will be on their way to planets around nearby stars.

Now, against my better judgment, having made some predictions, let me reiterate that I don't think anyone knows what the world will look like in one hundred years, let alone one thousand. A century or a millennium is nothing in the life of our species or our planet, but for reasons I have described, this century will be pivotal for both. Among the game changers that could completely alter the world of the next century we can list new energy technologies, new forms of connectivity, machine intelligence, or discoveries about alien life. Since I was a teenager in the 1970s, fusion power has always been twenty years off in the future, but one of these days it will arrive. Cheap, abundant power would change a lot. It would not only allow us to reduce our carbon effluence, but would have other beneficial environmental effects, perhaps permitting us to further intensify agriculture, reducing the amount of land we need to feed people, and to make freshwater from seawater wherever it is needed.

Global connectivity may be the biggest game changer of all,

one whose effects are just beginning to play out. A revolution is under way that may ultimately dwarf in impact the invention of the printing press. We are in the early days of this. If, as some think is likely, we eventually are able to connect our minds more directly to machines and to one another, the effects may be multiplied by some huge factor.[25] Machine intelligence might change things in unimaginable ways. Much has been written on this, so I won't elaborate here, but it is fair to say that the most extreme scenarios, which cannot be ruled out, would end our civilization completely, transforming us into something currently unimaginable. There are no guarantees. We may not be human beings at century's end—or we may be more than human. Is this something to fear, or mourn, or welcome? Who can say? Our individual cells resemble bacteria. Do they mind not being bacteria and instead being part of human beings? In the same way, we may not mind, or even fully notice, our transformation into something else that may be equipped with the skills for global self-management.

If inhabited worlds or radio-friendly civilizations are out there in large numbers, we may soon be able to find them. Nobody knows what the effect of such a discovery would be. It might lead us toward a less divided view of ourselves.

I don't know if all this sounds optimistic or pessimistic. People want to be one or the other. Either the world is all going completely to hell in an orgy of waste and destruction, or we are about to experience the techno-utopian Rapture and live in eternal post-human bliss. I don't buy either picture.

On much longer timescales, prediction is even more pointless and futile, except insofar as knowing that the laws of physics cannot be broken (although we will likely discover that our current understanding of them is pitifully incomplete). One galactic year from now, if sentient technological life exists on this planet, we can be sure that climate will be managed, that population and energy use will have stabilized, and that the economy, such as it is, will not depend upon continued growth in use of resources.

We are going through an awkward and destructive phase, but there is another role we could play on this planet: wise guardians, protectors of the biosphere against long-term threats and destructive changes. If we handle the Anthropocene well, it could be the first epoch of the Sapiezoic Eon. You and I will not know this in our lifetimes. It will not be clear within the lives of any of our immediate descendants. It may be millions of years before enough time has passed for an eon boundary to be obvious. In the proto-Anthropocene, we have, without knowing it, become a major geological force. Without self-awareness, such a force is dangerous and unstable, but the mature Anthropocene is when we realize what we are doing, and as that realization becomes incorporated into the way it functions, this human geology becomes a stabilizing force. If this self-conscious negative feedback becomes integral, in a sustained way, the planet will have entered its Sapiezoic Eon. Then we and our planet together will have become Terra Sapiens.

Eyes of the World

Here in Washington, DC, in the springtime, with bees flitting about in a riotous orgy of magnolia and cherry, it is obvious that flowering trees and their pollinators have evolved to need each other. In a way that was less obvious at first, and that becomes clearer the more we learn, life and the planet have coevolved to need each other as well. They are inseparable. That is the insight of Gaia. Now what we call intelligence is becoming a major influence on Earth. Can this last? Can it, can *we*, become a symbiotic part of the way the planet runs?

Thoughtful control of Earth is now a skill that, under threat of catastrophe, we are forced to learn and master. The hardest part of that, it seems, will be mastering ourselves. Rising to this occasion means accepting a new role on the planet. When we step back and examine current trends against the long sweep of planetary history, it becomes clear that our world is now being rapidly remade by changes we have inadvertently caused. How should we respond to this unsettling realization? Even if we wanted to, we couldn't just stop what we're doing. Because by now we've made our world of seven billion (and rising) people fully dependent upon and deeply desirous of world-changing technology. So we need to find a way to finish what we've started, to more fully embrace our role as world-changers and learn to get good at it. Then, safely beyond today's reckless adolescent planetary joyride, this self-destructive techno-tantrum, we'll find that we are no longer at the whim of deadly climate fluctuations and asteroid strikes. There may be a large rock that has Earth's name written on it, but we'll erase it, give the interloper a little nudge, and watch it hurl harmlessly past, on into the void.

The planetary perspective can help us. We now know that change, even catastrophe, is in the nature of planets. Habitability for even one species is fickle, and even for an entire biosphere there is a certain timescale over which, without intelligent technological intervention, life will be toast. On a much shorter timescale, without intervention, mass extinctions will occur. So, obviously, it is the responsibility of intelligent life to learn how to shape a planet. Yet that is not yet our task. First we need to learn to stop being a menace. As an initial step, we have to halt the wave of extinctions and climate havoc we are causing. When we consider the long history of catastrophic changes on Earth, we find the novelty of our situation jumps out. We are facing a challenge not remotely like what any other species has faced.

Accepting our role as planetary engineers means, most immediately, recognizing our obligation to reduce our ignorance of the Earth system. It's not just curiosity now that motivates us to study the workings of Earth and the other planets. It's survival. Planetary exploration itself, now entering its second half century, is a long-term, transnational, intergenerational technological project—exactly the kind of activity we need to move toward a world guided by planetary changes of the fourth kind, intentional global change.

Right now the subject of the future is rife with anxiety. Visions of apocalypse dance in our heads. The topic of the Anthropocene is often associated with doom and gloom, with an "Is Earth fucked?" mentality. This is understandable, but it's not the whole story. Let's not dwell on these prophecies to the point where they become self-fulfilling. I propose that, on the contrary, the true Anthropocene is something that should be welcomed. Though it is yet only in its infancy, it can be glimpsed. Don't fear it. Learn to shape it. It is the awareness of ourselves as geological change agents that, once propagated and integrated, will provide us with the capacity to avoid doom and take our future into our own hands. Understandably, we are uncomfortable with our role as reluctant planetary engineers. Discomfort sometimes manifests as self-loathing and denial, but this is our task, and we can't afford to wallow. It's time to human up. We have to stand and face it. Get up on our big bipedal frames and look in the mirror. Wake up to find out that we are the eyes of the world.

Earlier in this chapter, I described flavors of pessimism. There are flavors of optimism, too. There's cosmic optimism, stemming from a belief that the universe, in its vastness, bends toward life and intelligence and that what happens here doesn't really matter because "there's plenty more where we come from." There's data-driven and historically based optimism, which focuses on positive indicators, of which there are many. Poverty, malnutrition, and infant mortality are in retreat globally. Levels of education are on the rise. Communication continues to become cheaper and easier. Population is plausibly heading toward stability. Solar and wind energy are getting cheaper and will continue to do so. These are all trends toward human freedom and environmental sustainability.

There's pragmatic optimism: we really don't know what is going to happen, so why not spread hope and encourage engagement? Exponential technological innovation is transforming our world in surprising and accelerating ways. Possibilities that until recently seemed magical are now imminent, rendering the future frightening and exhilarating but, above all, unpredictable. Where there is uncertainty there is also hope—and choice, and room for faith in ourselves. I believe we're just getting started on this planet.

Nobody knows the odds of our being able to navigate the evolutionary obstacles before us, but there is a real hope, and it is this: that our evolving technological capacities can allow us to maximize our innate social prowess, equipping us to meet the novel threats we have accidentally created, and to become something new in the process. We have done this before.

In 1929, as a young man, British biologist J. D. Bernal wrote a book entitled "The World, The Flesh and the Devil" that Arthur C. Clarke called, "the most brilliant attempt at scientific prediction ever made." It closed with the following question:

> We hold the future still timidly, but perceive it for the first time as a function of our own action. Having seen it, are we to turn away from something that offends the very nature of our earliest desires? Or is the recognition of our new power sufficient to change those desires into the service of the future which they will have to bring about?

From this cosmic vantage point on our own evolutionary history and our current global situation, we have the option to choose the future we want and, with self-awareness, bravery, and humility, to reach for the wheel of history and steer a path toward infinite potential. Then the next Enlightenment can begin: when we learn how to live well within a finite world, with ourselves as its conscious shapers. We can embrace Earth as human beings: creative, cooperative, imaginative, storytelling, engineering problem solvers. We can care for our planet and begin to contemplate our galactic destiny.

Notes

Introduction: A Planetary Perspective on the Human Predicament

1. Several of these recorded conversations can be found on the Center's website https://www.loc.gov/loc/kluge/.

Chapter 1: Listening to the Planets

1. We call it planetary "*geo*logy" because the science of rocks, land-forms, and their evolution arose from the only planet we had access to until recently. Similarly, we talk about the "geography" of other worlds, avoiding the awkwardness of a separate prefix for each planet. Yet it does perhaps also speak to the difficulty of thinking of geology, biology, or any other -ology without some degree of Earth bias.

2. The rival volcanic hypothesis has not died off, however. There is a huge volcanic deposit in India, the Deccan Traps, that formed at the right time also to be implicated. A recent idea with some traction is that the impact event may have exacerbated the volcanic outpouring, so that some combination of the two is responsible. The timing is striking. Both the impact and the huge volcanic floods seem to be precisely timed to match the extinction event.

3. For this part of my thesis, Carl Sagan became my main adviser. He had worked on the climate effects of dust clouds on Mars, and nuclear winter

on Earth (which I'll describe shortly). Among the papers Carl asked me to read was a somewhat obscure 1967 PhD dissertation by a student at the University of Iowa entitled "The Atmosphere and Surface Temperature of Venus: A Dust Insulation Model." It explored the possibility that the high surface temperature observed on our sister planet might be caused not by a thick carbon dioxide greenhouse, but rather by a planet-wide blanket of atmospheric dust holding in heat. Like many ideas in science, this turned out to be wrong, but in a useful way. It's impressive that Carl remembered this work and saw that it would help me model the climate of dusty atmospheres. That's how his mind worked, and why he was such a valuable collaborator on so many planetary science projects. He remembered everything he read (which was a lot) and was able to make unlikely connections. Oh, and the author of that obscure dissertation? It was James Hansen, who went on to become one of the leading climate modelers studying global warming on Earth—and definitely one of the most visible and effective public communicators on the subject. His popular book *Storms of My Grandchildren* is one of the best on global warming: clearly written, scientifically accurate, and passionate. For those who have studied comparative planetology, it is no surprise that Jim Hansen cut his teeth on the Venus climate.

4. *Climatic Change: Evidence, Causes and Effects.* Ed. Harlow Shapley. (Harvard University Press, 1953).
5. C. P. McKay, J. B. Pollack, and R. Courtin, "Titan: Greenhouse and Anti-Greenhouse Effects on Titan," *Science* 253, no. 5024 (1991): 1118–21.
6. E. A. Petigura, A. W. Howard, and G. Marcy, "Prevalence of Earth-size Planets Orbiting Sun-like Stars," *Proceedings of the National Academy of Sciences* 110, no. 48 (2013): 19273–78.

Chapter 2: Can a Planet Be Alive?

1. Among his inventions was the electron capture detector, with which he made the first measurements showing that the chlorofluorocarbons we used for refrigeration were building up in Earth's atmosphere, a discovery that led directly to the realization that we were harming the ozone layer.
2. Carl Sagan, "The Long Winter Model of Martian Biology: A Speculation," *Icarus* 15 (1971): 511–14.

3. Margulis founded and managed NASA's Planetary Biology Internship Program, which for decades supported undergraduate and graduate students exploring the "bio" side of exobiology.

4. Given our fundamental ignorance of where teleology resides and how it arises (or seems to) within us, I always take those confident critiques of teleological Gaia with a grain of biologically mediated salt. At the very least, this controversial aspect of Gaia theory raises interesting questions about how both homeostasis and intentionality may result from the complex interplay of subsystems. These are questions that we need to consider anew as we grapple with what it means to be a species that is accidentally altering our planet and wondering how we should act, and even if we *can* act, on a planetary scale. A Darwinian approach suggests that teleology has not been a factor in Earth evolution for most of its history. Yet what about now, when we have entered the Anthropocene, this new time of supposed human control of Earth? Now people are making decisions that, whether they know it or not, are affecting the planet in major ways. So if we look at the Anthropocene as a transitional event in planetary evolution, it may be the time when teleology, intentionality, clearly becomes a part of the Earth story. Thus, the question of where, and at what level, a sense of identity and intentionality arises is quite relevant to the core predicament of humanity in the Anthropocene: Is our global human civilization, as a whole, capable of acting with intention? We will return to this question.

5. Norman H. Sleep, Dennis K. Bird, and Emily Pope, "Paleontology of Earth's Mantle," *Annual Review of Earth and Planetary Sciences* 40, no. 1 (2012): 277–300; see also Robert M. Hazen, *The Story of Earth: The First 4.5 Billion Years, from Stardust to Living Planet* (New York: Penguin Books, 2012).

6. Sleep, Bird, and Pope, "Paleontology of Earth's Mantle," p. 293.

7. If these little whiffs of methane recently reported by the *Curiosity* rover turn out to be real—this is still in question—and actually are revealed to be the signs of little pockets of Martian life on an otherwise generally dead world, I would consider that proof that my Living Worlds hypothesis is wrong, and that life can take on very non-Gaia-like forms elsewhere. I don't consider this likely, but nothing would make me happier.

8. Or, if you want to get nerdy, think of it in terms of nonequilibrium thermodynamics, as a self-perpetuating dissipative structure that maintains its form in a flow of material and energy.

Chapter 3: Monkey with the World

1. B. Werner, "Is Earth Fucked? Dynamical Futility of Global Environmental Management and Possibilities for Sustainability via Direct Action Activism," abstract for invited talk at the fall 2012 meeting of the American Geophysical Union.
2. By the way, there are some interesting parallels here to the astrobiological quest to identify biological processes on other planets. Recall in chapter 2 how I describe the *Viking* biological results that initially seemed to show evidence for life on Mars. These were ruled out largely because the gas release over time did *not* fit a biological pattern. It's the same kind of analysis that led Rothman and his colleagues to suspect biology over geology in the Great Dying. In many ways, the paleo-Earth is another planet, and our methods of investigation, including the search for biosignatures, reflect that.
3. Around that time, a kind of bacteria called *Methanosarcina* evolved the ability to much more efficiently produce methane from organic matter. This innovation seems to have come about not through traditional Darwinian evolution of mutation among competing organisms, but rather the kind of more fluid sharing, mixing, matching, and swapping of genes among organisms that Lynn Margulis and others championed for a long time and that has more recently been recognized as an important factor throughout evolutionary history. In a rather slick genetic move, rather than evolve this ability through their own mutations, these bacteria borrowed the necessary gene from another kind of bacteria, in something called "horizontal gene transfer." Rothman's colleagues performed a sophisticated phylogenetic analysis to derive the timing of this change to more efficient methane production, and though this method is imprecise, they showed that it plausibly happened at the time of the pulse of volcanism. Why do methane-producing bacteria need nickel to grow? The metabolism of organics into methane depends on a protein complex that is built around a nickel atom. In general, seawater is too deficient in nickel. Yet it is known that the Siberian Traps

magmas were very rich in nickel. In fact, our planet's most valuable economic concentrations of nickel are located in Siberia because of these deposits. Rothman and colleagues studied the history of nickel in the well-preserved sediments at Meishan, and they showed that, right before the extinctions, nickel increased up to seven times the background level, to an abundance that would indeed support rapid growth of these methanogenic bacteria.

4. Jan Zalasiewicz et al., "Scale and Diversity of the Physical Technosphere: A Geological Perspective," *The Anthropocene Review*, in press.

5. I say "seems" rather than "is" only because I have read too many flawed predictions written by the confident futurists of past decades and centuries. I am mindful of Arthur C. Clarke's observation that when a scientist states that something is impossible, he is probably wrong. I can at least imagine some bionanotechnological innovation that would prevent this seeming rush to inexorable acidification, but I would not bet on it.

6. See for example: J. Zalasiewicz, R. Kryza, and M. Williams "The Mineral Signature of the Anthropocene in Its Deep-Time Context," *Geological Society, London*, Special Publications, 2013.

7. An excellent book about this is *The Botany of Desire*, by Michael Pollan.

8. See, for example, E. S. Rice and J. Silverman, "Propagule Pressure and Climate Contribute to the Displacement of *Linepithema humile* by *Pachycondyla chinensis*," PLoS ONE 8, no. 2 (2013): e56281. doi: 10.1371/journal.pone.0056281, and references therein.

9. Prinn was also a grad student of John Lewis's, which makes him my academic brother and Lewis's adviser, Harold Urey, the Nobel Prize–winning founder of planetary chemistry who oversaw the famous Miller–Urey experiment into the organic origins of life, our academic grandfather. Tracing these nerd trees is fun, but it also reveals the heredity of ideas and modes of thought—in this case, more cross-fertilization between planetary exploration and earth environmental science.

Chapter 4: Planetary Changes of the Fourth Kind

1. A summary of this history is given in W. H. Brune, "The Ozone Story: A Model for Addressing Climate Change?" *Bulletin of the Atomic Scientists* 71 (2015): 75–84.

2. "Global Warming Has Begun, Expert Tells Senate," *New York Times*, June 24, 1988.
3. Lately some pundits have been decrying the notion of "peak oil" as a myth, insisting that as the price goes up because of scarcity, we will always find new sources, drill deeper, extract oil locked up in various kinds of sediments, etc. All this just delays the inevitable peak. We could scrape clean the crust of Earth, and at some point the oil will run out. There is no way around peak oil, unless we find a wormhole inside Earth that links to a filling station in another galaxy. The oil supply is finite.
4. Stapledon's book also describes a race of humans genetically engineered to survive on Neptune. That brings up an alternative to terraforming that has been explored in the speculative literature. As Pollack and Sagan put it, "At some time in the future, a much more elegant way to overcome our parochial habitat restrictions may be to genetically engineer humans for other worlds than to physically engineer other worlds for humans."
5. In fact, it occurred to me years ago that you might just use the same comet impacts that bring in the water also to raise the dust that would cause an anti-greenhouse and cool the planet. This could be tricky, though, because water is also a greenhouse gas that could work the opposite way and heat things up. Yet if the initial comet impacts were effective enough at raising dust without adding too much warming gas, then things might go in the right direction. I've never done the calculations, but this could make a fun little paper or at least a grad student project or term paper. You're welcome.
6. The quote is from *The Travels of Marco Polo*, circa 1300.
7. A great place to start is the two National Academies of Sciences, Engineering, Medicine Climate Intervention Reports issued in early 2015: "Climate Intervention: Carbon Dioxide Removal and Reliable Sequestration," and "Climate Intervention: Reflecting Sunlight to Cool Earth," at https://nas-sites.org/americasclimatechoices/other-reports-on-climate-change/2015-2/climate-intervention-reports/ (henceforth NRC reports). Note, the National Research Council is an arm of the National Academy of Sciences.
8. M. Bullock and D. H. Grinspoon, "The Stability of Climate on Venus," *Journal of Geophysical Research* 101 (1996): 2268.
9. NRC reports 2015.

10. There is also an expanding cloud of orbital debris from collisions, anti-satellite military tests, and abandoned hardware. This swarm threatens at some point to become self-sustaining (as colliding debris makes more debris) and hazardous to all future spaceflight. It is a looming potential tragedy of the orbital commons. Fortunately, some smart folks are on the case and some imaginative solutions are on the drawing board.

Chapter 5: Terra Sapiens

1. Though they did not invent the concept, chemist Paul Crutzen and paleontologist Eugene Stoermer are largely responsible for modern use of the term "Anthropocene" after the 2000 publication of a paper in *Global Change* magazine, in which they wrote, "It seems to us more than appropriate to emphasize the central role of mankind in geology and ecology by proposing to use the term 'anthropocene' for the current geological epoch." They proposed a start date in the late eighteenth century, to mark the beginning of the increase in greenhouse gases due to fossil fuel use.

2. Yes, for several of these agreed-upon geological transitions, there is an actual marker in the rock at a location that is designated as typical, but it is not always golden and not always an actual spike. For example, there is a bronze disk in a place called Klonk, in the Czech Republic, marking the Silurian–Devonian boundary.

3. R. J. Nevle and D. K. Bird, "Effects of Syn-pandemic Fire Reduction and Reforestation in the Tropical Americas on Atmospheric CO_2 During European Conquest," *Palaeogeography, Palaeoclimatology, Palaeoecology* 264 (2008): 25–38.

4. It was Harlow Shapley, Carl Sagan's predecessor at Harvard, whose climate change conference I discuss in chapter 1, who discovered this remarkable fact.

5. Our sun is also bobbing above and below the flattened plane of the galactic disk with a period of about sixty-five million years. From time to time there have been various proposals linking this to a cycle of extinction events on Earth because of comet impacts or dark matter or what have you, but earth scientists haven't taken these ideas too seriously. The evidence doesn't support the existence of any cycle that correlates with this motion.

6. David Christian, *Maps of Time: An Introduction to Big History*, 2005.

7. In part, the idea of the Anthropocene doesn't seem new to me because I grew up reading the "golden age" science fiction of the postwar years and early space age, which is full of techno-utopian visions in which human influence, guided by science and reason, took full hold of Earth and the other planets. When I should have been going out to play, I was reading Isaac Asimov's *Foundation* trilogy, about a planet called Trantor, a world covered by one continuous city, obviously a future version of Earth with mid-twentieth-century trends extrapolated. In my 2003 book, *Lonely Planets*, strongly influenced by Lem, I use "the psychozoic era" to describe what many are now calling the Anthropocene.

8. Cynics rightly point out that this can also be interpreted as having been motivated by their wish to keep others from joining the nuclear club, and that they were simply ready, at that point, to start testing underground. Still, the treaty was signed and it held, and as a result, the layer of radioisotopes from atmospheric testing shows a sharp peak and then a decline in the early 1960s, a lasting geologic signature of global cooperation on an Earth on the brink of disaster.

9. One random moment from later in that day stands out in my memory. After the symposium was over, a small group of us left the convention center together and walked to a somewhat decrepit sedan parked out in the blistering late afternoon LA sun. As we reached the car, I noticed that the interior was black and the windows rolled up. The car was hotter than Venus. In wiseguy grad student mode, I said, "Oh no, look out—low albedo interior!" Atmospheric scientist Richard Turco, whose car it was, looked at me as if to say, "Who the hell is this hairy kid spouting jargon?" Sagan promptly said, "Rich, have you met David Grinspoon? He's a grad student studying planetary atmospheres." And that was how I met the lead author of the TTAPS paper.

10. The presentations from this meeting were collected and published as *The Long Darkness: Psychological and Moral Perspectives on Nuclear Winter*, ed. Lester Grinspoon (New Haven, CT: Yale University Press, 1986).

11. "An Ecomodernist Manifesto" can be found at http://www.ecomod ernism.org/.

12. A recently published collection of essays, *Keeping the Wild: Against the Domestication of Earth*, ed. G. Wuerthner, E. Crist, and T. Butler

(Washington, DC: Island Press, 2014), captures the response of traditional conservationists to ecomodernism. It could almost be called the "Anti-Ecomodernist Manifesto." The publisher's release calls it a "critique of neo-environmentalists' attacks on traditional conservation."

13. This is a flare-up of an old fight. The status of the debate at the end of the twentieth century was summarized well in 2001 by environmental sociologists Dana Fisher and William Freudenburg, in "Ecological Modernization and Its Critics: Assessing the Past and Looking Toward the Future," *Society and Natural Resources* 14, no. 8 (2001). They describe ecomodernism as follows: "The lynchpin of the argument involves technological innovation...the authors see continued industrial development as offering the best option for escaping from the ecological crises of the developed world. Unlike theorists who see technological development as being generally problematic—pointing to a potential need to stop capitalism and/ or the process of industrialization to deal with ecological crises, [the ecological modernists] argue that environmental problems can best be solved through *further* advancement of technology and industrialization." This also builds on the debate within conservation that boiled up in the 1990s over how to reconcile the needs of indigenous and traditional peoples with the management of parks and conservation areas. Should locals be allowed to harvest animals and trees, or should the very notion of a protected area include a prohibition on all human exploitation? In many places, the strict "no harvesting allowed" approach has given way to one where local people who have a long-standing relationship with the land have been made partners in conservation, continuing a certain level of sustainable exploitation and improving overall land management.

14. Erle Ellis, one of the more thoughtful ecomodernists, despite his love of incendiary phrases like "used planet," wrote an op-ed in the *New York Times* in September 2013 entitled "Overpopulation Is Not the Problem," in which he stated that "there really is no such thing as a human carrying capacity on the Earth." When I pointed out to Erle, whom I count as a friend, that his statement was inconsistent with the laws of physics, he clarified that what he really meant was that these limits would not be reached in the coming century, and that our current challenges were more social in nature. I agreed with his point but still objected to his phraseology, as it implied the

complete lack of limits, which is a habit of thought we need to move beyond.

15. Actual quote from a string of tweets sent on January 16, 2015, by Alex Trembath (@atrembath), a senior analyst at the Breakthrough Institute, a leading ecomodernist think tank. The tweet storm began with, "I keep seeing this new climate communications message: 'Global warming will continue as long as we use the sky as a waste dump.'" and continued: "'the sky' may be the best waste dump we can imagine!! Nobody lives in the sky! If our shit has to go somewhere, why not there?"

16. For example, in "What's So New About the 'New Conservation'?" Curt Meine, senior fellow at the Aldo Leopold Foundation, writes, "We need to think of conservation in terms of whole landscapes, from the wildest places to the most urban places...We need to do more and better conservation work outside protected areas and sacred spaces; on our 'working' farms, ranches, and forests; and in the suburbs and cities where people increasingly live. We need to meet our needs for food, fiber, and fuel in ways that do not simplify and deplete but actively replenish, ecosystems close to home and around the world..." In describing his vision of where traditionalist conservation is going, he seems to be describing a more thoughtful strain of ecopragmatism, fortified with a healthy dose of the essential ingredient lacking in the most strident ecomodernists: humility. Meine writes, "We are engaged in a collective effort to understand and redirect the relationship between the human (and humanized) and the 'natural, wild, and free.' To do so, we need to understand, in ways we do not yet fully understand, the complicated history of humans and nature, and the evolution of what we now call conservation and environmentalism, over decades, centuries, and millennia...It is a vast task of intellectual and spiritual synthesis. It demands more than oversimplification and caricature. It requires, above all, humility. We have work to do."

17. Another fascinating area of disagreement is over the idea of "de-extinction." Many ecomodernists are smitten with it. For example, some scientists hope to resurrect woolly mammoths with DNA extracted from frozen remains. These new/old creatures would be born to elephant mothers. They (at least the portion of them that was not microbial) would be genetically similar to actual woolly mammoths but born into a world no mammoth ever walked, full of alien

creatures and strange diseases. Who knows how they would take, internally and externally, to this strange new world? Is this act of creation/ restoration benevolent or cruel? It is an interesting question, or pair of questions, whether we could, or should try to, re-create lost species. The first answer to "should we?" is another question: *why* should we? Is it to assuage our guilt? Or an attempt to return some part of the world to some semblance of what is "right" or "normal"? If extinction is bad, then is de-extinction good? When you look at the reality of what this would mean, it seems at most a freak show distraction from the larger questions we confront in owning up to the mass extinction we are in the process of starting. Because the answer to "could we?" is, strictly speaking, "no." While it is true that we may be able to *sort of* bring back certain hand-picked popular or charismatic species, or at least somewhat transmogrified simulacra of them (genetically similar, born to mothers of a related species), we will never be able to bring them back as they originally were. We certainly can't bring back all the lost species, or re-create the bygone worlds they inhabited. What about recently extinct species, who lived not in some vanished world but in ours? Or those still here but threatened? Would you oppose bringing them back if we lost them? This is more challenging. My instinct is to say this is not where our conservation efforts and attention should be focused. Yet I stumble if you present me with the following hypothetical: It's fifty years from now and there are no Siberian tigers, but we could create a family of them. Should we?

18. Tom Butler, in the introduction to *Keeping the Wild*.
19. A certain nostalgia for the Paleolithic appears from time to time in which we lament the change from forager to conqueror. It is, in fact, an interesting question: were we actually better off before we discovered agriculture and started the slide toward modernity? Yuval Noah Harari is eloquent on this question in the book *Sapiens: A Brief History of Humankind* (Harper, 2015). In some ways we may have been happier, though less safe and secure, when we were all hunter-gatherers. Perhaps, but this doesn't really help us now, does it?
20. For a thoughtful introduction to this subject, see Peter G. Brown and Peter Timmeran, eds., *Ecological Economics for the Anthropocene* (New York: Columbia University Press, 2015).
21. Around the time *Our Final Hour* came out, Rees also placed a bet on the Long Bets website, soliciting others to bet against his proposition that "By 2020, bioterror or bioerror will lead to one million

casualties in a single event." Sir Martin wrote his book in 2003, and by now, at the time of this writing, we have already survived 16 percent of the twenty-first century. There are only a few years to go until Rees loses his bet. Does that change the math? I am always wary of such sanguine arguments. People sometimes imply that since we survived the Cold War, it may not have been such an existential threat. Yet of course in all those possible histories where the unthinkable did happen, we would not have been here to have the conversation. So there is a strange selection effect, hard to factor, that might lead one to unwarranted optimism.

22. Caveat emptor: it's not a perfect system. This history is still being worked out and will, at some level, always be in revision. So if the boundaries are assigned to specific events, then the dates will have to shift. Alternatively, if we leave the dates set in (ahem) stone, the boundaries will soon no longer match the actual physical transitions they were meant to mark. So there is some tension between constancy and accuracy. It's like discussing a party that was supposed to start at 8:00 but for which people showed up at 8:35. When you later refer to the start time of the party, do you mean 8:00 p.m. or when the people actually showed up? The hour 8:00 p.m. is easier but less accurate. Who cares? Nobody—unless suddenly it becomes important to reconstruct events. Say you were visited by police investigating a crime that occurred at 8:15 and in which one of your party guests was a suspect? All of a sudden the actual, rather than the idealized, start time would be very important. So it is when we investigate the mysteries of Earth history—we need a common timescale.

    Science has not made up its mind, so there is minor confusion and debate about all these boundaries—but that's okay. They work well enough, and when you talk to geologists about the transition from the Archean to the Proterozoic, they generally know what you mean.

23. Zahnle et al., "Emergence of a Habitable Planet," *Space Science Reviews* 129 (2007): 35–78.

Chapter 6: Intelligent Worlds in the Universe

1. What we know about life elsewhere: it makes copies of itself. That's it, really. Must they be carbon copies? Or is there another chemical system that would do the trick? We don't know. So we search for what

we know, sometimes forgetting to retain the humility appropriate for our level of ignorance.

2 . C. H. Lineweaver, "Paleontological Tests: Human-Like Intelligence Is Not a Convergent Feature of Evolution," in J. Seckbach and M. Walsh, eds., *From Fossils to Astrobiology* (Netherlands: Springer, 2008).

3. After presenting his challenge to its conceptual basis, Charley wraps up his paper by stating his strong support for SETI. This would be a strange position to take if he fully embraced his own argument and was thoroughly convinced that the field was based on a wrong-headed (so to speak) assumption. This increases my suspicion that he is largely just arguing for the sheer joy of it. In this he differentiates himself from his predecessors Mayr and Simpson, who argued that it was foolish and wasteful to expend any resources on searching for extraterrestrial life or intelligence. Charley concludes: "We do not need to misinterpret the fossil record to justify this inspiring research." Quite true—and we do not need to overinterpret the fossil record to realize that the odds of success are not to be determined here on Earth but out among the numberless stars.

4. Phil cowrote and narrated (in his distinctive, soothing, raspy voice) the classic, somewhat psychedelic 1977 film *Powers of Ten*, which brilliantly synthesizes modern scientific understanding of where we stand within the scales and structures of the universe. It takes viewers on a cosmic journey beginning with a scene of a happy heterosexual couple picnicking on the Chicago lakeside, receding outward to a view of the whole Earth, past the orbit of the Moon, the planets, the other stars, on out to the farthest reaches of the known universe. Then it reverses and plunges inward, back again to the lakeside, only to keep going, zooming into the man's hand and his cells, molecules, atoms, down to the subatomic particles. At the end, a single proton fills the field, full of fuzzy, quivering quarks. Phil's final statement that "we have reached the edge of present understanding" carries a promise that in the future we would see farther, and continue the trip.

5. Shapley's estimates of the specific quantities involved are quite different from the numbers we would use today. Admitting that astronomers of 1952 knew little about how planets form, he said, "Let us be skeptical and suppose that, in whatever way they come, only one star out of a million is blessed with a family of planets." His estimate

for the fraction with planets in the habitable zone was similarly cautious: "Suppose we skeptically guess that only one star family out of a thousand has a planet with the happy requirements of a suitable distance from the star, near-circular orbit, proper mass, salubrious atmosphere, and reasonable rotation period—all of which are necessary for life as we know it on the Earth." This is pretty good astrobiology from about forty-five years before the birth of astrobiology.

6. A brouhaha about possible "alien megastructures" around the star KIC 8462852, or Tabby's Star, arose because the enigmatic changes in light from this star seem to reveal some strange configuration of orbiting material. Conceivably, this could be something like a Dyson swarm. Almost surely there will be a more prosaic explanation, but at the time of this writing, it is still a mystery. I wrote about this here: http://www.skyandtelescope.com/astronomy-blogs/cosmic-relief-david-grinspoon/could-it-be-possible-signs-of-e-t-intelligence/.

7. This dependence of the density of civilizations on L is very well illustrated with a visual analogy in Timothy Ferris's television film *Life Beyond Earth*. Ferris showed a Christmas tree with lights flashing on and off, and demonstrated that if you vary the length of time the lights stay on, then the appearance of the tree, at any instant in time, changes noticeably. If each light comes on for only a fraction of a second, then at any given time a snapshot shows only a few lights on an otherwise dark tree. If the lights all stay on for several seconds before blinking off, then the tree is always full of glowing lights. The longer the average "lifetime," the shorter the distance from one lit bulb to the next. The dark tree, with just a few lights, is analogous to the situation in the galaxy if civilizations generally don't last long. There may be others, somewhere out in the vast Milky Way, but the average distance between them would likely render two-way communication impossible. If the average civilization has a long lifetime, then the tree, or the galaxy, is densely lit up. Then the distance to our nearest communicating neighbor may be quite short.

8. Small world department: McNeill was invited by Sagan to broaden the generalized discussion of "civilizations" with a historian's perspective. He is the father of and occasional collaborator with environmental historian and Georgetown University history professor John McNeill, who today is a leading scholar of the Anthropocene. I had no idea of this connection when I met the younger McNeill, and

we both became founding members of our informal Washington Anthropocene Group.

9. Recently some scholars analyzing the detailed history of the escalations, and miscommunications of that episode, have concluded that it was a very close call. The commanders on the ground in Cuba were given the authority to launch nuclear missiles if the Americans attacked. The Americans were poised to attack with conventional weapons, assuming, wrongly, that this would not provoke a nuclear counterattack. Some analyses suggest there was perhaps a fifty-fifty chance that it could have led to the start of a nuclear war, a favorable coin toss allowing the rest of history after 1963.

10. It is also said that he didn't actually say this. If so, then he should have.

11. Note that for the purposes of this discussion I may use the term *intelligent life* or *intelligence* as shorthand for technological intelligence. Quite clearly there are other species, such as dolphins and perhaps elephants, some primates, and others possessing quite a bit of intelligence but no global technological civilization. I don't mean to assert any claim of superiority over such mindful, less manipulative creatures. There is a reasonable argument for the opposite claim. Here, however, I am discussing the possibility of planets where intelligence, for better or worse, has produced some kind of global technological civilization.

12. If there is a bifurcation in civilizational lifetimes, with one branch being quasi-immortal, it implies that the number of civilizations is not in steady state, as implicitly assumed in the Drake equation, but increasing as the universe ages. A further implication is that our own prospects for long-term survival and developing a sustainable civilization may be completely decoupled from the number of technical civilizations in the universe. We often conflate the two in our discussions of L. Given that ET civilizations, if they exist and we are able to find them, are assumed to represent some imagined, hoped-for, future state of our own evolution, we often identify L with ideas about our own future prospects with statements such as "If we are likely to destroy ourselves in a nuclear war, then it implies that L is short and there is nobody out there." This is wrong because L is not our own longevity but the average of all civilizations in the galaxy, and there is always the chance that we are highly atypical, lucky or unlucky. If there is a bifurcation of lifetimes, then our fate,

our future prospect, may have even less correspondence with the number of advanced civilizations out there. The universe could be populated with an increasing number of aged civilizations that have made it through the bottleneck, even if most civilizations at our stage typically don't survive for long. To see how both could be true, consider the reproductive strategy of dandelions or mice: organisms that make huge numbers of offspring so that just a few can survive to adulthood (what biologists call r-strategy, or r-selected, species, as opposed to k-strategy, or k-selected, species, which produce far fewer young and carefully nurture each individual toward adulthood). For creatures with this life history, the typical fate of any newborns is not to survive to maturity, but those few who do may have a very long life, even while there is not much hope for any young individual. Most acorns do not become tall trees, but the forest is dominated by trees. Our civilization may just be an acorn. In which case, we may have little chance of making it to maturity. Yet even if that is the case, there may still be plenty of mature civilizations, tall trees growing in the galactic forest. Even if it is highly unusual for a baby civilization like ours to achieve adulthood, the galaxy could still be full of adults. Although it matters very much to us how long we survive, it is quite possible that our longevity has very little connection to the overall chance that intelligence can thrive for very long timescales. Of course this picture is all wrong if the formation of an immortal is so improbable as never to happen within the age of the universe. The weakest link here is the assumption that any civilization ever makes it through the technological bottleneck to a state where they have the ability to use their technology for long-term survival. This will depend on how "the race between education and catastrophe" is won on other worlds. If catastrophe always wins, there will be no immortals.

13. The quote is from the beautiful first paragraph of H. G. Wells's *War of the Worlds.*

14. The use of this phrase has been criticized by some who point out that the Fermi paradox, strictly speaking, is neither Fermi's nor necessarily a paradox, since it's only a paradox if you accept the arguable premises, and others before and after Fermi have contributed to the idea. But it rolls off the tongue better than the "Fermi-Tsiolkovsky -Fontenelle-Lem-Bracewell-Hart question."

15. Philip Morrison, who worked with Fermi designing bombs at Los Alamos, before essentially founding SETI, described Fermi questions as follows: "the estimation of rough but quantitative answers to unexpected questions about many aspects of the natural world. The method was the common and frequently amusing practice of Enrico Fermi, perhaps the most widely creative physicist of our times. Fermi delighted to think up and at once to discuss and to answer questions which drew upon deep understanding of the world, upon everyday experience, and upon the ability to make rough approximations, inspired guesses, and statistical estimates from very little data." An example of a Fermi question is: how many piano tuners are there in the city of Chicago? You can make a decent estimate by starting with a rough idea of the population of Chicago, estimating what percentage of households has a piano, roughly how often those pianos need to get tuned, how long that takes, and how many people that could employ. The final answer you get, provided you've set it up correctly and made reasonable guestimates, is often surprisingly close to the actual number you would find if you solved the problem more carefully. Other Fermi questions described by Morrison include: What is the photon flux at the eye from a faint visible star? How far can a crow fly? How many atoms could be reasonably claimed to belong to the jurisdiction of the United States? What is the output power of a firefly, a French horn, an earthquake?

16. Start with David Brin's classic paper "The Great Silence—the Controversy Concerning Extraterrestrial Intelligent Life," *Quarterly Journal of the Royal Astronomical Society* 24, no. 3 (1983): 283–309. Then read *If the Universe Is Teeming with Aliens… WHERE IS EVERYBODY?: Fifty Solutions to the Fermi Paradox and the Problem of Extraterrestrial Life*, by Stephen Webb (New York: Copernicus Books, 2002). I also discuss many possible answers in my book *Lonely Planets: The Natural Philosophy of Alien Life* (New York: HarperCollins, 2003).

17. Fermi's paradox was discussed in 1933 by Tsiolkovsky (and even alluded to by French natural philosopher Bernard le Bovier de Fontenelle in his 1686 book, *Conversations on the Plurality of Worlds*). Tsiolkovsky's answer was that the advanced ETs were aware of us but were leaving us alone because we are not ready for contact. Just as today we are motivated to protect biodiversity in part to see what nifty unknown chemicals we can derive from wild, unknown species,

Tsiolkovsky also proposed that by allowing us to flourish, as long as we remained harmless to galactic civilization, advanced aliens could see what intellectual innovations we would develop on our own that might ultimately add to cosmic culture.

18. J. D. Haqq-Misra and S. D. Baum, "The Sustainability Solution to the Fermi Paradox," *Journal of the British Interplanetary Society* 62 (2009): 47–51.

Chapter 7: Finding Our Voice

1. In fact—now it can be told—the day Lewis introduced us at a Space Resources workshop at the Scripps Research Institute in the summer of 1984, the three of us went to see the then-new film *Revenge of the Nerds* in a San Diego multiplex.

2. The possibility of machine sentience also raises some moral questions. Asimov's laws, if enacted, would create robots whose only purpose was to do our bidding. Yet if these were truly sentient creatures, wouldn't they have rights? And wouldn't it be wrong, then, to design them to remain our slaves forever?

3. Referenced above. NRC Climate Intervention reports 2015.

4. This kind of planetary radar observation is how Sasha Zaitsev made a name for himself in astronomy long before he decided to send his controversial targeted messages.

5. The question of best practices seems tricky. With biotechnology and planetary protection, expert opinion was marshaled into a list of guidelines for how to proceed safely. Honestly, it's hard for me to think of a comparable set of guidelines for active SETI. I've tried to imagine what practicing "safe METI" would look like. Is there really any way to broadcast more safely? If we limit power or length of a broadcast, we are limiting our chances of success. I suppose we could insist that stars must be over a certain distance to be "contacted," thus ensuring that our descendants have plenty of time to think about how they would react to any response, but that is just kicking the can down the time line. The real question seems to be binary. We decide that it's either too risky or not. The discussion that needs to happen is "go or no-go." If we consider the very far future, there is a possible scenario for broadcasting with safety. An advanced species could send beacons from somewhere far out in space, distant

enough from their own world that they could avoid giving away their address. Someday, once we are a confident deep-spacefaring species, we might be able to accomplish this. To do so, we would need to be operating from beyond our own solar system. We are so far off from being able to do this that it doesn't seem a very helpful solution to consider.

6. See, for example, http://www.dearet.org/, https://www.newhorizons message.com/, and http://breakthroughinitiatives.org/Initiative/2.

Chapter 8: Embracing the Human Planet

1. This comes from a letter Jefferson wrote to Madison from Paris in September 1789.
2. Video online at http://www.loc.gov/loc/kluge/webcasts/grinspoon -julynov2013-feature.html.
3. J. F. Mouhot, "Thomas Jefferson and I," *Solutions* 5, no. 6 (March 2015): 74–78, http://www.thesolutionsjournal.org/node/237268.
4. Ibid.
5. Since moved on to the University of California, Davis.
6. The Longevity of Human Civilization: Will We Survive Our World-Changing Technologies? symposium held at the Library of Congress, September 12, 2013. Video and transcript at http://www .loc.gov/loc/kluge/news/nasa-program-2013.html.
7. The political process reinforces our tendency to take sides in order to win rather than seek a reasoned solution. Bertrand Russell captured this beautifully in an essay written in 1924: "A person imbued with the scientific spirit would hardly even examine these extreme positions. Some people think that we keep our rooms too hot for health, others that we keep them too cold. If this were a political question, one party would maintain that the best temperature is the absolute zero, the other that it is the melting point of iron. Those who maintained any intermediate position would be abused as timorous time-servers, concealed agents of the other side, men who ruined the enthusiasm of a sacred cause by tepid appeals to mere reason. Any man who had the courage to say that our rooms ought to be neither very hot nor very cold would be abused by both parties, and probably shot in No Man's Land" (from *Icarus, or the Future of Science*).

8. As Isaac Asimov, through his fictional scientist Hari Seldon, a "psychohistorian," describes brilliantly in his *Foundation* trilogy.
9. In the classic book *Extraordinary Popular Delusions and the Madness of Crowds*, Charles Mackay writes about the tulip mania that gripped the Netherlands in the early 1600s, when, as a result of a positive feedback between perception and price, some varieties of tulip briefly became the most valuable objects in the world. Tulips can be extraordinarily lovely, but there is magical thinking and shared delusion in believing a single tulip bulb is more valuable than any building, home, or artifact. There's an obvious continuum from this to today's speculative bubbles in real estate or tech stocks.
10. I don't wish to get sidetracked into the somewhat masturbatory philosophical debate over whether or not we have free will or it is just an illusion. Some sophists argue passionately that there is no such thing as free will. If so, then perhaps there really is no fundamental difference between us and a bacterium, or even an orbiting swarm of rocks. Maybe we're all just doing what we must do, following physical law, acting out predetermined scripts. Maybe we can't help anything and therefore bear no responsibility for the changes we are causing on this planet. I don't believe that. I believe we have responsibility for our actions. I declare that there *is* free will, and you cannot choose to disagree with me. Or if you do choose to disagree with me, you are proving my point. *Q.E.D.*
11. A vivid description of these dual forces at work in the Egyptian revolution of 2011 can be found in the transcript of the talk "Let's Design Social Media That Drives Real Change," by Egyptian activist and computer engineer Wael Ghonim, at https://www.ted.com/talks/wael_ghonim_let_s_design_social_media_that_drives_real_change/transcript?language=en.
12. Where I had the great good fortune to be as an undergraduate assistant at Jupiter, as a grad student at Uranus, and as a postdoc for the Neptune encounter.
13. In an article in *Earth Island Journal* entitled "Anthropocene Is the Wrong Word," environmental philosopher Kathleen Dean Moore rejected the term as too self-serving. After reiterating the oft-repeated (and undeniable) environmental sins of humankind, she suggested that a better term for the age of human influence would be the Unforgiveable–CrimeScene Epoch or, simply, the Obscene Epoch.

14. It must be noted, however—especially as an antidote to the tendency to romanticize the relationship that "primitive" humans had with nature—that forcing the extinction of megafauna, large mammals, is a human tradition that extends into the past long before most dates proposed for the beginning of the Anthropocene. Rhinos are among the few megafauna left in the human-altered Earth. If we change course and avoid causing the extinction of the remaining species of rhinos, elephants, hippos, and whales, it will be a sign that we have consciously evolved to transcend our deeply ingrained ancient "natural" tendencies.

15. As of this writing, there is only one male individual left alive on the planet: a forty-two-year-old named Sudan, living in the Ol Pejeta Conservancy in Kenya.

16. I've noticed that in discussions of our future, both optimism and pessimism are employed to get the moral upper hand. If you express a view of the human future that someone else thinks is unrealistically or annoyingly hopeful, that person can always accuse you of obviously not caring about this or that undeniable tragedy. If you are being too much of a downer, you can be accused of being too much of a downer.

17. http://www.vice.com/en_uk/read/what-do-we-do-when-we-know-the-world-is-ending.

18. Our world is finite in some dimensions and infinite in others: physically finite, but infinite in the inventiveness of its life. Similarly, although as individuals we come to recognize our physical limits, and our mortality, we are also infinite in other dimensions: spiritual, intellectual, creative, and loving.

19. Frederick Ferré, *Hellfire and Lightning Rods: Liberating Science, Technology, and Religion* (Maryknoll, NY: Orbis Books, 1993).

20. Lederberg also coined the term *exobiology* and helped found the field that later morphed into astrobiology. Often *microbiota* is used to refer to the organisms, and *microbiome* to the DNA within them.

21. Emma Marris, in her book *The Rambunctious Garden: Saving Nature in a Post-Wild World* (2011), a very good ecomodernist book that traditionalists love to hate, is quite eloquent on this point.

22. The subtitle of the anti-ecomodernist volume *Keeping the Wild* is *Against the Domestication of Earth*.

23. The very next year, British scientist/writer/futurist J. D. Bernal independently described a similar concept, writing, "Journeys would

have to last for hundreds and thousands of years, and it would be necessary—if man remains as he is—for colonies of ancestors to start out who might expect the arrival of remote descendants."

24. Peter Zahler and George Schaller, "Saving More than Just Snow Leopards," *New York Times*, February 1, 2014.

25. An excellent book that explores this haunting possibility is *World Wide Mind*, by Michael Chorost.

Acknowledgments

The opportunity for me to write this book was made possible by a partnership between two great institutions. I did the research and began writing while serving as the inaugural NASA/ Blumberg Chair of Astrobiology at the United States Library of Congress. This position, started in 2013, is meant to allow researchers to pursue questions at the intersection of Astrobiology and wider humanistic concerns. It is jointly supported by the NASA Astrobiology Institute and by the John W. Kluge Center at the Library. I am extremely appreciative of both for supporting my work. Mary Voytek, who directs NASA's Astrobiology program, has championed the Astrobiology Institute's efforts to engage deeply with the humanities and was as supportive and encouraging as you could want your NASA officials to be. Carl Pilcher, former director of the Astrobiology Institute, was instrumental in spearheading these connections and provided me with good advice and useful references. I'm grateful to the late Barry Blumberg who exemplified the scientist as humanitarian, whose discovery of the Hepatitis B vaccine literally saved millions of human lives, and who later became the first director of the Astrobiology Institute. His energetic efforts to forge ties between Astrobiology and the Library bore fruit with the chair founded in his name.

My time at the Kluge Center was immensely enriched by the kind and capable directorship of Carolyn Brown and the expert help of Mary Lou Rekker, Jason Steinhauer, Dan Turello, Travis Hensley, Camila Escobar, and JoAnne Kitching. During my tenure there I was fortunate to overlap with many brilliant scholars. In studying various aspects of the human presence on Earth, I frequently found myself in territory beyond the normal purview of a research scientist and it was extremely valuable to be embedded within a community of people studying a vast range of subjects, including literature, history, theology, political science, and ethnomusicology. In particular, I benefited from conversations with Steven Dick, Jean-Francois Mouhot, Jane McAuliffe, Charlotte Rogers, Matthias Klestil, and Nathaniel Comfort. The Library is a seemingly infinite trove with no possibility of a complete, systematic indexing system. Tapping into it effectively depends on reference staff who have cultivated and preserved the knowledge of its navigation. In the Science Reference section, I'm particularly grateful for the wizardry and generosity of Margaret Clifton and Jennifer Harbster.

For being a sounding board and for sharing references, ideas, and a few beers, I'm grateful to the regular and occasional members of our "Washington Anthropocene Group," including Scott Wing, Rick Potts, Odile Madden, John McNeill, Erle Ellis, Antoinette WinklerPrins, and Timothy Beach.

For enjoyable conversations, correspondence, or generously suggesting sources, I thank Jill Tarter, David Brin, Frank Drake, David Tatel, Vikki Meadows, Jeff Moore, John Spencer, Lindy Elkins-Tanton, David Catling, Colin Goldblatt, Dorion Sagan, Shawn Domagal-Goldman, Natalie Batalha, Andrew Revkin, Kim Stanley Robinson, Ursula Heise, David Biello, Seth Shostak, Ken Caldeira, Jacob Haqq-Misra, Gavin Schmidt,

Diane Ackerman, John Perry Barlow, Curtis Marean, Ariel Anbar, Kirk Johnson, Kevin Zahnle, Johannes Lunderhausen, Peter Swirski, Kate McKinnon, Martin Bohle, Brian Toon, Ray Pierrehumbert, Jim Hansen, Jan Zalasiewicz, Will Steffen, David Baker, David Christian, Pamela Engebretson, Elise Bohan, Jim Gates, Peter Brown, Jon Erickson, Connie Bertka, Melvin Konner, Robin Lovin, Will Storrar, Ann Kruger, Aaron Goldman, Sara Walker, Julaine Rossner, Lori Marino, Alan Stern, Ellen Stofan, Bill McKinnon, Damon Santostefano, Susan Schneider, Martin Bohle, and Clément Vidal.

Thanks to my writer buddies Peter Heller, Helen Thorpe, Rebecca Rowe, Juan Thompson, Janis Hallowell, Lisa Jones, Florence Williams, Juliet Eilperin, Joshua Horowitz, Jacki Lyden, Eric Weiner, Tim Zimmerman, Maarten Troost, George Musser, and Michael Chorost for tea and sympathy, advice, and solidarity.

I thank the editors and staff at *Sky & Telescope* magazine for many years of support and patience. For the last seven years I've been contributing a short column called "Cosmic Relief." Parts of chapters 3, 4, and 5 began as those columns, though none are reprinted here in whole. Similarly, chapter 6 incubated in a 2013 piece I wrote for *Slate* called "In Search of Planetary Intelligence" and chapter 7 in a December 2007 article I wrote for *SEED* magazine called "Who Speaks for Earth?"

I thank my editor, Lindsey Rose, for her excellent judgment, patience, encouragement, and good humor, all of which have made this book a pleasure to produce. Everyone I've worked with at Grand Central Publishing and Hachette has been kind, supportive, and extremely good at what they do. I also want to thank Mitch Hoffmann for his support, guidance, and key input into earlier drafts. I'm extremely grateful to my agent, Eric Lupfer, for his wise counsel and excellent instincts and for guiding me through the entire process, from proposal through publication.

For able assistance with all manner of tasks, I thank Julia DeMarines (aka T-Spoon) and Shana Hausman.

My brother Peter Grinspoon gave me detailed and valuable feedback on an earlier draft. I thank him and Liz, Josh, Lester, Betsy, Emma, Zach, Isabel and Audrey Grinspoon, and Jacob Leher for encouragement, advice, fun, love, and various forms of support.

Jennifer Goldsmith-Grinspoon has been an infinite source of encouragement, ideas, laughter, and inspiration throughout the process of writing this book. For all that and more, I am grateful.

Art Credits

Chapter 1
Jim Pollack: photo by NASA; page 30.

Chapter 2
Lynn Margulis: photo by Javier Pedreira; page 64.

Chapter 3
Keeling curve: Wikimedia Creative Commons; page 117.
Graphs: graphs by Will Steffen; pages 131 and 132.

Chapter 4
Mars rover tracks: photo by NASA; page 174.
Beach tire tracks: photo by Susan Bruce; page 175.
Fossilized animal tracks: photo by Grzegorz Niedźwiedzki; page 175.

Chapter 5
Moon footprint with fossilized ancient footprint; Right: NASA, Left: John Reader / Science Source; page 223.

Chapter 7
Dr. Grinspoon and Dr. Zaitsev: photo by H. Paul Shuch / SETI League Photo / SETILeague.org; page 363.

Bookshelf: photo by the author; page 364.

Photo Insert

Earth from space: photo by NASA; page 1.

Asteroid painting: artwork by Don Davis; page 2.

Earthrise: photo by NASA; page 3.

Korean Peninsula from space: photo by NASA; page 4.

Geological timescale: artwork by Aaron Gronstal; page 5.

"A Short History of America": copyright © 1979 by Robert Crumb; page 6.

"Epilogue": copyright © 1979 by Robert Crumb; page 7.

Global collaboration map: computed by Olivier H. Beauchesne & Scimago Lab, data from Scopus; page 8.

Index

Page numbers in italics refer to photographs and illustrations in the text.

About the Author

David Grinspoon is an astrobiologist, award-winning science communicator, and prize-winning author. He is a senior scientist at the Planetary Science Institute and adjunct professor of Astrophysical and Planetary Science at the University of Colorado. His research focuses on climate evolution on Earth-like planets and potential conditions for life elsewhere in the universe. He is involved with several interplanetary spacecraft missions for NASA, the European Space Agency, and the Japanese Space Agency. In 2013, he was appointed as the inaugural chair of Astrobiology at the U.S. Library of Congress, where he studied the human impact on Earth systems and organized a public symposium on the Longevity of Human Civilization. His technical papers have been published in *Nature, Science,* and numerous other journals, and he has given invited keynote talks at conferences around the world. Grinspoon's popular writing has appeared in *Slate, Scientific American, Natural History, Nautilus, Astronomy, SEED,* the *Boston Globe,* the *Los Angeles Times,* the *New York Times,* and *Sky & Telescope* magazine, where he is a contributing editor and writes the quasi-monthly "Cosmic Relief" column. He is the author and editor of several books, including

Lonely Planets: The Natural Philosophy of Alien Life, which won the PEN Center USA Literary Award for Nonfiction. Grinspoon has been a recipient of the Carl Sagan Medal for Public Communication of Planetary Science by the American Astronomical Society and has been honored with the title "Alpha Geek" by *Wired* magazine. He lectures widely and appears frequently as a science commentator on television, radio, and podcasts, including as a frequent guest on *StarTalk Radio* and host of the new spinoff *StarTalk All-Stars.* Also a musician, he currently leads the *House Band of the Universe.* He resides in Washington, DC, with his wife and an imaginary dog.